BIOPROCESS COMPUTATIONS IN BIOTECHNOLOGY
Volume 1

ELLIS HORWOOD SERIES IN BIOCHEMISTRY AND BIOTECHNOLOGY

Series Editor: Dr ALAN WISEMAN, Senior Lecturer in the Division of Biochemistry, University of Surrey, Guildford

Aitken, A.	**Identification of Protein Consensus Sequences**
Ambrose, E.J.	**The Nature and Origin of the Biological World**
Austin, B. & Brown, C.M.	**Microbial Biotechnology: Freshwater and Marine Environments***
Berkeley, R.C.W., *et al.*	**Microbial Adhesion to Surfaces**
Bertoncello, I.	**Human Cell Cultures for Screening Anti-Cancer Rays***
Blackburn, F. & Knapp, J.S.	**Agricultural Microbiology***
Bowen, W.R.	**Membrane Separation Processes***
Bubel, A. & Fitzsimons, C.	**Microstructure and Function of Cells**
Clark, C.R. & Moos, W. H.	**Drug Discovery Technologies**
Cook, N.	**Potassium Channels**
Corkill, J.A.	**Clinical Biochemistry: The Analysis of Biologically Important Compounds and Drugs***
Crabbe, M.J.C.	**Enzymes Biotechnology: Protein Engineering, Structure Prediction and Fermentation**
Crabbe, M.J.C.	**Kinetics of Enzymes***
Denyer, S. & Baird, R.	**Handbook of Microbiological Control***
Dolly, J.O.	**Neurotoxins in Neurochemistry**
Espinal, J.	**Understanding Insulin Action: Principles and Molecular Mechanisms**
Eyzaguirre, J.	**Chemical Modification of Enzymes**
Eyzaguirre, J.	**Human Biochemistry and Molecular Biology**
Ferencik, M.	**Immunochemistry***
Fish, N.M.	**Computer Applications in Fermentation Technology***
Francis, J.L.	**Haemostasis and Cancer***
Gacesa, P. & Russell, N.J.	**Pseudomonas Infection and Alginates: Structures, Properties and Functions in Pathogenesis***
Gemeiner, P. *et al.*	**Enzyme Engineering***
Ghose, T.K.	**Bioprocess Computations in Biotechnology: Vol. 1**
Ghose, T.K.	**Bioprocess Engineering: The First Generation**
Harding, J.	**Biochemistry and Pharmacology of Cataract Research: Drug Therapy Approaches***
Horobin, R.W.	**Understanding Histochemistry: Selection, Evaluation and Design of Biological Stains**
Hudson, M.J. & Pyle, P.L.	**Separations for Biotechnology, Vol 2***
Hughes, J.	**The Neuropeptide CCK**
Jordan, T.W.	**Fungal Toxins and Chemistry**
Junter, G.A.	**Electrochemical Detection Techniques in the Applied Biosciences: Vol. 1: Analysis and Clinical Applications; Vol. 2: Fermentation and Bioprocess Control, Hygiene and Environmental Sciences**
Kennedy, J.F. & White, C.A.	**Bioactive Carbohydrates**
Krcmery, V.	**Antibiotic and Chemotherapeutic Compounds***
Krstulovic, A. M.	**Chiral Separations by HPLC**
Lembeck, F.	**Scientific Alternatives to Animal Experiments**
Palmer, T.	**Understanding Enzymes, 2nd Edition**
Paterson, R.	**Biological Applications of Membranes**
Reizer, J. & Peterkofsky, A.	**Sugar Transport and Metabolism in Gram-positive Bacteria**
Russell, N. J.	**Microbes and Temperature**
Scragg, A.H.	**Biotechnology for Engineers: Biological Systems in Technological Processes**
Sikyta, B.	**Methods in Industrial Microbiology**
Sluyser, M.	**Molecular Biology of Cancer Genes**
Sluyser, M. & Voûte, P.A.	**Molecular Biology and Genetics of Childhood Cancers: Approaches to Neuroblastoma**
Verrall, M.S.	**Discovery and Isolation of Microbial Products**
Verrall, M.S. & Hudson, M.J.	**Separations for Biotechnology**
Walum, E., Stenberg, K. & Jenssen, D.	**Understanding Cell Toxicology: Principles and Practice**
Webb, C. & Mavituna, F.	**Plant and Animal Cells: Process Possibilities**
Winkler, M.	**Biochemical Process Engineering***
Wiseman, A.	**Handbook of Enzyme Biotechnology, 2nd Edition**
Wiseman, A.	**Topics in Enzyme and Fermentation Biotechnology Vols. 1, 2, 4, 6, 8**
Wiseman, A.	**Enzyme Induction, Mutagen Activation and Carcinogen Testing in Yeast**
Wiseman, A.	**Genetically Engineered Proteins and Enzymes from Yeasts**

* *In preparation*

BIOPROCESS COMPUTATIONS IN BIOTECHNOLOGY

Volume 1

Editor

T.K. GHOSE Ph.D.(ETH, Zurich)
Biochemical Engineering Research Centre
Indian Institute of Technology
New Delhi, India

ELLIS HORWOOD
NEW YORK LONDON TORONTO SYDNEY TOKYO SINGAPORE

First published in 1990 by
ELLIS HORWOOD LIMITED
Market Cross House, Cooper Street,
Chichester, West Sussex, PO19 1EB, England

A division of
Simon & Schuster International Group
A Paramount Communications Company

Typeset in Times by Ellis Horwood Limited
Printed and bound in Great Britain
by Hartnolls Limited, Bodmin, Cornwall

British Library Cataloguing in Publication Data

Ghose, T. K.
Bioprocess computations for biotechnology. Vol. 1
1. Biotechnology. Mathematical aspects
I. Title
660.0151

ISBN 0–13–084674–0 (Library Edn.)
ISBN 0–13–084658–9 (Student Edn.)

Library of Congress Cataloging-in-Publication Data

Bioprocess computations for biotechnology /
editor, T. K. Ghose.
p. cm. — (Ellis Horwood series in biochemistry and biotechnology)
ISBN 0–13–084674–0 (Library Edn.)
ISBN 0–13–084658–9 (Student Edn.)
1. Biotechnology — Mathematics. 2. Biotechnology — Problems, exercises, etc. 3. Biochemical engineering — Mathematics. 4. Biochemical engineering — Problems, exercises, etc. I. Ghose, T. K. II. Series: Ellis Horwood books in the biological sciences. Series in biochemistry and biotechnology.
TP248.24.B56 1990
660′.6–dc20

89–71624
CIP

Table of contents

Contributors

Tarun Ghose, Biochemical Engineering Research Centre, Indian Institute of Technology, Hauz Khas, New Delhi- 110 016, India.

K. Suga, Department of Fermentation Technology, Faculty of Engineering, Osaka University, Osaka 565, Japan.

T. Shioya, Department of Fermentation Technology, Faculty of Engineering, Osaka University, Osaka 565, Japan.

T. Yoshida, International Centre of Cooperative Research in Biotechnology, Faculty of Engineering, Osaka University, Osaka 565, Japan.

John Villadson, Institut of Biotechnology, Technical University of Denmark, DK-2800, Lynghy, Denmakr.

K. B. Ramachandran, Biochemical Engineering Research Centre, Indian Institute of Technology, Hauz Khas, New Delhi- 110 016, India.

S. N. Mukhopadhyay, Biochemical Engineering Research Centre, Indian Institute of Technology, Hauz Khas, New Delhi- 110 016, India.

Purnendu Ghosh, Biochemical Engineering Research Centre, Indian Institute of Technology, Hauz Khas, New Delhi- 110 016, India.

Subhash Chand, Biochemical Engineering Research Centre, Indian Institute of Technology, Hauz Khas, New Delhi- 110 016, India.

V. S. Bisaria, Biochemical Engineering Research Centre, Indian Institute of Technology, Hauz Khas, New Delhi- 110 016, India.

Foreword

Success in bioprocessing involves the combination of basic principles and methods from chemical engineering and the biological sciences. To design and operate bioprocesses efficiently, this integration must surpass conceptual and qualitative union. Rigorous quantitative description of the properties of the materials to be used and the characteristics of the processes employed are necessary for bioprocess engineering. One of the best ways for learning the intellectual and analytical strategies appropriate to the engineering description and analysis of biological systems is confrontation with well-designed problems which afford the opportunity to practise quantitative approaches to the complex materials and interactions found in bioprocesses. Further, such exercises provide a useful forum in which engineers can learn and become aware of certain biological principles and properties since, by their training, engineers are accustomed to working with systems in quantitative terms and to seeking central common properties. Problems which have been carefully prepared and selected for achieving these pedagogical purposes are few. This volume will greatly enrich the opportunities for biochemical engineers to learn about bioprocessing through excellent problems.

This collection of problems is organized into eleven chapters which cover fundamental principles as well as engineering analysis and design and bioproject engineering. The volume also includes glossary of nearly 450 different terms in biotechnology and a list of terms and symbols which many students and practitioners in the field will find extremely useful. This book is intended to be the first in a series of volumes providing instructive and informative problems to assist students in learning bioprocess engineering.

James E. Bailey
California Institute of Technology,
USA

Preface

Imparting knowledge in inter-disciplinary engineering sciences through good teaching is considered by many teachers more important than research done with empirical tools. Since learning bioprocesses principles cannot be equated with acquiring knowledge in philosophy a good teacher must be equipped with problem-solving abilities to make students appreciate what the special inputs in bioprocess engineering are that differ from teaching biology and chemical engineering separately.

At the time of my spending a year of Visiting Professorship in the Department of Chemical Engineering, University of Delaware during 1985–86 the need for a book dealing with problems in various areas of bioprocess engineering was keenly felt. Extension of the Coulson and Richardson series on Chemical Engineering by J. R. Backhurst and J. H. Harkar in separate volumes dealing with numerical problems has been very useful in the classroom teaching of chemical engineering courses. Another book entitled *Examples and Problems to the Course of Unit Operations of Chemical Engineering* by K. F. Peblov, P. G. Ramankov and A. N. Noskov, MIR Publishers, Moscow is also considered as a very useful text for first-degree students. A similar book to support the teaching of biochemical engineering was very much missed. It is believed that the application of the acquired knowledge of bioprocesses is likely to remain weak if the learners are not properly and quantitatively exposed to many and multiple questions that occur again and again at the delivery end of biotechnology. The compilation of contributions contained in this volume is therefore an attempt to reduce the gap.

The volume is designed essentially for students in a first-degree programme. Much of the material provided will drill them on fundamental principles. Graduate students and practising bioprocess engineers may find insight and satisfaction in many of these problems and the manner in which these are handled. The volume is organized in ten chapters each pertaining to the principles and application of concepts, mainly in the areas of conversion reactions and downstream processing, elaborated with the help of 125 solved problems. There are also 40 discrete problems, pertaining to most of these areas, without solutions. Nearly 250 terms and symbols used in the book are based on the recommendations of IUPAC Commission on Biotechnology as well as those prescribed by the American Institute of Chemical Engineers. These are consolidated and presented in a separate chapter provided with both SI and customary units.

In the course of preparation of this volume I have benefited from the experience and advice of many colleagues during the last three years. Assistance received from reviewers, friends, colleagues and some graduate students. In particular the late Professor H. Taguchi of Osaka University, Pradeep Roy Choudhury, T. R. Sreekrishan. Radhika Satsangee and Soma Chakraborty, and typing help from R. N. Shukla and Sunita Verma of the Centre is thankfully acknowledged. Critical reviews of many of the solved problems were made by Professors John Howell, University of Bath, UK, Prasad Dhurjati, University of Delaware, USA, Paul Peringer, Swiss Federal Institute of Technology, Lausanne, Switzerland, and T. Imanaka, University of Osaka, Japan. Special mention should be made to the dedicated work of deciphering of several hand written materials and proof reading done by Sreekrishan. My sincere thanks are due to all of them.

This volume, dedicated to my wife, Atreyee, who has been a constant source of inspiration and indulgence to what I have been doing in biochemical engineering all my life, will grow and prosper only with the help, advice and comments from the students, teachers and engineers associated with the design of bioprocesses. The second volume will provide integrated process problems handling both old and new biotechnology.

Tarun K. Ghose

1

Thermodynamics of biosystems

Contributors: Tarun Ghose, V. S. Bisaria, Subhash Chand

PROBLEM 1.1

The equilibrium constant (K_{eq}) for the reaction:

$$\text{Fumarate} + H_2O \rightleftharpoons \text{L-malate}$$

has been reported as a function of temperature by Scott and Powell (1948) (Table 1). From these data, calculate the values of $\Delta H°$, $\Delta G°$ and $\Delta S°$ for the above enzyme-catalysed reaction at 30°C, where $\Delta H°$ is the standard heat of reaction, $\Delta G°$ is the standard free energy change and $\Delta S°$ is the standard entropy change.

Table 1.1 — K_{eq} as a function of temperature for the hydrolysis of fumarate

Temp (°C)	15.0	20.2	25.0	30.0	34.6	40.0	44.4	49.6
K_{eq}	4.78	4.46	3.98	3.55	3.27	3.09	2.75	2.43

Gas constant $R = 8.314$ J mol^{-1} K^{-1}.

Reference

Scott, E. M. & Powell, R. (1948) *J. Amer. Chem. Soc.* **70**, 1104.

Solution

Second law of thermodynamics gives the relationship:

$$\Delta G° = \Delta H° - T\Delta S° \tag{1.1.1}$$

Also, standard free energy change ($\Delta G°$) is related to the equilibrium constant as

$$\Delta G° = -RT \ln K_{eq} \tag{1.1.2}$$

Therefore,

$$\frac{d(\ln K_{eq})}{dT} = -\frac{1}{R}\frac{d(\Delta G°/T)}{dT} \tag{1.1.3}$$

from equation (1.1.1),

$$\frac{\Delta G^\circ}{T} = \frac{\Delta H^\circ}{T} - \Delta S^\circ$$

substituting in equation (1.1.3), we get,

$$\frac{d(\ln K_{eq})}{dT} = \frac{\Delta H^\circ}{RT^2} \tag{1.1.4}$$

assuming that ΔH° is constant over the range of temperature used, or,

$$\ln K_{eq} = -\frac{\Delta H^\circ}{RT} + I \tag{1.1.5}$$

where, I is an integration constant. ΔH° can, therefore, be determined from a plot of ($\ln K_{eq}$) versus ($1/T$) the slope of which is equal to ($-\Delta H^\circ/R$). Such a plot for the data in the given problem is shown in Fig. 1.1.

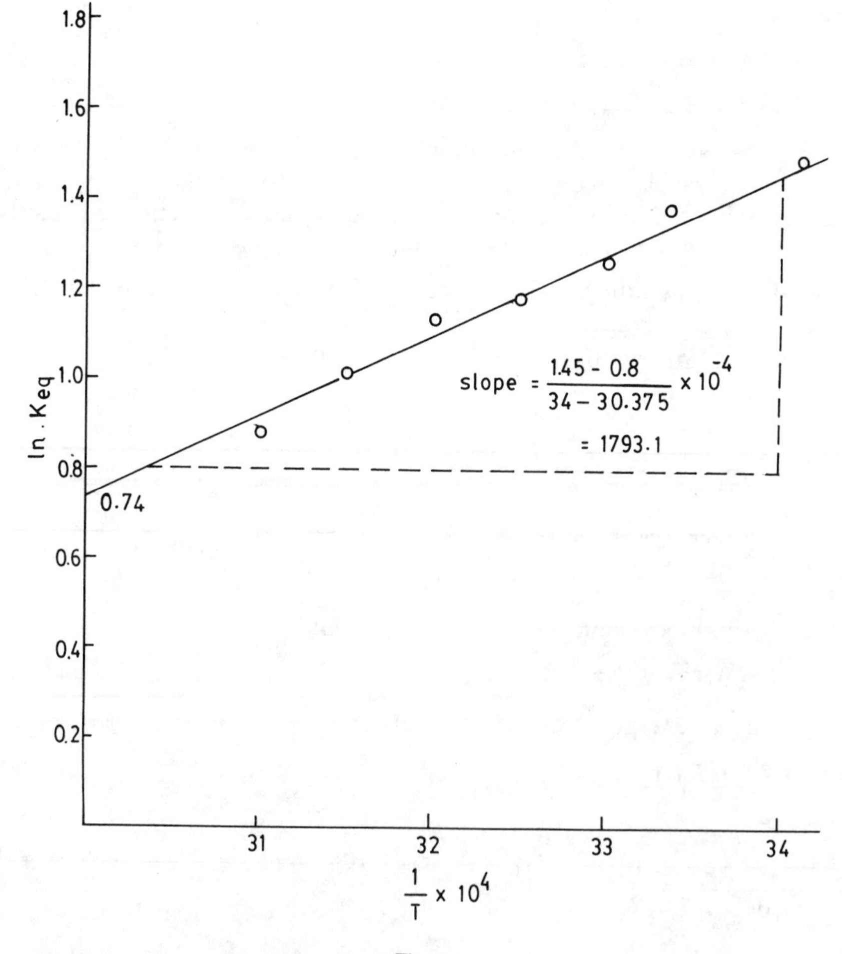

Fig. 1.1.

The slope of the straight line obtained $= 1793.1 = -\Delta H°/R$. Hence,

$$\Delta H° = -1793.1 \times 8.314$$
$$= -14907.833 \text{ J mol}^{-1}$$

To calculate $\Delta G°$ (at 30°C), substituting values of K_{eq} and T(K) in the equation (1.1.2), we get,

$$\Delta G° = -(8.314) \times (303) \times (1.267)$$
$$= -3191.753 \text{ J mol}^{-1}$$

The standard entropy change ($\Delta S°$), from the equation (1.1.1) will be,

$$(303)\ \Delta S° = -14907.833 + 3191.753$$

$$\therefore \Delta S° = -\frac{11716.08}{303}$$
$$= -38.667 \text{ J K}^{-1} \text{ mol}^{-1}$$

Thus, the effect of temperature on the equilibrium constant for an enzyme-catalysed reaction can give information on the thermodynamic parameters for the reaction (free energy change, enthalpy and entropy change).

PROBLEM 1.2

Association of ATP–ADP cycle is essential in almost all biochemical reactions catalysed by enzyme *in vitro* and *in vivo*. It is therefore necessary to understand this participation in terms of both rate and equilibrium. In a typical situation in which intracellular ATP is hydrolysed by ATPase, the hydrolysis rate, Q_r is related to intracellular concentration of ATP as

$$\frac{Q_r}{Q_m} = \frac{C_{ATP}}{K_m + C_{ATP}}$$

where Q_m = maximum hydrolysis rate, C_{ATP} = intracellular concentration of ATP, and K_m = Michaelis–Menten constant.

Cultivation of a bacterium gave the data shown in Table 1.2 at different concentrations of pantothenic acid in two different media.

Table 1.2

Culture medium	Pantothenic acid (mg ml^{-1})	$Y_{x/s}$ (g cell mol^{-1} glucose)	μ (h^{-1})	ATP pool (mg ATP g^{-1} cell)
Complex	—	7	0.37	1.54
Synthetic	5×10^{-3}	6.4	0.39	1.55
Synthetic	1×10^{-7}	2.8	0.20	3.15
Mineral	5×10^{-3}	4.5	0.28	3.55
Mineral	1×10^{-4}	2.9	0.16	4.52

Regardless of the culture medium, the specific rate of glucose uptake (v) was 0.06 mole glucose g^{-1} cell h^{-1}. Taking $K_m = 10^{-3}$ mol ATP l^{-1} and water content of the cell as 75%, discuss the bioenergetic role of pantothenic acid from the aspect of bacterial growth. Molecular weight of ATP is 507.

Solution

$$C_{ATP} = \frac{\text{ATP pool} \times 10^{-3}}{507} \times \frac{\text{moles ATP}}{\text{g cell}}$$

$$= \frac{\text{ATP pool} \times 10^{-3}}{507} \times \frac{0.25}{0.75} = 6.57 \times 10^{-4}\ \text{ATP}_{\text{pool}}\ \text{mol ATP l}^{-1}\ \text{of cell fluid}$$

Table 1.3

Medium	Panthothenic acid (mg.ml^{-1})	$Y_{X/S}$ (g cell mol^{-1} glucose)	ATP pool (mg ATP g^{-1} cell)	C_{ATP} (mol ATP l^{-1})	Q_r/Q_m
Complex	—	7	1.54	1.01×10^{-3}	0.50
Synthetic	5×10^{-3}	6.4	1.55	1.02×10^{-3}	0.50
Synthetic	1×10^{-7}	2.8	3.15	2.07×10^{-3}	0.67
Mineral	5×10^{-3}	4.5	3.55	2.33×10^{-3}	0.70
Mineral	1×10^{-4}	2.0	4.52	2.97×10^{-3}	0.75

Addition of pantothenic acid in synthetic or mineral media decreases the hydrolysis rate and $Y_{X/S}$ is thus increased.

PROBLEM 1.3

An anaerobic culture of *Lactobacillus casei* in a complex medium of peptone and meat extract gave the following values of growth yield $Y_{X/S} = 40.5$ g cell. mol^{-1} of mannitol and $Y_{X/S} = 62$ g cell. mol^{-1} of glucose when mannitol and glucose were used as energy source respectively. Metabolites other than the cells were, per mol mannitol, when used as sole carbon source:

lactate	0.4 mol
acetate	0.22 mol
ethanol	1.29 mol
formate	1.6 mol

When glucose was used instead of mannitol, the metabolites were, per mol glucose:

lactate	0.05 mol
acetate	1.05 mol
ethanol	0.94 mol
formate	1.76 mol

Assuming that enthalpy change of dry cell is 22.175 J g^{-1} cell, assess the value of $Y_{\Delta H}$ for each energy source. Enthalpy change of available electron that takes part in oxidation of each metabolite is assumed to be 110.876 kJ l^{-1} (av.).

Solution

$\Delta H_{av} = -22.175$ kJ g^{-1} cell
$\Delta H_g = -110.876 \times 4 \times 6 = -2661.024$ kJ mol^{-1} glucose
$\Delta H_m = -110.876 \times 4 \times 13/2 = -2882.776$ kJ mol^{-1} mannitol
$\Delta H_l = -110.876 \times 4 \times 3 = -1330.512$ kJ mol^{-1} lactate
$\Delta H_a = -110.876 \times 4 \times 2 = -887.008$ kJ mol^{-1} acetate
$\Delta H_e = -110.876 \times 4 \times 3 = -1330.512$ kJ mol^{-1} ethanol
$\Delta H_b = -110.876 \times 4 \times 1/2 = -221.752$ kJ mol^{-1} formate

Where subscripts denote: av = average; g = glucose; m = mannitol; l = lactate; a = acetate; e = ethanol and f = formate.

For mannitol

$$\Sigma(-H_p)\,Y_{p/s} = 1330.512 \times 0.4 + 887.008 \times 0.22 + 1330.512 \times 1.29 + 221.752 \times 1.6$$
$$= 2798.51 \text{ kJ mol}^{-1} \text{ mannitol}$$

$$\therefore Y_{\Delta H} = \frac{40.5}{(22.175 \times 40.5) + 2882.776 - 2798.51}$$
$$= 0.0412 \text{ g cell kJ}^{-1}$$

For glucose

$$\Sigma(-H_p)\,Y_{p/s} = 1330.512 \times 0.05 + 887.008 \times 1.05 + 1330.512 \times 0.94 + 221.752 \times 1.76$$
$$= 2638.849 \text{ kJ mol}^{-1} \text{ glucose}$$

$$\therefore Y_{\Delta H} = \frac{62}{22.175 \times 62 + 2661.024 - 2638.849}$$
$$= \underline{0.044 \text{ g cell kJ}^{-1}}$$

PROBLEM 1.4

The reaction to produce isocitric acid from α-keto glutarate is shown by the following formula:

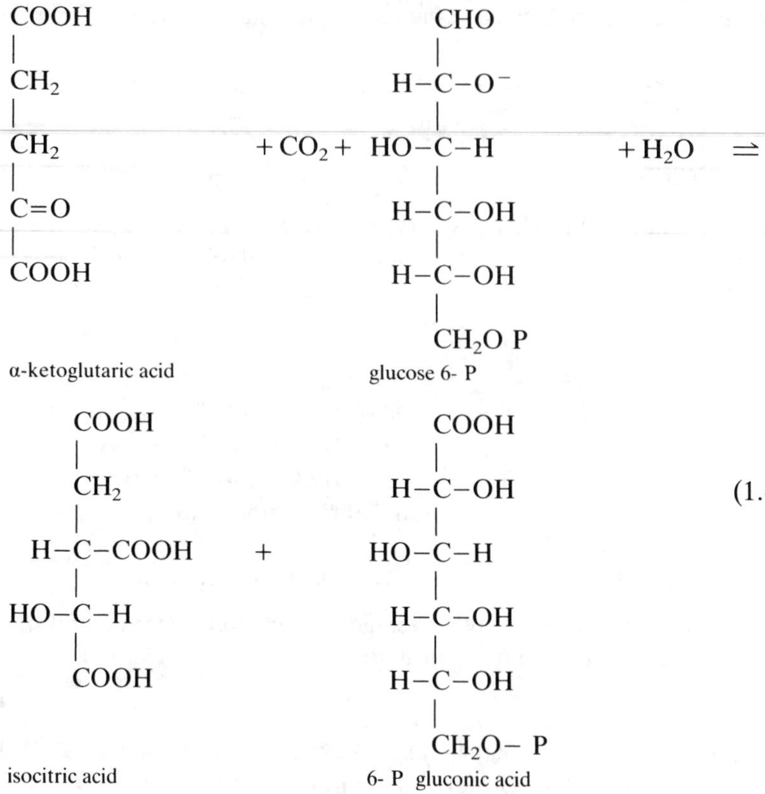

$$
\begin{array}{c}
\mathrm{COOH}\\ | \\ \mathrm{CH_2}\\ | \\ \mathrm{CH_2}\\ | \\ \mathrm{C{=}O}\\ | \\ \mathrm{COOH}\\ \\ \alpha\text{-ketoglutaric acid}
\end{array}
+\mathrm{CO_2}+
\begin{array}{c}
\mathrm{CHO}\\ | \\ \mathrm{H{-}C{-}O^-}\\ | \\ \mathrm{HO{-}C{-}H}\\ | \\ \mathrm{H{-}C{-}OH}\\ | \\ \mathrm{H{-}C{-}OH}\\ | \\ \mathrm{CH_2O\ P}\\ \text{glucose 6- P}
\end{array}
+\mathrm{H_2O} \rightleftharpoons
$$

$$
\begin{array}{c}
\mathrm{COOH}\\ | \\ \mathrm{CH_2}\\ | \\ \mathrm{H{-}C{-}COOH}\\ | \\ \mathrm{HO{-}C{-}H}\\ | \\ \mathrm{COOH}\\ \\ \text{isocitric acid}
\end{array}
+
\begin{array}{c}
\mathrm{COOH}\\ | \\ \mathrm{H{-}C{-}OH}\\ | \\ \mathrm{HO{-}C{-}H}\\ | \\ \mathrm{H{-}C{-}OH}\\ | \\ \mathrm{H{-}C{-}OH}\\ | \\ \mathrm{CH_2O{-}\ P}\\ \text{6- P gluconic acid}
\end{array}
\qquad (1.4.1)
$$

$$\Delta G_1^\circ = -6{,}694.4 \text{ J mol}^{-1}$$

On the other hand, the following reactions are considered for the formation of isocitric acid from α-keto glutaric acid.

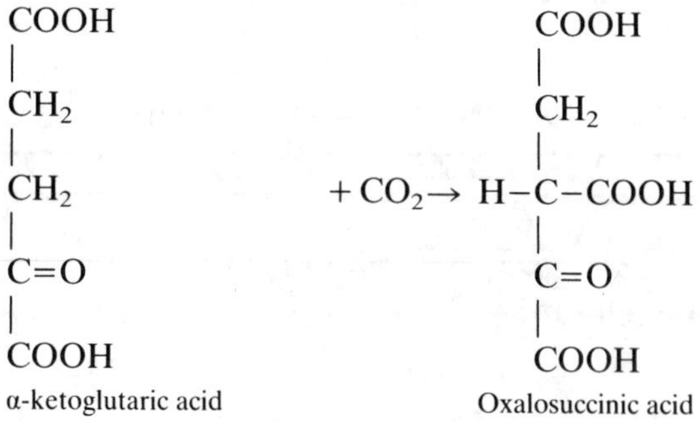

$$
\begin{array}{c}
\mathrm{COOH}\\ | \\ \mathrm{CH_2}\\ | \\ \mathrm{CH_2}\\ | \\ \mathrm{C{=}O}\\ | \\ \mathrm{COOH}\\ \alpha\text{-ketoglutaric acid}
\end{array}
+\mathrm{CO_2}\rightarrow
\begin{array}{c}
\mathrm{COOH}\\ | \\ \mathrm{CH_2}\\ | \\ \mathrm{H{-}C{-}COOH}\\ | \\ \mathrm{C{=}O}\\ | \\ \mathrm{COOH}\\ \text{Oxalosuccinic acid}
\end{array}
\qquad (1.4.2)
$$

$$\Delta G_2^\circ = 19{,}246.4 \text{ J mol}^{-1}$$

and,

$$\text{Oxalosuccinic acid} + \mathrm{NADH} + \mathrm{H^+} \rightarrow \text{isocitric acid} + \mathrm{NAD^+} \qquad (1.4.3)$$

$$\Delta G_3^\circ = 2{,}928.8 \text{ J mol}^{-1}$$

In order to realize the oxidation of glucose-6-phosphate to 6-phosphogluconic acid, what type of coupling mechanism is required amongst (1.4.1), (1.4.2) and (1.4.3)? Calculate the value of ΔG° for the coupled reaction.

Solution
Given,

$$\alpha\text{-ketoglutaric acid} + CO_2 + \text{glucose} - 6\ P\ + H_2O \rightarrow \text{isocitric acid} + 6 -\ P\ \text{gluconic acid} \quad (1.4.1)$$

$$\Delta G_1^\circ = -6{,}694.4 = \text{J mol}^{-1}$$

$$\alpha\text{-ketoglutaric acid} + CO_2 \rightarrow \text{oxalosuccinic acid} \quad (1.4.2)$$

$$\Delta G_2^\circ = 19{,}246.4\ \text{J mol}^{-1}$$

$$\text{oxalosuccinic acid} + NADH + H^+ \rightarrow \text{isocitric acid} + NAD^+ \quad (1.4.3)$$

$$\Delta G_3^\circ = 2{,}928.8\ \text{J mol}^{-1}$$

The coupled reaction for formation of 6-phospho gluconic acid from glucose-6-phosphate will be given by [(1.4.1) – {(1.4.2) + (1.4.3)}], which is:

$$\text{Glucose} - 6\ P\ + H_2O + NAD^+ \rightarrow 6 -\ P\ \text{gluconic acid} + NADH + H^+$$

$$\Delta G_4^\circ = \underline{-28{,}869.6\ \text{J mol}^{-1}}$$

PROBLEM 1.5

The overall reaction of biological oxidation may be represented by:

$$H_2 + \tfrac{1}{2}O_2 \rightleftharpoons H_2O$$

The standard free energy change of this reaction is 236,396 J mol^{-1}. Deduce from this information, the oxidation-reduction potential of the oxygen-activating system that catalyses this final reaction at pH = 7.0 and 30°C. Take the pressure of atmospheric oxygen as 20,265 N m^{-2}. In aqueous solutions the concentration of H_2O is taken as unity.

$$R = 8.314\ \text{J mol}^{-1}\ \text{K}^{-1};\ F = 96{,}487.224\ \text{J V}^{-1}\ \text{mol}^{-1}$$

Solution
The standard free energy change, ΔG°, for the reaction is

$$\Delta G^\circ = -RT \ln K_{eq}$$

$$\therefore K_{eq} = \exp(-\Delta G^\circ/RT) = \exp\left[-\frac{(236{,}396)}{(8.314).(303)}\right] = 5.6773 \times 10^{40}$$

$$K_{eq} = \frac{[H_2O]}{[H_2][O_2]^{1/2}} = \frac{1}{[H_2][O_2]^{1/2}} = \frac{1}{p_{H_2}\, p_{O_2}^{1/2}}$$

$$\therefore p_{H_2} = \frac{1}{p_{O_2}^{1/2}} \times \frac{1}{K_{eq}} = \frac{1}{(20{,}265)^{1/2}} \times \frac{1}{(5.6773 \times 10^{40})}$$

$$= 1.23732 \times 10^{-43}\ \text{N m}^{-2}$$

At pH = 7.0

$[H^+] = 10^{-7}$ mol.l^{-1}

The thermodynamic relationship of redox potential, E_h, is given by:

$$\begin{aligned} E_h &= \frac{RT}{nF} \ln \frac{[H^+]}{[H_2]^{1/2}} \\ &= \frac{8.314 \times 303}{1 \times 96{,}487.224} \times \ln \frac{10^{-7}}{(123{,}732 \times 10^{-43})^{1/2}} \\ &= 0.87\ \text{V} \\ &= \underline{870\ \text{mV}} \end{aligned}$$

PROBLEM 1.6

Assuming that the composition of yeast cell can be given by the following formula, calculate the heat of formation of 100 g of yeast from the heat of combustion of each element.

chemical formula of the cells = $C_{3.92}\,H_{6.5}\,O_{1.94}$

heat of combustion

C = 394.133 kJ mol^{-1}

H_2 = 285.767 kJ mol^{-1}

cell = 1,518.792 kJ mol^{-1}

Solution

Mol. wt. $C_{3.92}\,H_{6.5}\,O_{1.94} = 84.58$ g

Moles of each element in 100 g yeast cell are:

$$\begin{aligned} C &= (3.92 \times 100)/84.58 = 4.63 \text{ moles} \\ H_2 &= (6.5/2 \times 100)/84.58 = 3.84 \text{ moles} \\ O_2 &= (1.94/2 \times 100)/84.58 = 1.15 \text{ moles} \end{aligned}$$

$$4.63C + 3.84H_2 + 1.15O_2 \xrightarrow{Q} \underset{(100\text{ g})}{\text{yeast cells}}$$

The heat of formation of yeast cells becomes therefore:

$$\begin{aligned} \Delta H_f &= 4.63 \times 394.133 + 3.84 \times 285.767 - 1{,}518.792 \\ &= \underline{1{,}403.389\ \text{kJ 100 g}^{-1}\ \text{cell}} \end{aligned}$$

PROBLEM 1.7

A cell possesses phosphorylated high energy compounds which can transfer their phosphate group to ADP to produce ATP. The ATP, which is an intermediate energy compound, in turn is used to phosphorylate a number of substrates like glucose to activate and prepare them for catabolic and anabolic reactions. ATP/ADP cycle is therefore central to meet the energy requirements of a cell. The phosphate group transfer potential of a compound has been defined as '$\Delta G^{\circ\prime}$', which is the standard free energy change which a compound experiences upon transfer of its phosphate group to water or an acceptor compound.

(a) The $\Delta G^{\circ\prime}$ of hydrolysis of ATP to ADP and P_i, at pH 7.0 and 25°C, is $-32{,}216.8$ J mol^{-1}. The $\Delta G^{\circ\prime}$ of hydrolysis of glucose-6-phosphate to glucose and P_i is $-13{,}129.39$ J mol^{-1}. From this information, calculate the value of $\Delta G^{\circ\prime}$ for the reaction between glucose and ATP which is catalysed by hexokinase to produce glucose-6- P and ADP.

$$\text{glucose} + \text{ATP} \rightleftharpoons \text{glucose-6- P} + \text{ADP} \qquad (1.7.1)$$

Also, find out the ratio of ATP/ADP at equilibrium if the ratio of glucose-6- P / glucose is 10^4 at equilibrium.

(b) If the ADP, produced in the above reaction, is to be rephosphorylated to ATP involving phosphoenol pyruvate in the following reaction:

$$\text{PEP} + \text{ADP} \rightleftharpoons \text{ATP} + \text{pyruvate} \qquad (1.7.2)$$

Find the ratio of PEP/pyruvate that would be required to maintain a steady state ratio of ATP/ADP at level determined by equation (1.7.1). $\Delta G^{\circ\prime}$ for PEP is $-61{,}923.2$ J mol^{-1}.

Solution

(a) The overall reaction,

$$\text{glucose} + \text{ATP} \rightleftharpoons \text{glucose-6- P} + \text{ADP} \qquad (1.7.1)$$

can be broken to

$$\text{glucose} + P_i \rightleftharpoons \text{glucose-6- P} + H_2O \qquad \Delta G^{\circ\prime}_1 = 13{,}129.39 \text{ J mol}^{-1}$$

and,

$$\text{ATP} + H_2O \rightleftharpoons \text{ADP} + P_i \qquad \Delta G^{\circ\prime}_2 = -32{,}216.8 \text{ J mol}^{-1}$$

Thus the $\Delta G^{\circ\prime}$ for reaction (1.7.1) is

$$\begin{aligned}\Delta G^{\circ\prime} &= \Delta G^{\circ\prime}_1 + \Delta G^{\circ\prime}_2\\ &= +13{,}129.39 - 32{,}216.8 \text{ J mol}^{-1}\\ &= -19{,}087.41 \text{ J mol}^{-1}\end{aligned}$$

since,

$$\Delta G^{\circ\prime} = -RT \ln K_{eq}$$

$\therefore$

$$\begin{aligned}K_{eq} &= 10^{-\Delta G^{\circ\prime}/RT(2.303)}\\ &= 10^{\{+(19{,}087.41)/(8.314)(298)(2.303)\}}\\ &= 10^{3.345} = 2{,}213.1\end{aligned}$$

or,

$$\frac{[\text{glucose-6- P}][\text{ADP}]}{[\text{glucose}][\text{ATP}]} = 2{,}213.1$$

or,

$$\frac{\text{ADP}}{\text{ATP}} = \frac{2{,}213.1\ [\text{glucose}]}{[\text{glucose-6- P}\ ]}$$
$$= 2.2131 \times 10^{3} \times 10^{-4}$$
$$= 0.22131$$

or,

$$\text{ATP/ADP} = \underline{4.5}$$

(b) For the reaction,

$$\text{PEP} + \text{ADP} \rightleftharpoons \text{pyruvate} + \text{ATP} \qquad (1.7.2)$$

the standard free energy change is

$$\Delta G^{\circ\prime} = -61{,}923.2 + 32{,}216.8$$
$$= -29{,}706.4\ \text{J mol}^{-1}$$

The equilibrium constant for reaction (1.7.2) is:

$$K_{eq} = 10^{(-\Delta G^{\circ\prime}/(2.303)RT)}$$
$$= 10^{((+29{,}706.4)/(2.303)(8.314)(298))}$$
$$= 10^{5.2}$$

or,

$$\frac{[\text{pyruvate}][\text{ATP}]}{[\text{PEP}][\text{ADP}]} = 1.585 \times 10^{5}$$

To maintain the same steady state level of ATP/ADP at 4.5, the ratio of (pyruvate/PEP) will be:

$$\frac{\text{pyruvate}}{\text{PEP}} = \frac{1.585 \times 10^{5}}{4.5}$$
$$= 3.522 \times 10^{4}$$

or,

$$\frac{\text{PEP}}{\text{pyruvate}} = \underline{2.84 \times 10^{-5}}$$

PROBLEM 1.8

The conversion of glucose to lactic acid has an overall $\Delta G^{\circ\prime}$ of $-217{,}568$ J mol^{-1}. In an anaerobic cell, this conversion is coupled to the formation of 2 ATP per mole of glucose

(a) Calculate the $\Delta G^{\circ\prime}$ of the overall reaction. ($\Delta G^{\circ\prime}$ of ATP hydrolysis is $-$ 30,543.2 J mol^{-1}).

(b) What is the efficiency of energy conversion to ATP?

(c) How many moles of ATP per mole of glucose can be obtained in an aerobic organism at 40% efficiency of energy conversion to ATP?

$$\Delta G^{\circ\prime} \text{ of glucose} \rightleftharpoons CO_2 + H_2O \text{ is } -2{,}870{,}224\ \text{J mol}^{-1}.$$

Solution

Given,

$$\text{Glucose} \rightleftharpoons 2 \text{ lactic acid } \Delta G^{\circ\prime}_1 = -217{,}568 \text{ J mol}^{-1}$$
$$2 \text{ ADP} + 2 \text{ P}_i \rightleftharpoons 2 \text{ ATP } \Delta G^{\circ\prime}_2 = +2 \times 30{,}543.2$$
$$= +61{,}086.4 \text{ J mol}^{-1}$$

(a) Overall reaction is:

$$\text{glucose} + 2 \text{ ADP} + 2 \text{ P}_i \rightleftharpoons 2 \text{ lactic acid} + 2\text{ATP}$$

$$\begin{aligned}\Delta G^{\circ\prime} &= \Delta G^{\circ\prime}_1 + \Delta G^{\circ\prime}_2\\ &= -217{,}568 + 61{,}086.4\\ &= \underline{-156{,}481.6 \text{ J mol}^{-1}}\end{aligned}$$

(b) Efficiency of energy conservation $= \dfrac{61{,}086.4}{217{,}568} \times 100$

$$= \underline{28.1\%}$$

(c) For complete combustion of glucose,

$$\text{glucose} \rightleftharpoons 6 \text{ CO}_2 + 6\text{H}_2\text{O}, \qquad \Delta G^{\circ\prime} = -2{,}870{,}224 \text{ J mol}^{-1}$$

Maximum number of ATP which can be synthesized at 100% efficiency

$$= \frac{2{,}870{,}224}{30{,}543{,}2} = 93.97 \text{ moles}$$

At 40% efficiency, ATP synthesized

$$\begin{aligned}&= 0.4 \times 93.97\\ &= 37.58\\ &= \underline{38 \text{ moles.}}\end{aligned}$$

PROBLEM 1.9

The tendency of the electrons to be transferred from a compound depends on its half-cell potential. Under standard conditions of 1M concentration of reactants and products, it depends on the difference in the standard half-cell potentials of the participating redox couples.

(a) Indicate if the following redox reaction will proceed in the direction of acetoacetate formation if the initial concentrations of reactants and products are equal at 25°C.

$$\beta\text{-hydroxybutyrate} + \text{NAD}^+ \rightleftharpoons \text{acetoacetate} + \text{NADH} + \text{H}^+ \quad (1.9.1)$$

Given are the standard half-cell potentials,

$$E^{\circ\prime}\left(\frac{\text{acetoacetate}}{\beta - \text{OH butyrate}}\right) = -0.27 \text{ V}$$

$$E^{o\prime}\left(\frac{NAD^+}{NADH}\right) = -0.32\ V$$

(b) Since the electron transport chain in an organism is coupled to ADP phosphorylation, it is theoretically possible to drive electrons backward up the chain against the potential gradient at the expense of ATP. This type of mechanism has been observed in some organisms such as *Nitrobacter*.

Consider a bacterium at 25°C that can transfer electrons from β-hydroxybutyrate to NAD^+ by means of reverse electron transport (reaction (1.9.1)). If intracellular concentration of P_i is constant at 15 mmol l^{-1}, calculate the ATP/ADP ratio that would be required to maintain the steady-state ratio of $NADH/NAD^+$ and acetoacetate/β-hydroxybutyrate at 100 mmol l^{-1} each.

$\Delta G^{o\prime}$ for ATP hydrolysis $= -30.5432$ kJ mol^{-1}
Faraday constant, F $= 96.232$ kJ V^{-1} mol^{-1}.

Solution

When a reaction takes place under non-standard conditions of temperature and concentration, the potential difference ($\Delta E'$) is related to standard potential difference ($\Delta E^{o\prime}$) by the Nernst equation:

$$\Delta E' = \Delta E^{o\prime} - \frac{RT}{nF}\ln\frac{[\text{acetoacetate}][\text{NADH}]}{[\beta\text{-OH butyrate}][\text{NAD}^+]}$$

(a) At equal concentrations of reactants and products,

$$\begin{aligned}\Delta E' &= \Delta E^{o\prime}\\ &= E^{o\prime}\left(\frac{NAD^+}{NADH}\right) - E^{o\prime}\left(\frac{\text{acetoacetate}}{\beta\text{-OH butyrate}}\right)\\ &= -0.32-(-0.27)\\ &= -0.05\ V\end{aligned}$$

since standard free energy change $\Delta G^{o\prime}$ and standard half-cell potential difference, $\Delta E^{o\prime}$ are related by the relation

$$\begin{aligned}\Delta G^{o\prime} &= -n_e F\,\Delta E^{o\prime}\\ &= -2\times\frac{96.232\ \text{kJ}}{\text{V. mol}}(-0.05\ \text{V})\\ &= +9.6232\ \text{kJ mol}^{-1}\end{aligned}$$

since $\Delta G^{o\prime}$ is +ve, it does not proceed in the direction of acetoacetate formation under these conditions.

(b) The actual free energy change under the intracellular steady-state conditions is given by

$$\Delta G' = \Delta G^{o\prime} + RT\ln\frac{[\text{acetoacetate}][\text{NADH}]}{[\beta\text{-OH butyrate}][\text{NAD}^+]}$$

$$= +9.6232 + \frac{8.314 \times 298}{1000} \times 2.303 \log [(0.1)(0.1)]$$
$$= 1.7885 \text{ kJ mol}^{-1}$$

To balance this reaction, ATP hydrolysis must yield 1.7885 kJ mol^{-1}.

For ATP hydrolysis,

$$\text{ATP} \rightleftharpoons \text{ADP} + \text{P}_i$$

actual free energy change is

$$\Delta G' = \Delta G^{\circ\prime} + RT \ln \frac{[\text{ADP}][\text{P}_i]}{[\text{ATP}]}$$

or,

$$-1.7885 = -30.5432 + \left(\frac{8.314}{1000}\right)(298)(2.303) \log \frac{[\text{ADP}](0.015)}{[\text{ATP}]}$$

or,

$$+28.7547 = 5.7058 \log \frac{\text{ADP}}{\text{ATP}} + 5.7058 \log (1.5 \times 10^{-2})$$

or,

$$\frac{\text{ATP}}{\text{ADP}} = 1.3 \times 10^{-7}$$

Thus, the ATP/ADP ratio should be maintained at 1.3×10^{-7} to get the steady-state values of reactants and products.

PROBLEM 1.10

The actual redox potential of a redox couple depends on the concentrations of the reductant and oxidant. The redox potential is given by $E_h = E_0' + \frac{RT}{nF} \ln \frac{[\bar{e}\text{ acceptor}]}{[\bar{e}\text{ donor}]}$ where E_h is the actual redox potential, E_0' is the standard redox potential, R is the gas constant, F is the Faraday constant and n is the number of electrons transferred. Consider the transfer of electrons from succinate to NAD^+ and FAD to give fumarate and reduced forms of NADH and $FADH_2$ (E-FAD represents enzyme-bound FAD).

$$\text{succinate} + \text{E-FAD} \rightleftharpoons \text{fumarate} + \text{E-FADH}_2 \quad (1.10.1)$$

$$\text{succinate} + \text{NAD}^+ \rightleftharpoons \text{fumarate} + \text{NADH} + \text{H}^+ \quad (1.10.2)$$

The standard redox potentials of the three couples are:

$$E_0'\left(\frac{\text{fumarate}}{\text{succinate}}\right) = +0.03 \text{ V}$$

$$E'_0\left(\frac{\text{E-FAD}}{\text{E-FADH}_2}\right) = -0.06\ \text{V}$$

$$E'_0\left(\frac{\text{NAD}^+}{\text{NADH}}\right) = -0.32\ \text{V}$$

Under actual intracellular conditions, assume that the initial concentrations of NAD, NADH, FAD, $FADH_2$ are each at 0.01M, and fumarate and succinate are present at a level at which E_h is -0.4 V.

(a) Calculate the ratio of initial concentration of fumarate/succinate.

(b) Which of the two reactions will have the greater tendency towards fumarate formation?

(c) Find how many fold will the fumarate/succinate ratio be for reaction (1.10.1) compared to that of reaction (1.10.2) at equilibrium.

Solution

(a) The Nernst equation for a redox reaction of general type is given by

$$A_{red} \rightleftharpoons A_{ox} + e^-$$

is

$$E_h = E'_0 + \frac{RT}{nF} \ln \frac{[A_{ox}]}{[A_{red}]}$$

where, A_{ox} and A_{red} represent oxidized (electron-accepting) and reduced (electron donating) forms of substrate, A.

At 25°C, $T = 298$ K, $n = 2$, as two electrons are transferred in the reactions (1.10.1) or (1.10.2) $F = 96{,}232$ J V^{-1} mol^{-1} and $R = 8.314$ J mol^{-1} K^{-1}

$$E_h = E'_0 + \frac{8.314 \times 298 \times 2.303}{2 \times 96232} \log \frac{A_{ox}}{A_{red}}$$

or,

$$E_h = E'_0 + 0.03 \log \frac{A_{ox}}{A_{red}} \qquad (1.10.3)$$

The actual potential can be calculated from equation (1.10.3).

Initially,

$$E_h \text{ for NAD}^+/\text{NADH} = E'_0 = -0.32\ \text{V}$$

$$E_h \text{ for FAD/FADH}_2 = E'_0 = -0.06\ \text{V}$$

For the fumarate/succinate couple,

$$E_h = E'_0 + 0.03 \log \frac{[\text{fumarate}]}{[\text{succinate}]}$$

or,

$$-0.4 = +0.03 + 0.03 \log \frac{[\text{fumarate}]}{[\text{succinate}]}$$

or,

$$\log \frac{[\text{fumarate}]}{[\text{succinate}]} = -\frac{0.43}{0.03} = -14.33$$

or,

$$\frac{\text{succinate}}{\text{fumarate}} = 2.15 \times 10^{14}$$

or,

$$\frac{\text{fumarate}}{\text{succinate}} = \underline{4.65 \times 10^{-15}}$$

(i.e. for each molecule of fumarate, 2.15×10^{14} molecules of succinate must be present so as to have $E_h = -0.4$ V.

(b) The tendency of electron transfer in an overall reaction,

$$A_{ox} + B_{red} \rightleftharpoons A_{red} + B_{ox}$$

is given by $\Delta E_0'$ which is computed by subtracting the half-cell potential ($\Delta E_0'$) of the half-cell undergoing oxidation from the half-cell undergoing reduction.

For reaction (1.10.1),

$$\Delta E_0' = \Delta E_0' \left(\frac{\text{E}-\text{FAD}}{\text{E}-\text{FADH}_2}\right) - \Delta E_0' \left(\frac{\text{fumarate}}{\text{succinate}}\right)$$
$$= -0.06 - 0.03$$
$$= -0.09 \text{ V}$$

For reaction (1.10.2),

$$\Delta E_0' = \Delta E_0' \left(\frac{\text{NAD}^+}{\text{NADH}}\right) - \Delta E_0' \left(\frac{\text{fumarate}}{\text{succinate}}\right)$$
$$= -0.32 - 0.03$$
$$= \underline{-0.35 \text{ V}}$$

since $\Delta E_0'$ is negative (meaning $\Delta G_0'$ is positive), none of the reactions is likely to proceed in the direction of fumarate formation at these conditions. However, as $\Delta E_0'$ for reaction (1.10.1) is less negative, it has a greater tendency to proceed in the direction of fumarate formation than reaction (1.10.2).

(c) At equilibrium, the observed potential, E_h, of the two couples will be the same, i.e. mid-point of their initial potential.

For reaction (1.10.1),

$$E_h = \frac{-0.4 - 0.06}{2} = -0.23$$

From the Nernst equation,

$$-0.23 = +0.03 + 0.03 \log \frac{[\text{fumarate}]}{[\text{succinate}]}$$

or,

$$\log \frac{[\text{fumarate}]}{[\text{succinate}]} = -8.67$$

$$\frac{\text{succinate}}{\text{fumarate}} = 10^{8.67} = 4.67 \times 10^8$$

or,

$$\frac{\text{fumarate}}{\text{succinate}} = 2.14 \times 10^{-9}$$

For reaction (1.10.2),

$$E_h = \frac{-0.4 - 0.32}{2} = -0.36$$

From the Nernst equation,

$$-0.36 = +0.03 + 0.03 \log \frac{[\text{fumarate}]}{(\text{succinate}]}$$

or,

$$\log \frac{[\text{fumarate}]}{[\text{succinate}]} = -13$$

or,

$$\frac{\text{fumarate}}{\text{succinate}} = 10^{-13}$$

At equilibrium, the fold increase in fumarate/succinate concentration is, for reaction (1.10.1) $= \frac{2.14 \times 10^{-9}}{4.65 \times 10^{-15}} = 4.6 \times 10^5$ and for reaction (1.10.2) $= \frac{10^{-13}}{4.65 \times 10^{-15}} = 21.5$.

Thus the magnitude of increase in fumarate/succinate ratio for reaction (1.10.1) compared to that of reaction (1.10.2) is $= \frac{4.6 \times 10^5}{21.5} = 2.14 \times 10^4$ fold at equilibrium.

PROBLEM 1.11

The distribution of calcium ions across the membrane of a cell was found experimentally to give a concentration 80-fold greater inside than outside at 25°C. The measured potential difference across the membrane is -76.5 mV. Ascertain from these measurements whether active transport of calcium ions is likely to occur and calculate the amount of energy required to be expended for this transport to occur.

Solution

The transport of ionic nutrients across biological membranes depends upon physical forces determined by concentration gradient, and by electrical forces as well. The interior of most living cells is found to be electrically negative relative to the surrounding medium. Consequently, there will be a tendency for positively charged ions (cations) to move down the electric potential gradient into the cell. If this potential difference across the cell membrane is sufficiently great, it may overcome a concentration gradient and thereby permit an ion to move from an external dilute solution to the concentrated cell cytoplasm without increasing its free energy.

The electrochemical potential of the calcium ions is a function of both its chemical and electrical potentials, i.e.

$$\bar{E} = E^* + RT\,c + ZF\psi \tag{1.11.1}$$

where, E^* is the electrochemical potential at its standard state and c is the concentration (equated to its activity), Z is the valency, F is Faradays constant and ψ is the electric potential of the system containing the ions.

When the ions are at same $\bar{E}$ on either side of the cell membrane, i.e. $\bar{E}^{\mathrm{o}} = \bar{E}^{\mathrm{i}}$, (the superscripts o and i refer to outside and inside of the membrane), the system is under passive flux equlibrium.

At

$$\bar{E}^{\mathrm{o}} = \bar{E}^{\mathrm{i}} \tag{1.11.2}$$

$$RT\,c^{\mathrm{o}} + ZF\psi^{\mathrm{o}} = RT\,c^{\mathrm{i}} + ZF\psi^{\mathrm{i}} \tag{1.11.3}$$

and

$$\psi^{\mathrm{i}} - \psi^{\mathrm{o}} = \frac{RT}{ZF} \ln \frac{c^{\mathrm{o}}}{c^{\mathrm{i}}} = E^{\mathrm{N}} \text{ (Nernst potential)} \tag{1.11.4}$$

The *Nernst potential* is the electric potential difference required to maintain an asymmetric distribution of an ion across a membrane at equilibrium.

When the system is not under passive flux equilibrium,

$$\Delta\bar{E} = \bar{E}^{\mathrm{i}} - \bar{E}^{\mathrm{o}} \tag{1.11.5}$$

$$= (RT \ln c^{\mathrm{i}} + ZF\psi^{\mathrm{i}}) - (RT \ln c^{\mathrm{o}} + ZF\psi^{\mathrm{o}}) \tag{1.11.6}$$

$$= ZF(\psi^{\mathrm{i}} - \psi^{\mathrm{o}}) - RT \ln \frac{c^{\mathrm{o}}}{c^{\mathrm{i}}} \tag{1.11.7}$$

$$= ZF(\psi^{\mathrm{i}} - \psi^{\mathrm{o}}) - ZFE^{\mathrm{N}} \tag{1.11.8}$$

In equation (1.11.4) $\psi^{\mathrm{i}} - \psi^{\mathrm{o}}$ is the electric potential difference (E_{M}) across the membrane.

$$\bar{E}^{i} - \bar{E}^{o} = ZF(E_M - E^N) \quad (1.11.9)$$

The difference between E_M and E^N (ΔE) ascertains the operation of an active transport process. If the measured potential is the same as the predicted E^N, then a passive equilibrium situation exists. For a significant difference between E_M and E^N, energy must be expended to maintain the non-equilibrium state and active transport will occur.

Now, in the present case:

$$E^N = \frac{2.303\ RT}{ZF} \log \frac{[\mathrm{Ca}]^o}{[\mathrm{Ca}]^i}$$

$T = 298$ K and $Z = 2$

$$E^N = \frac{2.303 \times 8.314 \times \times 298}{2 \times 96.5} \log \frac{1}{80}$$
$$= -56.25 \text{ mV}$$

Since the measured potential difference; (E_M), across the membrane is -76.5 mV, active transport is occurring and the energy requirement is proportional to the difference $(E_M - E^N) = -20.25$ mV. Converting the ΔE into energy required ($1\text{ mV} = 96.5\text{ J mol}^{-1}$), one gets

$$\text{Energy requirement} = 20.25 \times 96.5$$
$$= \underline{1954\text{ J mol}^{-1}}.$$

2

Mass and energy balance

Contributors: Tarun Ghose, Purnendu Ghosh, V. S. Bisaria

PROBLEM 2.1

Yeast cells were cultivated by using glucose as the sole carbon source. An overall yield of the yeast cells (dry matter) harvested was 50% based on glucose consumed. The following empirical formula may be assumed to deal with the cell production:

$$6.67\ CH_2O + 2.10\ O_2 \rightarrow C_{3.92}H_{6.5}O_{1.94} + 2.75\ CO_2 + 3.42\ H_2O$$

Calculate the heat of reaction from the formula shown above provided

$$\text{heat of combustion of dry cells} = 1.517 \times 10^3\ \text{kJ}\ 100\ \text{g}^{-1}$$
$$\text{heat of combustion of glucose} = 2.817 \times 10^3\ \text{kJ mol}^{-1}.$$

Solution

For one mole of yeast cell

$$\text{glucose required} = 6.67 \times 30$$
$$= 200\ \text{g} = 1.11\ \text{moles}$$

$$\text{Yield of cells} = \frac{\Delta x}{\Delta s} = 0.5$$

$$\therefore \text{Yeast cell mass} = 200 \times 0.5 = 100\ \text{g}$$

$$\therefore \Delta H_r = 2.817 \times 10^3 \times \frac{6.67}{6} - 1.517 \times 10^3 \times \frac{100}{100}$$

$$= \underline{1.613 \times 10^3\ \text{kJ}\ 200\ \text{g}^{-1}\ \text{glucose (100 g cells)}}$$

PROBLEM 2.2

Pseudomonas ovalis is used to convert glucose to gluconic acid via gluconolactone in batch culture, following first order consecutive reactions. Designating k_1 and k_2 to the reaction rate constants, derive the time, θ_{max}, at which the concentration of the intermediate (gluconolactone) is maximized and also, derive the maximum concentration $(c_L)_{max}$ of gluconolactone, provided that concentrations of both the intermediate product and the final product (gluconic acid) at $\theta = 0$ are assumed to be zero.

Solution

Consider the consecutive, unimolecular, first order reaction,

$$\underset{\text{glucose}}{G} \xrightarrow{k_1} \underset{\text{gluconolactone}}{L} \xrightarrow{k_2} \underset{\text{gluconic acid}}{P}$$

Rate equation for gluconolactone

$$\frac{dc_L}{d\theta} = k_1 c_G - k_2 c_L$$

$$= k_1 C_{G_0} e^{-k_1\theta} - k_2 c_L$$

or,

$$\frac{dc_L}{d\theta} + k_2 c_L - k_1 c_{G_0} e^{-k_1\theta} = 0$$

$$C_L = \frac{k_1}{k_2 - k_1} c_{G_0} (e^{-k_1\theta} - e^{-k_2\theta})$$

Values of $k_1$and k_2 govern the maximum concentration of L. This may be found at $\frac{dc_L}{d\theta} = 0$; thus,

$$\theta_{max} = \frac{\ln (k_1/k_2)}{k_1 - k_2}$$

At $\theta = \theta_{max}$,

$$(c_L)_{max} = \frac{k_1}{k_2 - k_1} c_{G_0} (e^{-k_1\theta_{max}} - e^{-k_1\theta_{max}})$$

$$= \left(\frac{k^1}{k_2}\right) \frac{k_2}{k_2 - k_1}$$

PROBLEM 2.3

Clostridium acetobutylicum carries out anaerobic fermentation and converts glucose into acetone, butanol along with smaller concentrations of butyrate, acetate, etc. In a fermentation, the following products were obtained from 100 moles of glucose and 11.2 moles of NH_3 as nitrogen source.

Products formed	*Moles*
cells =	13
butanol =	56
acetone =	22
butyric acid =	0.4
acetic acid =	14
CO_2 =	221
H_2 =	135
ethanol =	0.7

(a) By performing a carbon, nitrogen, hydrogen and oxygen balance, determine the elemental composition of the cells.

(b) Find out the redox status of the fermentation using available electron balance principles.

(c) *Example of unbalanced fermentation*
By calculating the available electron balance, it is possible to find if the fermentation is balanced with respect to its redox status. Consider an acetone–butanol fermentation using the same *Clostridium* sp. which yields the following products from 100 moles of glucose and 11.2 moles of ammonia.

Products	*Moles formed*
cells =	13
butanol =	68
acetone =	22
butyric acid =	0.4
acetic acid =	14
CO_2 =	221
H_2 =	135
ethanol =	0.7

Calculate the excess or shortfall of available electrons in the products for 100 moles of glucose.

(d) *Calculation of products in homofermentative pathway*
The concept of available electron balance can also be used to find out the amount

of products which are falling short or are in excess in a fermentation yielding a single major product, such as in lactic acid fermentation.

In a homolactic acid fermentation using *Streptococcus* sp., 70% of the available electrons contained in 100 moles of glucose were found in the main product lactic acid and 10% in other minor products. Assuming that the error has been made in the estimation of the lactic acid only, calculate the moles of lactic acid which have been underestimated to balance the oxidation–reduction status of the fermentation.

Solution

(a) The fermentation balance can be writen as

$$100\ C_6H_{12}O_6 + 11.2\ NH_3 \rightarrow 13\ C_aH_bO_cN_d + 56\ C_4H_{10}O_{(butanol)}$$
$$+ 22\ C_3H_6O_{\ (acetone)} + 0.4\ C_4H_8O_{2\ (butyrate)}$$
$$14\ C_2H_4O_{2\ (acetate)}$$
$$+ 221\ CO_2 + 135\ H_2 + 0.7\ C_2H_6O_{\ (ethanol)} \qquad (2.3.1)$$

Where $C_aH_bO_cN_d$ represents the elemental composition of the *Clostridium* cells. Taking balances on C, H, O and N, the following relationships are obtained which yield the values of a, b, c and d.

(i) *Carbon balance*

$$100 \times 6 = 13a + (56 \times 4) + 22 \times 3) + (0.4 \times 4) + (14 \times 2) + 221 + (0.7 \times 2)$$
$$600 = 13a + 542$$
$$\therefore a = 4.46$$

(ii) *Hydrogen balance*

$$(11.2 \times 3) + (100 \times 12) = 13b + 560 + 132 + 3.2 + 56 + 270 + 4.2$$
$$\therefore b = 16.01$$

(iii) *Oxygen balance*

$$600 = 13c + 56 + 22 + 0.8 + 28 + 442 + 0.7$$
$$\therefore c = 3.88$$

(iv) *Nitrogen balance*

$$11.2 = 13d \qquad \therefore d = 0.86$$

Thus the elemental composition of the cells is

$$\underline{C_{4.46}H_{16.01}O_{3.88}N_{0.86}}$$

(b) The degree of reductance (γ) of a compound $C_aH_bO_cN_d$ can be calculated from following relation

$$\gamma = 4a + b - 2c - 3d \qquad (2.3.2)$$

Thus for the fermentation equation (2.3.1), υ values for the reactants and products are shown in Table 2.1.

Table 2.1

Compound	γ mol^{-1} compound	γ 100 moles^{-1} glucose
$C_6H_{12}O_6$	24	2,400
NH_3	0	0
$C_4H_{10}O$	24	1,344
C_3H_6O	16	352
$C_4H_8O_2$	20	8
$C_2H_4O_2$	8	112
CO_2	0	0
H_2	2	270
C_2H_6O	12	8.4
$C_{4.46}H_{16.01}O_{3.88}N_{0.86}$	23.51	305.63

Total available electrons in products per 100 moles glucose = 2400.03
Total available electrons in reactants = 2400

Thus the fermentation is balanced with respect to its oxidation–reduction status.

(c) Calculation of the degree of reductance (v) of the products and the reactants yield the values shown in Table 2.2 for 100 moles glucose.

Table 2.2

Compound	γ mol^{-1} glucose	γ 100 mol^{-1} glucose
Reactants		
Glucose	24	2,400
NH_3	0	0
Products		
cells	23.51	305.63
butanol	24	1,632
acetone	16	352
butyrate	20	8
acetate	8	112
CO_2	0	0
H_2	2	270
ethanol	12	8.4

Total available electrons in products per 100 moles glucose = 2,688.03
Total available electrons in reactants per 100 moles glucose = 2,400
Excess electrons in products per 100 moles glucose = 288.03

Thus the fermentation is unbalanced. Since the fermentation yields a mixture of products, it is theoretically not possible to pinpoint the product(s) which have been under- or over-estimated.

(d) The degree of reductance (v) of the substrate, glucose and the main product, lactic acid are as follows:

$$\gamma_{\text{glucose}} = 24$$
$$\gamma_{\text{lactate}} = 12$$

Available electrons in 100 moles glucose = 2400
Available electrons in lactic acid per 100 moles glucose = $0.7 \times 2400 = 1680$
Since 10% of the electrons are contained in other minor products, the shortfall of electrons in lactic acid

$$= 0.9(2400) - 1680 = 480$$

Thus the moles of lactic acid which can contain these available electrons

$$= \frac{480}{\gamma_{\text{lactate}}} = \frac{480}{12} = 40 \text{ moles}$$

Thus, 20% of the available electrons contained in glucose can be accounted for by 40 moles of lactic acid per 100 moles of glucose. This amount of lactic acid was in shortfall to balance the oxidation–reduction status of the fermentation.

PROBLEM 2.4

Brevibacterium ($C_8H_{13}O_4N$) is grown in a medium containing glucose for the production of lysine ($C_6H_{14}O_2N_2HCl$). The biomass growth rate and lysine production rate are 0.26 kg m^{-3} h^{-1} and 0.375 kg m^{-3} h^{-1} respectively. 60% of the substrate energy is released as heat. Estimate the fraction of substrate used for biomass and lysine production, yield of cell mass based on glucose and oxygen ($Y_{X/S}$, $Y_{X/O}$) and heat evolution rate.

Solution

The general balance equation for microbial growth can be written as (Erickson *et al.*, 1978):

$$CH_mO_l + aNH_3 + bO_2 \rightarrow y_c CH_pO_nN_q + zCH_rO_sN_t + (1 - y_c - z)\, CO_2 + CH_2O \tag{2.4.1}$$

where CH_mO_l is organic substrate, $CH_pO_nN_q$ is biomass and $CH_rO_sN_t$ is product, y_c is biomass carbon yield (fraction of organic substrate carbon in biomass).

Taking the balance of available electrons in this equation gives:

$$\nu_s + b(-4) = y_c \nu_b + z\nu_p \tag{2.4.2}$$

or,

$$\frac{4b}{\nu_s} + y_c\frac{\nu_b}{\nu_s} + \frac{z\nu_p}{\nu_s} = 1.0 \tag{2.4.3}$$

$$\varepsilon + \eta + \xi_p = 1.0 \tag{2.4.4}$$

where ν_s, ν_b and ν_p are the number of equivalents of available electrons per g atom carbon of substrate, biomass and product respectively.

In equation (2.4.3), the first term (ε) is the fraction of available electrons transferred to O_2, the second term (η) is the fraction of available electrons transferred to biomass and the third term (ξ_p) is the fraction of available electrons transferred to product.

Now growth rate $\frac{\Delta x}{\Delta t} = 0.26$ kg m^{-3} h^{-1}. In terms of available electrons,

$$1000 \frac{\Delta x}{\Delta t} \frac{\sigma_b \nu_b}{12} \frac{\text{available electrons for growth}}{\text{m}^3 \text{ h}}$$

where σ_b is the weight fraction of carbon in the cell

$$\sigma_b = \frac{8 \times 12}{8 \times 12 + 13 + 16 \times 4 + 14} = 0.51$$

$$\nu_b = \frac{8 \times 4 + 13 \times 1 - 4 \times 2 - 3 \times 1}{8} = 4.25$$

Available electron for

$$\text{growth} = \frac{0.26 \times 0.51 \times 4.25}{12} \times 1000 = 47$$

similarly, available electron for product

$$= \frac{\Delta c_p}{\Delta t} \cdot \frac{\sigma_p . \nu_p}{12}$$

$$= \frac{0.375 \times 0.395 \times 4.67}{12} \times 1000$$

$$= 57.6.$$

Since 60% of the available electrons in substrate is released as heat, the remaining 40% is used for growth and product formation.

Fraction of substrate used for growth

$$= \frac{47}{47 + 57.6} \times 0.40$$

$$= 0.18$$

Fraction of substrate used for product

$$= (0.40 - 0.18)$$

$$= 0.22$$

The energetic yield coefficient, η, is related to yield based on substrate, $Y_{X/S}$ as,

$$\eta = \left(\frac{\sigma_b \nu_b}{\sigma_s \nu_s}\right) Y_{X/S}$$

$$0.18 = \left(\frac{0.51 \times 4.25}{0.4 \times 4}\right) Y_{X/S}$$

$$Y_{X/S} = 0.133 \text{ g biomass g}^{-1} \text{ glucose}$$
$$= 23.94 \text{ g mol}^{-1}$$

Similarly, the yield of biomass based on oxygen, $Y_{X/O}$, is related to η as,

$$Y_{X/O} = \frac{3\eta}{2\sigma_b \nu_b (1 - \eta - \xi_p)}$$

$$Y_{X/O} = \frac{3 \times 0.18}{2 \times 0.51 \times 4.25\ (1 - 0.18 - 0.22)}$$

$$= 0.20 \text{ g cell g}^{-1}\ O_2$$

$$= 6.4 \text{ g mol}^{-1}$$

Equivalent electron released $= \dfrac{47}{0.18} \times 0.60$
as heat (60% is released as heat) = 157.

Experimentally oxygen consumption and heat evolution are observed to be the same. The heat of reaction per electron transferred to oxygen (Q_O) is assumed to be constant for all organic molecules. Q_O is approximately 112.968 kJ/g equivalent available electrons transferred to oxygen. Thus heat evolution rate

$$= 157 \times 112.968$$

$$= \underline{17{,}736 \text{kJ m}^{-3} \text{ h}^{-1}}$$

Reference

Erickson, L. E., Selga, S. E. and Viesturs, U. E. (1978), *Biotechnol, Bioeng.*, **20**, 1623.

PROBLEM 2.5

Aerobic culture of *Saccharomyces agalactiae* using pyruvic acid as energy source resulted in the following scheme of metabolites.

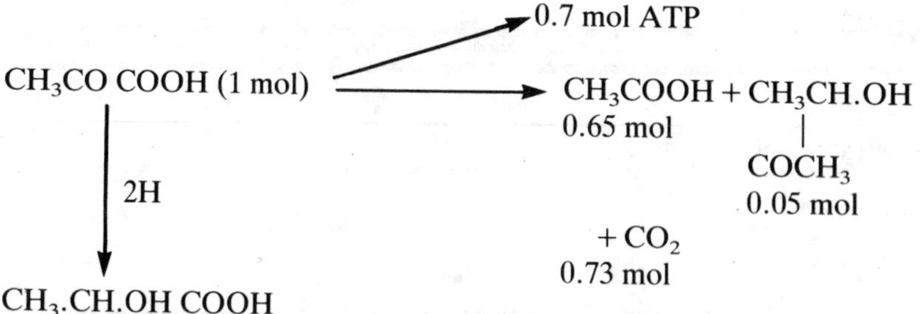

From the computed value of $Y_{X/S} = 12.4$ g dry cell mass mol^{-1} pyruvic acid and oxygen consumption = 0.2 mol O_2 mol^{-1} pyruvic acid, estimate the net generation of ATP per mol of O_2 (P/O). Given $Y_{ATP} = 10$ g dry cell mass mol^{-1} ATP.

Solution

0.7 mol ATP mol^{-1} pyruvate

or,

$$0.7 \times 10 = 7 \text{ g dry cell mol}^{-1} \text{ pyruvate}$$

Also given

$$Y_{X/S} = 12.4 \text{ g dry cell mol}^{-1} \text{ pyruvate}$$

Thus, the net yield of cell per mol of pyruvate = (12.4 − 7.0) = 5.4 g dry cell mol^{-1} pyruvate and $Y_{X/O} = \frac{5.4}{0.2} = 27 \text{ g dry cell mol}^{-1} O_2$

$$Y_{ATP} = 10 \text{ g dry cell mol}^{-1} \text{ ATP}$$

or,

$$\frac{27}{10} = 2.7 \text{ mol ATP mol}^{-1} O_2$$

$$\therefore P/O = \frac{2.7}{2} = \underline{1.35 \text{ mol ATP g atom}^{-1} O_2}.$$

PROBLEM 2.6

Aerobic culture of *Saccharomyces cerevisae* in a synthetic medium produced the following:

$$\mu = 0.2 \text{ h}^{-1}; \qquad Y_{X/S} = 90 \text{ g dry cell (g. glucose)}^{-1}$$

$$RQ = 1.0$$

$$\nu_{glutamate} = 0.043 \text{ g glutamate (g cell)}^{-1} \text{ h}^{-1}$$

Assuming that the carbon content of the cell is 45%, check carbon and oxygen balance with respect to the given culture.

Solution

$$aC_6H_{12}O_6 + bC_5H_9O_4N + cO_2 \rightarrow dCH_\alpha O_\beta N_\gamma + eCO_2 + fH_2O$$
(glutamic acid)

$$RQ = 1.0 \Rightarrow Q_{O_2} = Q_{CO_2} \Rightarrow \underline{c = e}.$$

$\mu = 0.2 \text{ h}^{-1}$ and 45% C in biomass.

$$\therefore d = \frac{0.2 \times 0.45}{12} = 7.5 \times 10^{-3} \text{ mol C (g dry cell)}^{-1} \text{ h}^{-1}$$

One molecular weight cell biomass contains 12 g C

100 g cell biomass contains 45 g C (45% C)

$$\therefore \text{Mol. weight} = \frac{12 \times 100}{45} = \underline{26.67\text{g}}$$

Again,

$$MW = 26.67 = 12 + \alpha(1) + \beta(16) + \gamma(14)$$

$$14.67 = \alpha + 16\beta + 14\gamma \qquad (2.6.1)$$

$$a = q_{glucose} = \frac{\mu}{Y_{X/S}} = \frac{0.2}{90} = 2.22 \times 10^{-3} \text{ mol glucose g}^{-1} \text{ (dry cell) h}^{-1}$$

$$b = q_{glut} = \frac{0.043}{147} = 0.29 \times 10^{-3} \text{ mol glut g}^{-1} \text{ (dry cell) h}^{-1}$$

Elemental balance:

C:	$6a + 5b = d + e$	(2.6.2)
H:	$12a + 9b = \alpha d + 2f$	(2.6.3)
O:	$6a + 4b = \beta d + f$ (because $2c = 2e$)	(2.6.4)
N:	$b = \gamma d$	(2.6.5)

From (2.6.2)

$$e = 6a + 5b - d = 6 \times (2.2 \times 10^{-3}) + 5 \times 0.29 \times 10^{-3} - (7.15 \times 10^{-3})$$

From (2.6.5)

$$\gamma = \frac{b}{d} = \frac{0.29 \times 10^{-3}}{7.5 \times 10^{-3}} = 0.0387$$

Combining (2.6.3) and (2.6.4):

$$b = \alpha d - 2\beta d = 0.29 \times 10^{-3} \qquad (2.6.6)$$

From (2.6.1), $14.66d = \alpha d + 16\beta d + 14\gamma d = 110 \times 10^{-3}$

$$106 \times 10^{-3} = \alpha d + 16\beta d \qquad (2.6.7)$$

Combining (2.6.6) and (2.6.7) and solving:

$$\beta = \frac{105.71 \times 10^{-3}}{18 \times 7.5 \times 10^{-3}} = 0.78$$

From (2.6.6),

$$\alpha ds = b + 2\beta d \Rightarrow \alpha = \frac{b + 2\beta d}{d} = \frac{0.29 \times 10^{-3} + 2 \times 0.78 \times 7.5 \times 10^{-3}}{7.5 \times 10^{-3}}$$

$$= \underline{1.6}$$

From (2.6.4),

$$f = 6a + 4b - \beta d = 6 \times 2.22 \times 10^{-3} + 4 \times 0.29 \times 10^{-3} - 0.78 \times 7.5 \times 10^{-3}$$

$$= \underline{8.6 \times 10^{-3}}$$

The value of the coefficient $c(=e)$ is computed from (2.6.2).

Therefore the final balance equation becomes:

$$2.22\ C_6H_{12}O_6 + 0.29\ C_5H_9O_4N + 7.27\ O_2 \rightarrow 7.5\ CH_{1.6}O_{0.78}N_{0.0387}$$
$$+ 7.27\ CO_2 + 8.6\ H_20$$

$$\frac{c \text{ in products}}{c \text{ in reactants}} = \frac{7.5 + 7.27}{(2.22 \times 6) + (0.29 \times 5)} = \frac{14.77}{14.77} = 1.$$

PROBLEM 2.7

The composition of macromolecules and their ATP requirement in a bacterial cell is shown in Table 2.3. What will be the distribution of biosynthetic energy for macromolecule synthesis? If the bacteria has a generation time of 20 minutes, estimate the specific rate of ATP production and efficiency of ATP utilization if Y_{ATP} (g cells mol^{-1} ATP generated) is 10.5.

Table 2.3

Macromolecule	wt%	ATP requirement (mol g^{-1} macromolecule)
DNA	5	330×10^{-4}
RNA	10	373.2×10^{-4}
Protein	70	391.1×10^{-4}
Polysaccharides	5	123.6×10^{-4}
Lipids	10	114.8×10^{-4}

Solution

Considering 100 g of bacterial cells, ATP requirement will be as given in Table 2.4.

Table 2.4

Polysaccharide	=	$5 \times 123.6 \times 10^{-4} =$	618×10^{-4} mol
Protein	=	$70 \times 391.1 \times 10^{-4} =$	$27{,}377 \times 10^{-4}$ mol
Lipid	=	$10 \times 114.8 \times 10^{-4} =$	$1{,}148 \times 10^{-4}$ mol
RNA	=	$10 \times 373.2 \times 10^{-4} =$	$3{,}732 \times 10^{-4}$ mol
DNA	=	$5 \times 330.0 \times 10^{-4} =$	$1{,}650 \times 10^{-4}$ mol
		Total	$= 34{,}525 \times 10^{-4}$ mol

Table 2.5 gives the percent of biosynthetic energy used for individual macromolecules. It indicates that nearly 80% of the total energy is used for protein synthesis and 20% is used for other macromolecular synthesis activities.

Table 2.5

Polysaccharide	1.8
Protein	79.3
Lipid	3.3
RNA	10.8
DNA	4.7

Moles of ATP used per g of bacterial cell

$$= \frac{34{,}525 \times 10^{-4}}{100} = 345 \times 10^{-4}$$

Moles of ATP generated per g of bacterial cell synthesized $= \dfrac{1}{10.5}$

Efficiency of ATP utilization

$$= 345 \times 10^{-4} \times 10.5 \times 100$$

$$= \underline{36.225\%}$$

PROBLEM 2.8

Saccharomyces cerevisiae is grown in a 100-litre bioreactor at a dilution rate such that glucose and cell concentration at the reactor outlet are 8 and 20 kg m^{-3} respectively. The inlet feed glucose concentration is 50 kg m^{-3}. In order to design the cooling arrangement for the bioreactor, estimate the rate of heat generation from the system. Based on the available data suggest a balance equation for microbial growth and compute RQ (respiratory quotient) and cell yield based on oxygen consumed. Assume that the organism follows Monod's growth kinetics and $\mu_m = 0.5\,\text{h}^{-1}$; $K_s = 2$ kg m^{-3}.

Solution

The rate of heat generation, Q is

$$Q = VDX \frac{1}{Y_\Delta} \times 1{,}000 \tag{2.8.1}$$

where, V is bioreactor volume (m^3), D is dilution rate (h^{-1}), X is cell mass concentration (kg m^{-3}) and Y_Δ is cell yield based on heat generated (g J^{-1}).

Considering heat balance in substrate consumption, Y_Δ is represented as (Bailey and Ollis, 1977)

$$Y_\Delta = \frac{Y_{X/S}}{\Delta H_s - Y_{X/S}\Delta H_c} \tag{2.8.2}$$

where, ΔH_s is heat of combustion of substrate (J g^{-1}), ΔH_c is heat of combustion of cell material (J g^{-1}), $Y_{X/S}$ is cell yield based on glucose (g g^{-1}). Considering Monod's growth model for cell concentration, $X = 20$ kg m^{-3} and glucose concentration, $c_s = 8$ kg m^{-3}, the dilution rate D becomes

$$D = \mu = \frac{\mu_m\, c_s}{K_m + c_s} = 0.4\ \text{h}^{-1}$$

At $D = 0.4\ h^{-1}$, $Y_{X/S} = 20/(50-8) = 0.476$ g cells g^{-1} glucose consumed.

$= 85.68$ g cells mol^{-1} glucose

In order to estimate ΔH_c, the balance equation for cell combustion is required.

Assuming microbial cell with empirical formula $C_8H_{13}O_4N$, the balance equation for cell combustion is

$$C_8H_{13}O_4N + 9.25O_2 \rightarrow 8CO_2 + 0.5\ N_2 + 6.5\ H_2O \tag{2.8.3}$$

It is known that the heat of combustion per electron transferred is 112.968 kJ and since O_2 has four such electrons, it is 451.87 kJ per mole of O_2. Thus

$$\Delta H_c = \frac{9.25 \times 451.87 \times 1000}{8 \times 12 + 13 \times 1 + 4 \times 16 + 1 \times 14}$$

$$= 22{,}351\ \text{J g}^{-1}$$

ΔH_s for glucose $= 15{,}062.4$ J g^{-1}
From equation (2.8.2),

$$Y_\Delta = \frac{0.476}{15{,}062.4 - 0.476 \times 20{,}795}$$

$$= 10.7 \times 10^{-5}\ \text{gJ}^{-1}$$

From equation (2.8.1),

$$Q = \frac{100 \times 0.4 \times 20}{10.7 \times 10^{-5}} = 74.7 \times 10^5\ \text{J h}^{-1}$$

using ammonia as the nitrogen source, the general balance equation for microbial growth is,

$$CH_mO_l + aNH_3 + bO_2 \rightarrow y_C CH_pO_nN_q + (1 - y_C)CO_2 + cH_2O \tag{2.8.4}$$

For this system, $m = 2$, $l = 1$, $p = 13/8$, $n = 4/8$, $q = 1/8$

$$Y_{X/S} = 0.476 = \frac{y_c 72/180}{96/(96 + 13 + 64 + 14)}$$

Carbon yield, $y_C = 0.61$ means 61% of the organic substrate carbon goes for cell synthesis and 39% is released as heat.

Taking a balance of equation (2.8.4),

$$14a = 0.61 \times q \times 14 \Rightarrow a = 0.076$$

$$a \times 3 + m = 0.61 \times p + c \times 2 \Rightarrow c = 0.618$$

$$b \times 32 + l \times 16 = 0.61 \times 16 \times n + (1 - 0.61) \times 32$$
$$+ c \times 16 \Rightarrow b = 0.351$$

Thus the balance equation for microbial growth becomes

$$CH_2O + 0.076NH_3 + 0.351O_2 \rightarrow 0.61CH_{1.625}O_{0.5}N_{0.125} + 0.39CO_2 + 0.61H_2O \tag{2.8.5}$$

Thus respiratory quotient $(RQ) = \dfrac{0.39}{0.351} = 1.11$ mol CO_2 evolved per mol O_2 consumed.

Cell yield based on O_2 consumed

$$= \frac{0.61(12 + 1.625 + 0.5 \times 16 + 0.125 \times 14)}{0.351 \times 32}$$

$$= \underline{1.27 \text{ g cell per g } O_2 \text{ consumed}}.$$

Reference

J. E. Bailey and D. F. Ollis (1977) *Biochemical Engineering Fundamentals*, McGraw-Hill Kogakusha, Tokyo.

3

Kinetics of growth and product formation

Contributors: Tarun Ghose, T. Shioya, John Villadsen, T. Yoshida, Purnendu Ghosh, V. S. Bisaria, K. B. Ramachandran

PROBLEM 3.1

An organism is grown in a CSTR at a dilution rate of $0.1\ \text{h}^{-1}$. The organism follows substrate inhibition kinetics of the type:

$$\mu = \frac{\mu_\text{m}\, c_\text{s}}{K_\text{s} + c_\text{s} + (c_\text{s}^2/K_\text{i})}$$

The kinetic parameters are as follows: $\mu_\text{m} = 0.2\ \text{h}^{-1}$; $K_\text{s} = 2\times 10^{-3}\ \text{kg m}^{-3}$; $K_\text{i} = 150\times 10^{-3}\ \text{kg m}^{-3}$; $c_{\text{s}_0} = 0.2\ \text{kg m}^{-3}$; $Y_\text{x/s} = 0.5$.
How many steady states are possible? Determine the stability of each steady state.

Solution

General cell (x) and substrate (c_s) balance equations are as follows:

$$\frac{\text{d}x}{\text{d}t} = \mu x - Dx \tag{3.1.1}$$

$$\frac{\text{d}c_\text{s}}{\text{d}t} = D(c_{\text{s}_0} - c_\text{s}) - \frac{1}{Y_\text{x/s}}\mu x \tag{3.1.2}$$

$$\mu = \frac{\mu_\text{m}\, c_\text{s}}{K_\text{s} + c_\text{s} + (c_\text{s}^2/K_\text{i})} \tag{3.1.3}$$

Making the equations (3.1.1) and (3.1.2) dimensionless by introducing

$$x^* = \frac{x}{K_\text{s} Y_\text{x/s}};\ y^* = \frac{c_\text{s}}{K_\text{s}};\ \theta = \mu_\text{m} t;\ m = \frac{D}{\mu_\text{m}};\ k = \frac{K_\text{s}}{K_\text{i}};\ y_0 = \frac{c_{\text{s}_0}}{K_\text{s}}$$

$$\frac{\text{d}x^*}{\text{d}\theta} = m(y_0^* - y^*) - \frac{x^* y^*}{1 + y^* + ky^{*2}} \tag{3.1.4}$$

$$\frac{\text{d}y^*}{\text{d}\theta} = \frac{x^* y^*}{1 + y^* + ky^{*2}} - mx^* \tag{3.1.5}$$

The solutions of the equations (3.1.4) and (3.1.5) at $\frac{dx^*}{d\theta}=\frac{dy^*}{d\theta}=0$, will give steady states. Three steady states are possible

$$\bar{x}_1^* = 0 \text{ and } \bar{y}_1^* = y_0^* \tag{3.1.6}$$

$$km\bar{y}^{*2} + (m-1)\bar{y}^* + m = 0 \tag{3.1.7}$$

$$\bar{x}^* = y_0^* - \bar{y}^* \tag{3.1.8}$$

The equation (3.1.6) indicates wash out when $x=0$ and $c_s = c_{s_0}$ and is a stable steady state. In order to study the stability of other steady states, the equations (3.1.4) and (3.1.7) are linearized around the steady state $(\bar{x}^*, \bar{y}^*)$ as

$$\frac{dx^{*1}}{d\theta} = (-m+a)x^{*1} + by^{*1} \tag{3.1.9}$$

$$\frac{dy^{*1}}{d\theta} = -ax^{*1} - (m-b)y^{*1} \tag{3.1.10}$$

where,

$$x^* = \bar{x}^* + x^{*1}$$

$$y^* = \bar{y}^* + y^{*1}$$

$$a = \frac{\bar{y}^*}{(1+\bar{y}^*+k\bar{y}^{*2})}$$

$$b = \frac{1-k\bar{y}^{*2}}{(1+\bar{y}^*+k\bar{y}^{*2})^2}$$

The characteristic equation for this linearized system is given by:

$$\begin{vmatrix} -m+a-\lambda & b \\ -a & -m+b-\lambda \end{vmatrix} = 0$$

The roots of which are:

$$\lambda_1 = -m$$

$$\lambda_2 = b - a - m$$

If the eigenvalues λ_1 and λ_2 are both real and negative, the steady state is stable. Instability will occur if $\lambda_2 > 0$ or $b - a - m - > 0$.

Using the parameter values, the roots of equations (3.1.7) and (3.1.8) are obtained as

$$\bar{y}_2^* = 0.5 \qquad \bar{y}_3^* = 76.4$$
$$\bar{x}_2^* = 99.5 \qquad \bar{x}_3^* = 23.6$$

For, $\bar{y}_2^* = 0.5$, the value of $b - a < m$, so the steady state is stable. For $\bar{y}_3^* = 76.4$, the value of $b - a > m$, and so the steady state is unstable.

In summary,

Steady state I $\underline{\bar{x}_1^* = 0}$, $\underline{\bar{y}_1^* = 100}$ wash out
$\underline{x = 0}$ $\underline{c_s = 0.2}$ steady state stable

Steady state II $\underline{\bar{x}_2^* = 99.5}$, $\underline{\bar{y}_2^* = 0.5}$ } steady state stable
$\underline{x = 0.1}$ $\underline{c_s = 1 \times 10^{-3}}$ }

Steady state III $\underline{\bar{x}_3^* = 23.6}$, $\underline{\bar{y}_3^* = 76.4}$ unstable
$\underline{x = 0.02}$ $\underline{c_s = 152.8}$ steady state

PROBLEM 3.2

Choose the reactor configuration from Fig. 3.1 which will give maximum cell productivity. The systems are operated at a feed rate of $15 \times 10^{-3}\ \mathrm{m^3\ h^{-1}}$ with initial

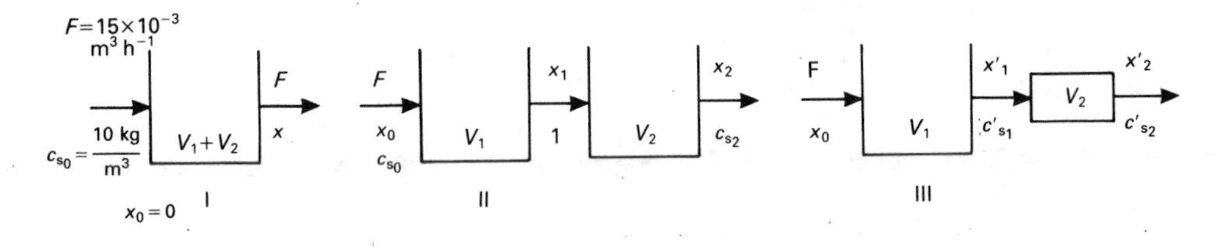

Fig. 3.1.

glucose concentration, $c_{s_0} = 10\ \mathrm{kg\ m^{-3}}$. Total bioreactor volume $(V_1 + V_2)$ is $0.10\ \mathrm{m^3}$ and $V_1 = V_2$. The organism is assumed to follow Monod's growth kinetics with:

$$\mu_m = 0.4\ \mathrm{h^{-1}},\ K_s = 2\ \mathrm{kg\ m^{-3}}\ \text{and}\ Y_{x/s} = 0.5\ \mathrm{g.g^{-1}}$$

Solution

At steady state, mass balance equations for I are:

$$D = \mu = \frac{\mu_m c_s}{K_s + c_s} \tag{3.2.1}$$

$$x = Y_{x/s}(c_{s_0} - c_s) \tag{3.2.2}$$

Using the constants at $D = 0.15\ \mathrm{h^{-1}}$ this gives $x = 4.4\ \mathrm{kg\ m^{-3}}$ and $c_s = 1.2\ \mathrm{kg\ m^{-3}}$. For the second system using (3.2.1) and (3.2.2), the values of x_1 and c_{s_1} at $D = 0.30\ \mathrm{h^{-1}}$ are $x_1 = 2\ \mathrm{kg\ m^{-3}}$, $c_{s_1} = 6\ \mathrm{kg\ m^{-3}}$.

For the second stage of II, the mass balance equations are:

$$D(x_2 - x_1) = \frac{\mu_m c_{s_2} x_2}{K_s + c_{s_2}} \tag{3.2.3}$$

$$x_2 - x_1 = Y_{x/s}(c_{s_1} - c_{s_2}) \tag{3.2.4}$$

Replacing x_2 by c_{s_2} in (3.2.3), a quadratic equation in c_{s_2} is obtained as follows:

$$0.05c_{s_2}^2 - 1.4c_{s_2} + 0.18 = 0 \tag{3.2.5}$$

solving (3.2.5), two values of c_{s_2} are obtained as

$$c_{s_2} = 0.2 \text{ or } 27.8 \text{ kg m}^{-3}$$

Taking $c_{s_2} = 0.2$ (as 27.8 kg m^{-3} is not possible)

$$x_2 = 4.90 \text{ kg m}^{-3}$$

In the third system, $x_1' = x_1$ and $c_{s_1}' = c_{s_1}$. Second stage is a plug-flow reactor, for which the kinetic expression in integrated form is represented as:

$$(K_s Y_{x/s} + c_{s_1}' Y_{x/s} + x_1')\ln\left(\frac{x_2'}{x_1'}\right) - (K_s Y_{x/s})\ln\left(\frac{Y_{x/s}c_{s_1}' + x_1' - x_2'}{Y_{x/s}c_{s_1}'}\right)$$

$$= \mu_m t(Y_{x/s}c_{s_1}' + x_1') \tag{3.2.6}$$

For

$$t = \frac{1}{D} = 3.33 \text{ h}$$

$$x_2' = 4.5 \text{ kg m}^{-3}$$

$$c_{s_2}' = 1.0 \text{ kg m}^{-3}$$

The maximum cell productivity (xD) is obtained from configuration II. (Fig. 3.1.)

PROBLEM 3.3

An organism is growing in a CSTR at a dilution rate $D = 0.2\,h^{-1}$ with an initial feed sugar concentration $c_{s_0} = 50$ kg m^{-3}. If dilution rates are changed to 0.4, 0.6 and 0.8 h^{-1}, what will be the transient response due to change in dilution rate? Draw your inference from the transient behaviour. Assume Monod's growth kinetics ($\mu_m = 1.0$ h^{-1}, $K_s = 2$ kg m^{-3}, $Y_{x/s} = 0.5$) is followed.

Solution

The following unsteady state mass balance equation will give the dynamic behaviour of the system due to change in dilution rate:

$$\frac{dx}{dt} = \frac{\mu_m c_s x}{K_s + c_s} - Dx \tag{3.3.1}$$

$$\frac{dc_s}{dt} = \frac{-\mu_m c_s x}{K_s + c_s}\frac{1}{Y_{x/s}} + D(c_{s_0} - c_s) \tag{3.3.2}$$

Solving equations (3.3.1) and (3.3.2) numerically by the Runge–Kutta method (Lapidus. 1962; Mickley *et al.*, 1959), using the kinetic parameters $\mu_m = 1.0$ h^{-1},

$k_s = 2\ \mathrm{kg\ m^{-3}}$, $Y_{x/s} = 0.5$, the transient behaviour of the system due to change in dilution rate from 0.2 to 0.4, 0.6 and $0.8\ \mathrm{h^{-1}}$ is obtained.

The transient response is given in Fig. 3.2. The figure clearly indicates that the

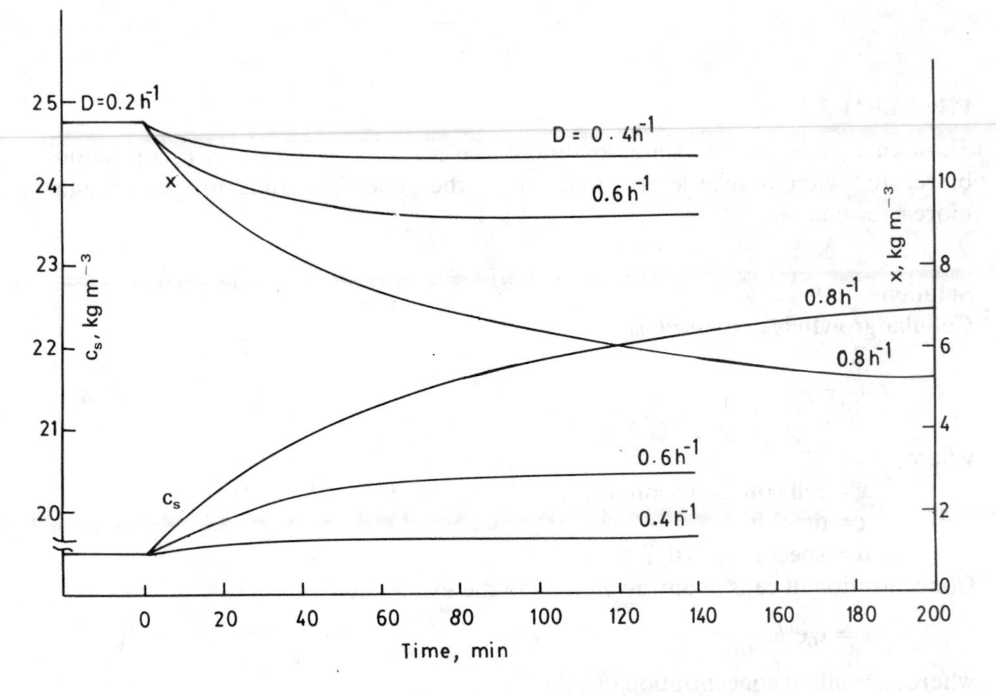

Fig. 3.2 — Transient behaviour with change in dilution rate.

response time to reach new steady state increases with an increase in ΔD. The response time to reach steady state is given in Table 3.1.

Table 3.1

Change in dilution rate ($\mathrm{h^{-1}}$)	Response time to reach steady state (min)
0.2–0.4	90
0.2–0.6	120
0.2–0.8	450

From Table 3.1 it can be observed that as dilution rate approaches μ_m, the response time to reach equilibrium is very large. This observation was pointed by Perram (1973). This analysis indicates that it is difficult to carry out transient experiments at high dilution rate (near washout conditions).

References
Lapidus, L. (1962) *Digital Computation for Chemical Engineers*, McGraw-Hill, New York.
Mickley, H. S., Sherwood, T K., & Reed, C. E. (1959) *Applied Mathematics in Chemical Engineering*, McGraw-Hill, New York.
Perram, J. W. (1973) *J. Theoret. Biology*, **38**, 571.

PROBLEM 3.4

Show that under steady state conditions the medium flow rate in a continuous bioreactor without recycle is a function of the generation time for a particular bioreactor volume.

Solution

Cellular growth is expressed as

$$\frac{\mathrm{d}x}{\mathrm{d}t} = \mu x \tag{3.4.1}$$

where,

x = cell concentration, kg m^{-3}
t = time, h
μ = specific growth rate, h^{-1}

Concentration of cells at any given time in the exponential phase is

$$x = x_0 \mathrm{e}^{\mu t} \tag{3.4.2}$$

where x_0 = initial concentration of cells.

The change in the microbial population during growth phase follows as:

$$1 \rightarrow 2^1 \rightarrow 2^2 \rightarrow 2^3 \rightarrow 2^4 \ldots 2^{N_\mathrm{g}} \tag{3.4.3}$$

where

N_g number of generations taking place by a process of binary fission and geometric progression.

We can thus write

$$x = x_0 2^{N_\mathrm{g}} \tag{3.4.4}$$

or,

$$N_\mathrm{g} = \frac{\ln(x) - \ln(x_0)}{\ln 2} \tag{3.4.5}$$

Combining equations (3.4.2) and (3.4.4) we get,

$$x_0 2^{N_\mathrm{g}} = x_0 . e^{\mu t}$$

or,

$$2^{N_\mathrm{g}} = \mathrm{e}^{\mu t}$$

or,

$$N_g = \frac{\mu t}{\ln 2} \tag{3.4.6}$$

If t_g is generation time, i.e. the time required for doubling the population, then,

$$t_g = \frac{t}{N_g} = \frac{\text{total time allowed for generations}}{\text{number of generations}}$$

From equation (3.4.6) we get,

$$\frac{t}{t_g} = \frac{t}{\mu t/\ln 2}$$

or,

$$t_g = \frac{\ln 2}{\mu} \tag{3.4.7}$$

The continuous culture mass balance gives (Fig. 3.3),

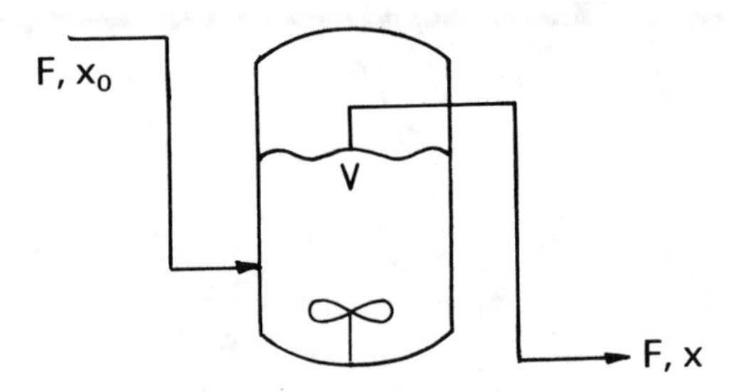

Fig. 3.3.

input + growth = output + accumulation

or,

$$Fx_0 + \mu x V = Fx + V\frac{dx}{dt} \tag{3.4.8}$$

where

F = feed (= effluent) flow rate
V = constant working volume
x_0 = cell concentration in feed
x = cell concentration in the reactor
cell concentration in the effluent
μ = growth rate constant

For steady state continuous bioreactor (chemostat) we can write,

$$\frac{dx}{dt} = D(x_0 - x) + \mu x \tag{3.4.9}$$

where, D = dilution rate $= \frac{F}{V}\,h^{-1}$

or,

$$\frac{dx}{dt} = \frac{F}{V}(x_0 - x) + \mu x$$

$$= Dx_0 + x(\mu - D) \tag{3.4.10}$$

If the initial cell concentration in the feed is zero, i.e. $x_0 = 0$, then at steady state, the equation (3.4.10) becomes:

$$\frac{dx}{dt} = (\mu - D)x \tag{3.4.11}$$

Since the system is at steady state,

$$\frac{dx}{dt} = 0$$

Therefore, $(\mu - D)x = 0$
or,

$$\mu = D = \frac{F}{V} \tag{3.4.12}$$

which implies that at steady state, $\mu = D$.

Combining equations (3.4.7) and (3.4.12), we get,

$$\frac{F}{V} = D = \mu = \frac{\ln 2}{t_g}$$

or,

$$F = V\frac{\ln 2}{t_g}$$

since,

V = constant in continuous culture

$$F = \frac{\text{a constant}}{t_g} = \underline{f(t_g)}$$

"To those who work with continuous culture, the preoccupation of microbial geneticists and molecular biologists with batch culture is both foolish and puzzling." This statement, made by S. B. Primrose some time ago, conveys a deep sense of urgency to undertake both the rate and dynamics of continuous culture studies. To the biochemical engineer in the class or in practice, this remains an essential part of his training or profession. Studies in this area are generally carried out in chemostats with various organisms, in particular with single-cell microbes.

PROBLEM 3.5

A chemostat culture of *Candida tropicalis* growing on hexadecane was carried out at a dilution rate $D = 0.18\,h^{-1}$. The following data were reported:

working volume	$= 0.015\,m^3$
aeration rate	$= 1.8\,m^3\,h^{-1}$(20°C, 9599.2 $N\,m^{-2}$)
feed substrate concentration	$= 10\,kg\,m^{-3}$
reactor substrate concentration	$= 0.032\,kg\,m^{-3}$
dry weight concentration	$= 10.2\,kg\,m^{-3}$
% CO_2 in the exit gas	= 1.0 v/v
% O_2 in the exit gas	= 2.5 v/v
μ_m	$= 0.27\,h^{-1}$

From the available data, calculate the following parameters:
$Y_{x/s}$, K_s, Q_s, Q_{O_2}, Q_{CO_2} and the carbon balance of the steady state system.

Solution

$$Y_{x/s} = \frac{10.2}{10} = 1.02$$

$$K_s = c_s\left(\frac{\mu_m}{D} - 1\right) = 0.032\,kg\,m^{-3}\left(\frac{0.27}{0.18} - 1\right) = 0.016\,kg\,m^{-3}$$

$$Q_s = \frac{\mu}{Y_{x/s}} = \frac{D(c_{s_0} - c_s)}{x} - \frac{1}{x}\frac{dc_s}{dt} \equiv 0$$

Under steady state conditions

$$= \frac{0.18\,(10.0 - 0.032)}{10.2} = 0.18\,kg.kg^{-1}h^{-1}$$

$$Q_{CO_2} = \frac{\text{air} \times K_{sc} \times \Delta\%CO_2 \times 10}{V \times x \times 22.4}$$

$$= \frac{1.8 \times 0.8864 \times 10 \times 10}{0.015 \times 10.2 \times 22.4}$$

$$Q_{CO_2} = 4.65\text{ moles }CO_2\,kg^{-1}(DW)\,h^{-1}$$

Analogously, Q_{O_2} can be assayed to be

$$Q_{O_2} = 11.64\text{ moles }O_2\,kg^{-1}(DW)\,h^{-1}$$

The RQ will then become

$$RQ = \frac{Q_{CO_2}}{Q_{O_2}} = \frac{4.65}{11.64} = 0.4$$

Carbon recovery:

$$C_{\text{substrate}} = C_{\text{DW}} + C_{\text{CO}_2}$$

$$\text{Carbon balance} = \frac{C_{\text{DW}} + C_{\text{CO}_2}}{\text{substrate}}$$

where c_{DW} and c_{CO_2} stand for carbon in cells (dry weight) and in carbon dioxide. Assuming that cell mass has 50% of its dry weight as carbon,

$$C_{\text{DW}} = 50\% \text{ of the DW}$$
$$= \frac{10.2}{2} = 5.1 \text{ kg m}^{-3}$$

CO_2 produced per hour

$$VDC_{\text{DW}} = 0.015 \times 0.18 \times 5.1$$
$$= 0.0138 \text{ kg h}^{-1}$$

$$C_{\text{CO}_2} = 1\% \text{ of } 1.8 \text{ m}^3 \text{ h}^{-1} = 0.018 \text{ m}^3 \text{ h}$$
$$= 0.8 \text{ mol h}^{-1} = 0.0096 \text{ kg CO}_2 \text{ h}^{-1}$$

$$C_{\text{substrate}} = VD\frac{192}{226} \; 10 \text{ kg m}^{-3}$$

[Hexadecane mol. wt. = 226, C = 192.]

$$= 22.9 \times 10^{-3} \text{ kg C h}^{-1}$$

$$c_{\text{recovery}} = \frac{0.0138 + 0.0096}{0.0229} \times 100$$
$$= \underline{102\%}$$

PROBLEM 3.6

In an experiment on the propagation of *Saccharomyces cerevisiae* yeast, uptake of 0.2 kg of sugar was found to produce 0.0746 kg of yeast cells, liberating 0.121 kg of CO_2(g). Oxygen requirement during the cell synthesis was determined as 0.0672 kg. Write the mass balance equation and compute:

(a) the value of yield of yeast cells, and
(b) the value of RQ.

Solution

$$CH_2O + O_2 \rightarrow \text{yeast cells} + CO_2 + H_20$$

$$\underset{0.2\text{ kg}}{n_1(CH_2O)} + \underset{0.0672\text{ kg}}{n_2O_2} = \underset{0.0746\text{ kg}}{\text{yeast}} + n_3\, CO_2 + \underset{0.121\text{ kg}}{n_4\, H_2O}$$

So, the number of moles of carbohydrate (n_1) involved in the reaction

$$\frac{0.2}{(12+2+16)} = \frac{0.2}{30} = 6.67 \times 10^{-3} \text{ kg mol}$$

and the number of moles of oxygen (n_2) involved in the reaction

$$\frac{0.0672}{32} = 2.1 \times 10^{-3}\ \text{kg mol}$$

Number of moles of carbon dioxide (n_3) produced $= \dfrac{0.121}{44} = 2.75 \times 10^{-3}$ kg mol

Amount of water produced

$$= (0.2 + 0.0672) - (0.0746 + 0.121)$$

$$= 0.2672 - 0.1956 = 0.0716\ \text{kg}$$

Number of moles of water (n_4) produced

$$= \frac{0.0716}{18} = 3.97 \times 10^{-3}\ \text{kg mol}$$

Mass balance of the steady state system can be written as

	CH_2O +	O_2	= yeast cells +	CO_2	+ H_2O
kg mol	6.67×10^{-3}	2.1×10^{-3}		2.75×10^{-3}	3.97×10^{-3}
weight (kg)	0.2	0.0672	0.0746	0.121	0.0716

(a) Yield of yeast cells

$$= \frac{0.0746}{0.2} = \underline{0.373\ \text{kg yeast kg}^{-1}\ \text{sugar}}$$

(b)
$$RQ = \frac{CO_2\ \text{produced}}{O_2\ \text{utilized}} = \frac{2.75 \times 10^{-3}}{2.1 \times 10^{-3}} = \underline{1.3}$$

PROBLEM 3.7

Prove that in a two-stage continuous bioreactor (CSTB) system the concentration of microorganisms in the second reactor cannot be greater than 1.718 times that in the first. See Fig. 3.4.

Solution

Mass balance of the first vessel of a multistage bioreactor is given by

$$F_1 x_0 + \mu_1 V_1 x_1 = F_1 x_1 + V_1 \frac{dx_1}{dt} \tag{3.7.1}$$

where

F_1 = equal flow rates to and from bioreactor
x_0, x_1, x_2 = concentrations of cells entering bioreactor 1, 2, and exit of bioreactor 2 respectively.
V = volume of bioreactor 1 or 2
μ = growth rate constant

For the second vessel, the mass balance can be written as

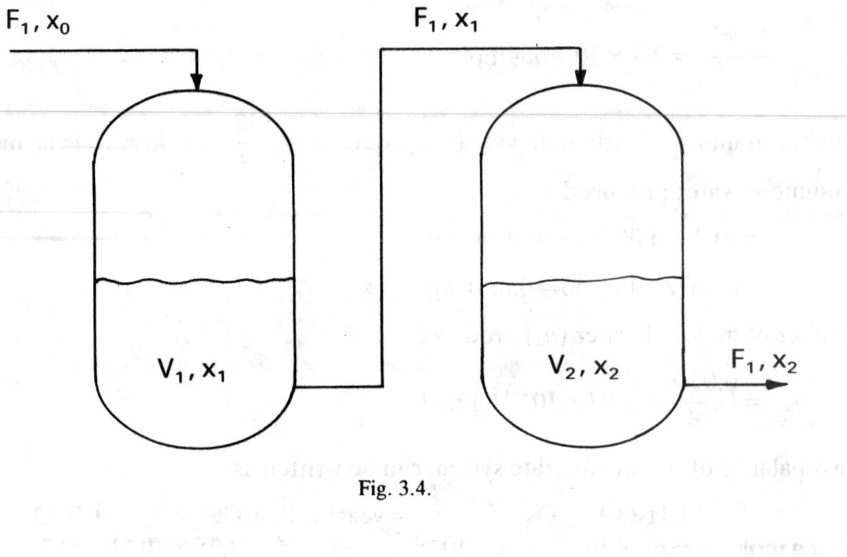

Fig. 3.4.

$$F_1x_1 + \mu_2V_2x_2 = V\frac{dx_2}{dt} + f_1x_1 \qquad (3.7.2)$$

where μ_2 = cell growth rate constant in the second vessel.

From equation (3.7.2) we get

$$\int dt = \int \frac{d(Vx_2)}{F_1x_1 + \mu_2Vx_2}$$

or

$$t = \frac{1}{\mu_2}\left\{\ln\frac{F_1x_1 + \mu_2.Vx_2}{F_1x_1}\right\} \qquad (3.7.3)$$

when $t = t_r$, that is when the vessel 2 begins to overflow at V_2/F_1 we have

$$V_1x_1 = V_2x_2$$

Thus, from equation (3.7.3),

$$\mu_2t = \ln\frac{F_1x_1 + \mu_2V_2x_2}{Fx_1}$$

or,

$$e^{\mu_2t} = \frac{F_1x_1 + \mu_2V_2x_2}{F_1x_1}$$

or,

$$x_1\, e^{\mu_2t} = x_1 + \mu_2\frac{V_2}{F_1}x_2 \qquad (3.7.4)$$

At steady state

$$V_1 = V_2, \frac{F_1}{V_2} = \mu = \frac{1}{t}$$

Thus, from equation (3.7.4)

$$x_1\, e^{\mu_2/\mu_1} = x_1 + \mu_2 \frac{1}{\mu_1} x_2$$

or

$$\mu_2 x_2 = \{x_1\, e^{\mu_2/\mu_1} - x_1\}\mu_1$$
$$= x_1 \mu_1 \{e^{\mu_2/\mu_1} - 1\}$$

or

$$x_2 = \frac{\mu_1}{\mu_2} x_1 \{e^{\mu_2/\mu_1} - 1\}$$

Now, x becomes maximum when $\mu_1 = \mu_2$ (since μ_1 cannot exceed μ_2); then,

$$x_2 = x_1(e^{-1})$$

or

$$\underline{x_2 = 1.718\, x_1}$$

PROBLEM 3.8

A specific bacterium is cultivated by using ammonium sulphate as sole nitrogen source. The chemical analysis of the cells harvested shows that the nitrogen content was 8% (by weight). What is the value of pH of the broth, if the cells are grown to an extent of 10 kg m^{-3}. It is assumed that no buffering action of the medium is available and that the initial concentration of hydrogen ion (10^{-7} M) could be disregarded.

Solution

$$(NH_4)_2SO_4 \rightleftharpoons NH_4^+ + SO_4^{2-}$$

$$10 \text{ kg cell m}^{-3} \times \frac{8}{100} \times \frac{1}{14} = 0.0571 \text{ kg.mol N m}^{-3} \ (= \text{ g mol. l}^{-1})$$

H^+ is equal to moles of NH_3 consumed by the cells ($NH_4^+ \rightleftharpoons NH_3 + H^+$) which in turn is equal to moles of N present in the cells

$$[H^+] = 0.0571$$

$$pH = -\log[H^+] = \underline{1.24}$$

PROBLEM 3.9

An aerobic culture of *Azobacter vinelandii* gave the following relationship between the overall growth yield $Y_{x/s}(= \Delta x/\Delta c_s)$ and the dilution rate $D(= \mu)$

$$\frac{1}{Y_{x/s}} = \frac{m}{\mu} + \frac{1}{Y_G} \tag{3.9.1}$$

where

m = specific rate of substrate uptake for maintenance
Y_G = true growth yield ($= \Delta x/(\Delta c_s)_G$), (growth yield constant)

Designating c_s and x to substrate (limiting, glucose) and cell mass concentration, show that

$$-\frac{dc_s}{dt} = -mx + \frac{1}{Y_G}\frac{dx}{dt} \tag{3.9.2}$$

Transforming both sides of equation (3.9.2) with Laplace transform, derive the tansfer function $G(p)$ with respect to the culture, provided:

$$G(p) = \frac{x(p)}{-c_s(p)}$$

where

p = Laplacian
$x(p), c_s(p)$ = transforms of $x(t)$, $c_s(t)$ respectively

See Fig. 3.5.

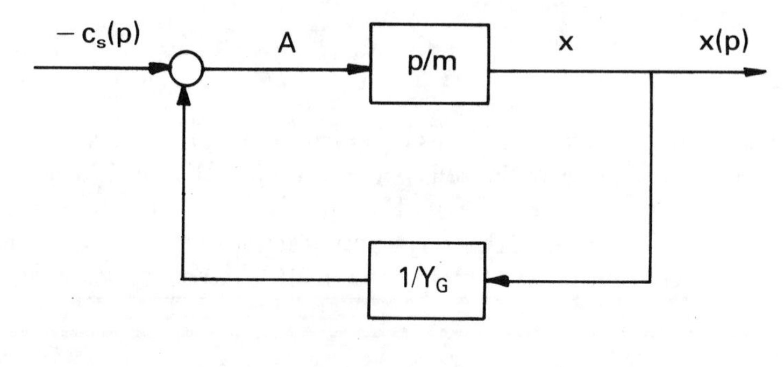

Fig. 3.5.

Solution

$$\frac{1}{Y_{x/s}} = \frac{m}{\mu} + \frac{1}{Y_G}$$

or,

$$\frac{\mu x}{Y_{x/s}} = mx + \frac{\mu x}{Y_G}$$

or,

$$\left(\frac{-\Delta c_s}{\Delta x}\right)\left(\frac{dx}{dt}\right) = mx + \frac{1}{Y_G}\frac{dx}{dt}$$

$$\therefore -\frac{dc_s}{dt} = mx + \frac{1}{Y_G}\frac{dx}{dt}$$

Taking the Laplace transform:

$$-pc_s(p) = mx(p) + \frac{1}{Y_G}p \times (p)$$

$$= \left(m + \frac{1}{Y_G}\right)x(p)$$

$$\therefore G(p) = \frac{x(p)}{-c_s(p)} = \frac{p}{m + \frac{1}{Y_G}p}$$

$$= \frac{p/m}{1 + \frac{1}{Y_G}\frac{p}{m}} \quad \ldots \text{ transfer function}$$

$$A\ p/m = x \qquad -c_s = x\left(\frac{m}{p} + \frac{1}{Y_G}\right)$$

$$x\frac{1}{Y_G} - c_s(p) = A$$

$$\left(-x\frac{1}{Y_G} - c_s\right) = x\frac{m}{p} \qquad \frac{Y_G}{-c_s} = \frac{1}{\frac{m}{p} + \frac{1}{Y_G}}$$

PROBLEM 3.10 Ex.

A pilot plant fermenter is used to cultivate aerobically a specific bacterium which shows the doubling time of 40 min when the following medium is employed:

glucose 20 kg m^{-3}
$(NH_4)_2SO_4$ 10 kg m^{-3}

Salts and growth factors are also present in the mediums.

The aeration and agitation conditions used corresponded to the volumetric oxygen transfer coefficient $k_L a = 4.9346 \times 10^{-6}$ kg mol O_2 m^{-3} h^{-1}:

Data given:

air flow rate at STP = 0.70 vvm
agitation speed of impeller = 300 rpm

Taking the inoculum size and Q_{O_2} value of this organism as 0.05 kg cell m^{-3} and 0.01 kg mol O_2/kg cell h, estimate the time when the cell ceases to grow logarithmically. No lag time is assumed in this bacterial culture.

Solution

$$\mu = \frac{\ln 2}{t_g} = \frac{\ln 2}{40/60} = 1.04\ h^{-1}$$

$$O_2 \text{ transfer} = k_L a(c^* - \bar{c}) = k_L aH(\bar{p} - p^*) = k_v(\bar{p} - p^*)$$

driving force

$$p_{O_2} = px_{O_2} = 1.01325 \times 10^5 = 21{,}278.25\ N\ m^{-2} \Rightarrow p_{in}$$
$$p_{out} = 0$$

$$\Delta p_{ave} = \left(\frac{21{,}278.25 - 0}{2}\right) = 10{,}639.125$$

O_2 transfer $= 4.9346 \times 10^{-6} \times 10639.125$
O_2 supply $= 0.0525$ kg moles/m^3 h
O_2 demand $= Q_{O_2}x$ where, $x = x_0 e^{\mu t}$
and $x_0 = 0.05$ kg/m^3
Thus, O_2 demand $= 0.05 \times 0.01\ e^{1.04t}$

$$= 5 \times 10^{-4}\ e^{1.04t}\ \frac{\text{kg mole } O_2}{m^3\ h}$$

Let t_{growth} = time when cell growth ceases
then at any $t < t_{growth}$
supply of $O_2 >$ demand of O_2 until $t = t_{growth}$
supply O_2 = demand O_2
$0.0525 = 5 \times 10^{-4}\ e^{1.04t_g}$

$$\frac{\ln \dfrac{0.0525}{5 \times 10^{-4}}}{1.04} = t_{growth} \Rightarrow t_{growth} = \underline{4.475\ h}$$

Check on glucose supply

Cell formation $= x = x_0\ e^{1.04t}$ when $t = t_{growth} = 4.475$

$$x = 0.05\ e^{1.04 \times 4.475}$$
$$= 5.223\ \text{kg cells } m^{-3}$$

Usually $Y_{x/s} = 0.5$ (50% yield)

Or, Carbon in cell = 50% of cell weight

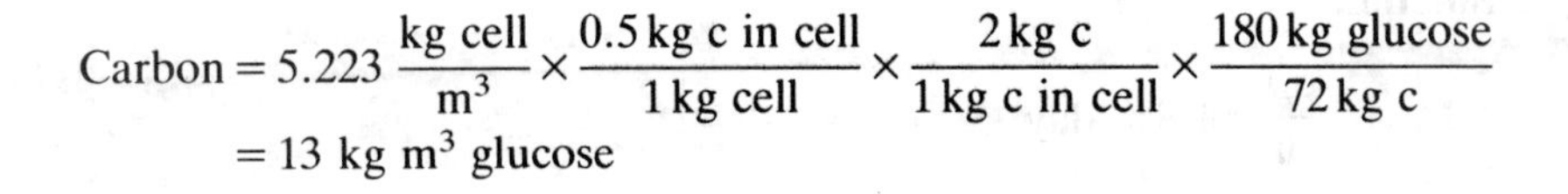

$$\text{Carbon} = 5.223\ \frac{\text{kg cell}}{\text{m}^3} \times \frac{0.5\,\text{kg c in cell}}{1\,\text{kg cell}} \times \frac{2\,\text{kg c}}{1\,\text{kg c in cell}} \times \frac{180\,\text{kg glucose}}{72\,\text{kg c}}$$

$$= 13\ \text{kg m}^3\ \text{glucose}$$

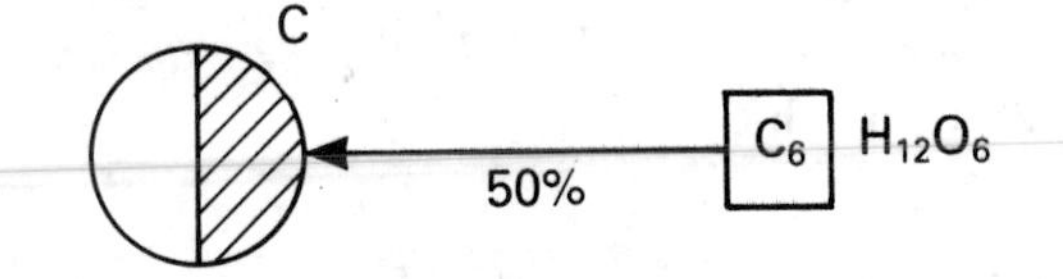

Initial glucose conc. $= 20\ \text{kg m}^{-3}$
$\therefore$ Glucose unutilized $= 7\ \text{kg m}^{-3}$

Sake is a common drink in Japan based on rice fermentation. The *sake* industry is very extensive and it is, next to green tea, perhaps the most important in Japan. There are some specific problems with sake brewing and some of these are dealt with in the following problems.

PROBLEM 3.11

It is claimed in the *sake* brewing that yeast cells do not grow beyond the concentration of around 10^8 cells ml^{-1}. It is difficult to measure the cell concentration owing to solid ingredients in the mash. The following formula is available to predict the rate of ethanol production as affected by ethanol concentration in the later phase of this brewing.

$$\frac{dp}{dt} = \nu_0 x\, e^{-k_2 p}$$

where

ν_0 = specific rate of ethanol production at $p = 0$
x = cell mass concentration
p = ethanol concentration
k_2 = empirical exponent
t = time

A run of this brewing gave the following data:

$p = 10\%$ v/v

$$\frac{dp}{dt} = 1.8\%/\text{day}$$

Taking $k_2 = 0.148\%\ \text{day}^{-1}$ and $\nu_0 = 1.0\ \text{h}^{-1}$, check the cell concentration by using the above empirical formula. Data given:

Ethanol density $= 790\ \text{kg m}^{-3}$
Av. diam. of yeast cells $= 5\ \mu\text{m}$.

Solution

$$\frac{dp}{dt} = 1.8\%\ \text{day}^{-1}$$

$$= \frac{1.8}{100} \times 790 \times \frac{1}{24}$$

$$= 0.5925\ \text{kg m}^{-3}\ \text{h}$$

$$\frac{dp}{dt} = \nu_0 x\, e^{-k_2 p}$$

$$x = \frac{e^{k_2 p}}{\nu_0}\frac{dp}{dt}$$

$$= \exp\left(\frac{0.148 \times 10}{1.0}\right) \times 0.5925 = 2.6\ \text{kg m}^{-3}$$

Volume of a yeast cell

$$= \frac{4}{3}\pi\left(\frac{5 \times 10^{-6}}{2}\right)^3 = 6.54 \times 10^{-17}\ \text{m}^3\ \text{cell}^{-1}$$

Assuming cell density as 1000 kg m^{-3}, the number of cells

$$= 2.6 \times \frac{1}{1000} \times \frac{1}{(6.54 \times 10^{-17})}$$

$$= \underline{3.98 \times 10^{13}\ \text{cells m}^{-3}}$$

PROBLEM 3.12

A continuous system for yeast propagation and/or alcohol production is to be designed. The data for the batch fermentation of molasses under conditions similar to the continuous fermentation are given in Table 3.2. For a feed volume of 0.5 m^3 h^{-1} of molasses, compute:

(a) primary fermenter volume if the fermenter is operated at the maximum rate of yeast propagation,
(b) the volume of secondary fermenter in series with the first for operating at the maximum rate of glucose utilization,
(c) volume of the secondary fermenter, if it is to be operated at the maximum rate of alcohol production.

Reference

Bailey, J. & Ollis, D. F. (1985) *Biochemical Engineering Fundamentals*, 2nd edn, McGraw-Hill, New York, Chapter 9.

Table 3.2

Time (h)	Yeast count (10^6 cell ml^{-1})	Glucose utilization (kg mol m^{-3})	Alcohol production (kg mol m^{-3})
2	6	2	1
4	10	4	5
6	20	7	11
8	38	12	20
10	180	18	30
12	117	26	41
14	140	34	55
16	155	44	70
18	160	53	86
20	163	58	101
22	164	61	110
24	165	63	116

Solution

The maximum cell productivity means dx/dt is maximum. The rate data are given (see Table 3.3).

Table 3.3

Time (h)	x (cell conc.) (10^6 cell ml^{-1})	dx/dt (growth rate) (10^6 cell ml^{-1} h^{-1})	$c_0 - c_s$ (substrate conc.)	$-$ dc_s/dt (glucose uptake rate) (kg mol m^{-3} h^{-1})
2	6	1	2	0.5
4	10	3.5	4	1.25
6	20	6.92	7	1.96
8	38	14.25	12	2.75
10	80	20.29	18	3.5
12	117	15.84	26	4.86
14	140	9.75	34	5.65
16	155	4.8	44	4.96
18	160	1.75	53	3.54
20	163	1.36	58	2.04
22	164	0.58	61	1.125
24	165	0.37	63	0.375

(a) From Fig. 3.6 $D_1 = F/V = 0.25\ h^{-1}$

$F = 0.5\ m^3\ h^{-1}$

$\therefore V_1 = \underline{2.0\ m^3}$

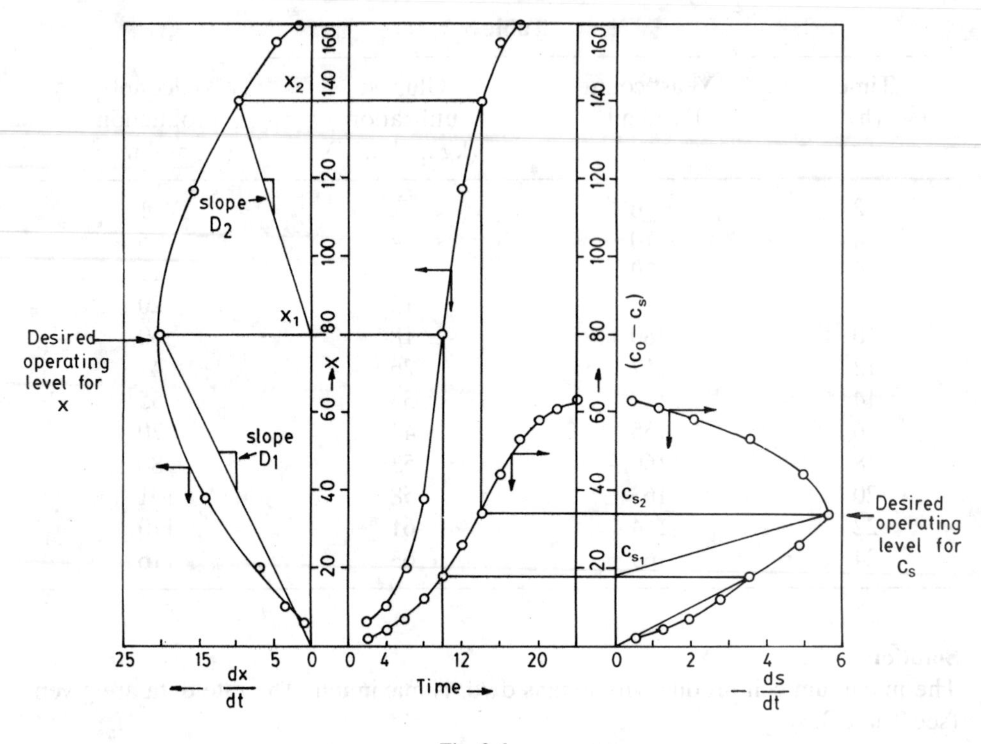

Fig. 3.6.

Table 3.4

t (h)	x (10^6 cells.ml^{-1})	dx/dt (10^6 cells.ml^{-1}h^{-1})	c_p (kg.mol m^{-3})	dc_p/dt (kg.mol m^{-3}h^{-1})
2	6	1	1	1
4	10	3.5	5	2.5
6	20	6.92	11	3.79
8	38	14.25	20	4.79
10	80	20.29	30	5.24
12	117	15.84	41	6.21
14	140	9.75	55	7.21
16	155	4.8	70	7.83
18	160	1.75	86	7.83
20	163	1.36	101	6.125
22	164	0.58	110	3.9
24	165	0.37	116	1.58

(b) From Fig. 3.6 $D_2 = \frac{1}{6}\,\text{h}^{-1}$

$\because F = 0.5\ \text{m}^3\,\text{h}^{-1}$

$\therefore V_2 = \underline{3.0\ \text{m}^3}$

(c) dc_p/dt can be obtained similarly by differentiation (see Table 3.4).
From Fig. 3.7

$$D_2 = \frac{5}{75} = \frac{1}{15} h$$

$$\therefore V_2 = \underline{15 \times 0.5 = 7.5\ \text{m}^3}$$

PROBLEM 3.13

Continuous culture has been used for the selection of mutant organisms having desirable properties such as resistance to extreme environments, biocides, salts, etc. Most publications have dealt with the selection of constitutive mutants, hyper strains producing large amounts of intracellular enzymes and selection of microorganisms which have lost their ability to synthesize substances that are not needed under the selection conditions (Harder *et al.*, 1977). A more recent theoretical approach has appeared (Lelieveld, 1982) on the selection of microorganisms in continuous culture which produce special extracellular enzymes. This problem considers selection of a strain producing extracellular enzymes capable of degrading insoluble substrates such as cellulose.

The growth rate of a microorganism is determined by the concentration of the growth-limiting substrate (c_s) and the maximum specific growth rate (μ_{max}) of the microorganism in the presence of excess substrate. It is expressed by:

$$\mu = \mu_{max} \frac{c_s}{K_s + c_s} \tag{3.13.1}$$

where, K_s is a constant representing the substrate concentration at which $\mu = 0.5\mu_{max}$. The value of μ_{max} depends on the efficiency with which the microbial cells convert the *available substrate* into biomass.

From equation (3.13.1) it is clear that the growth rate will increase with increase in c_s. The growth limiting substrate might be a molecule which is not freely available but which has to be obtained by breaking down molecules (e.g. cellulose, proteins, lignin, etc.). In such cases, the growth of the microorganism will depend on the efficiency of the extracellular enzyme. Fig. 3.8 shows the schematic representation of a microbial cell in an aqueous suspension (Lelieveld, 1982).

Around any particle in a fluid, there is a layer where diffusion rather than turbulence determines the rate of transport of molecules. The concentration of an extracellular enzyme is always highest near the cell producing it and decreases with the distance from the cell. As a result, the breakdown of a macromolecular, non-assimilable substrate and the production of small assimilable molecules is fastest near the surface of a microbial cell, which results in the highest concentration of small molecules near the cell surface. It is this concentration which actually determines the growth rate of the cell. Thus, if during growth, there is a mutation which takes place

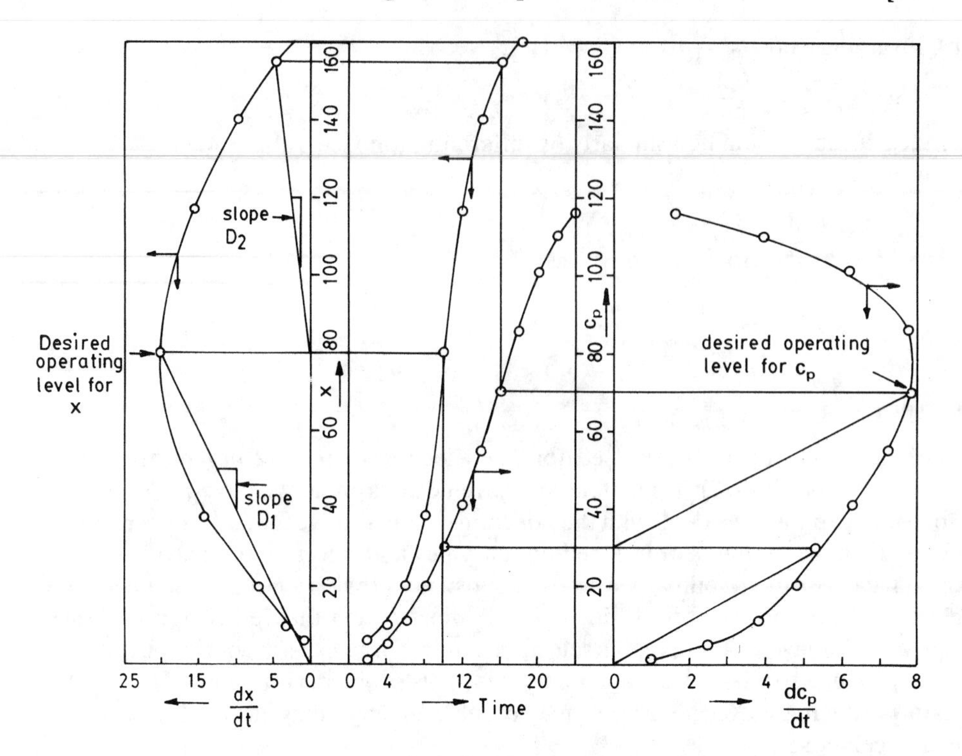

Fig. 3.7.

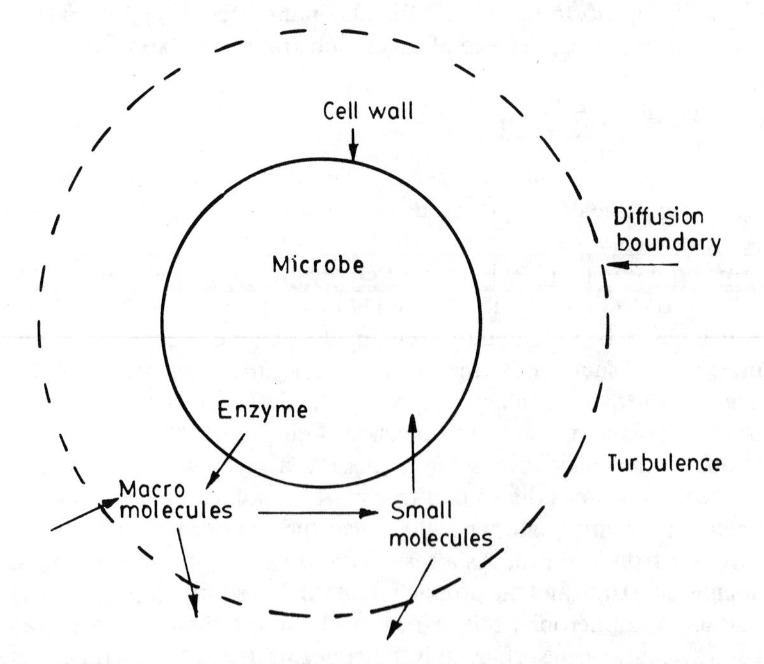

Fig. 3.8 — Schematic representation of a microbial cell in an aqueous suspension.

in one of the cells which leads to production of a more active enzyme under prevailing conditions, there will be higher concentration of assimilable substrate (c_s') around the mutant than around the cells of parent strain (c_s).

Assume that in a continuous culture (1×10^{-4} m^3 volume) a microbial culture is grown on cellulose at a dilution rate (D) of 0.6 μ_{max}. The cell concentration under steady state is 10^8 cells ml^{-1}. At any point, when the culture undergoes mutation, the concentration of assimilable substrate for it becomes 20% higher than that for the parent cell owing to synthesis of more effective cellulase enzyme by the mutant.

The detection of the mutant is possible only if its concentration in the vessel is 8% of the concentration of the parent cells. Calculate the time that will be required to detect the presence of a mutant in the culture assuming that the μ_{max} value is 0.8 h^{-1}.

References
Harder, W., Kuenen, J. G. & Matin, A. (1977) *J. Appl. Bacteriol.*, **43**, 1.
Lelieveld, H. L. M. (1982) *Biotechnol. Bioeng.*, **24**, 1419.

Solution

In continuous culture, $\mu = D$. From equation (3.13.1),

$$\mu = \mu_{max} \frac{c_s}{K_s + c_s}$$

or,

$$c_s = \frac{K_s \mu}{\mu_{max} - \mu} \tag{3.13.2}$$

For parent culture, $\mu = D = 0.6\,\mu_{max}$

$$c_s = \frac{K_s\, 0.6\mu_{max}}{\mu_{max} - 0.6\mu_{max}}$$

$$= \frac{0.6}{0.4} K_s = 1.5 K_s$$

or

$$K_s = 0.67 c_s$$

For the mutant organism, $c_s' = 1.2c_s$, therefore,

$$\mu' = \mu_{max} \frac{1.2c_s}{0.67c_s + 1.2c_s}$$

$$= 0.64\mu_{max}$$

Because of the difference in growth rates of the parent and the mutant, the concentration of the mutant will reach a level at which the isolation will become possible. The time required for this can be calculated by using the relationship.

$$\ln x = \ln x_0 + (\mu - D)t \tag{3.13.3}$$

For a mutant cell, let us write x' and μ' for x and μ.

$$\ln x' = \ln x'_0 + (\mu' - D)t \qquad (3.13.4)$$

For the detection to be possible x' must be $= 8\%$ of 10^8 cells ml^{-1}

$$x' = 8 \times 10^6 \text{ cells ml}^{-1}$$

when a mutant arises (one cell),

$$x'_0 = 1$$

From relation (3.13.4),

$$\ln(8 \times 10^6) = \ln(1) + (\mu' - D)t$$

or,

$$2.303\ [\log 8 + 6] = 0 + t(0.64\mu_{max} - 0.6\mu_{max})$$

or,

$$2.303\ [6.9] = (0.04)\ \mu_{max}\ t$$

or,

$$15.9 = (0.04)\ (0.8)t$$

or,

$$t = \frac{15.9}{0.032} = 496 \text{ h}$$

It will therefore be possible to detect the mutant having more efficient cellulase enzyme system after the continuous culture has run for 496 hours.

PROBLEM 3.14

A dilute suspension of *Chlorella* (cell concentration, $N_1 = 4.1 \times 10^{13}$ cells m^{-3}; turbidity, $-\log \mathcal{T}_1 = 0.5$) was subjected to agitation after adding Al^{3+} ions at a concentration of 10^{-3} kg mol m^{-3} to cause accelerated flocculation.

(i) In steady state when the turbidity of the suspension is expressed as $-\log \mathcal{T}_2 = 0.25$ due to flocculation, what should be the size of the flocs? The following empirical equation is given in which the diameter of single cells of *Chlorella* is taken as $dp = 3.2 \times 10^{-6}\ m$.

$$\frac{\log \mathcal{T}_1}{-\log \mathcal{T}_2} = (\overline{df}/dp)^{0.267}$$

(ii) Assuming the suspension being dilute, i.e.,

$$\begin{aligned} -\log \mathcal{T}_1 &= \log(I_0/I_1) \\ &= k_1(\pi/4)\ dp^2\ N_1, \\ -\log \mathcal{T}_2 &= \log(I_0/I_2) \\ &= k_2(\pi/4)\ d^2f\ N_2. \end{aligned}$$

derive the following equation:

$$1 - a' = (\overline{\mathrm{d}f}/\mathrm{d}p)^{-0.733}$$

where a' = water content of floc.

(*Hint*. The mass balance of the cell material per unit volume of the suspension is taken.)

(iii) What is the average value of a' in particular floc?

Solution

(i)

$$\frac{-\log \mathcal{T}_1}{-\log \mathcal{T}_2} = \left(\frac{\overline{\mathrm{d}f}}{\mathrm{d}p}\right)^{0.267}$$

$$\overline{\mathrm{d}f} = \mathrm{d}p \left(\frac{-\log \mathcal{T}_1}{-\log \mathcal{T}_2}\right)^{1/0.267}$$

$$= 3.2 \times \left(\frac{0.5}{0.25}\right)^{1/0.267}$$

$$= \underline{42.9\ \mu\mathrm{m}}$$

(ii)

$$(1 - a')\frac{\pi}{6}\overline{\mathrm{d}f}^3 N_2 = \frac{\pi}{6}\mathrm{d}p^3 N_1$$

$$\therefore \quad 1 - a' = \left(\frac{\mathrm{d}p}{\overline{\mathrm{d}f}}\right)^3 \frac{N_1}{N_2}$$

$$= \left(\frac{\mathrm{d}p}{\overline{\mathrm{d}f}}\right)^3 \left(\frac{-\log \mathcal{T}_1}{-\log \mathcal{T}_2}\right) \frac{k_2(\pi/_4)\overline{\mathrm{d}f}^2}{k_1(\pi/_4)\mathrm{d}p^2}$$

$$= \left(\frac{\mathrm{d}p}{\overline{\mathrm{d}f}}\right)\left(\frac{\overline{\mathrm{d}f}}{\mathrm{d}p}\right)^{0.267} \frac{k_2}{k_1}$$

$$= \frac{k_2}{k_1}\left(\frac{\overline{\mathrm{d}f}}{\mathrm{d}p}\right)^{-0.733}$$

$$\because \quad k_1 = k_2$$

$$\therefore \quad \underline{1 - a' = \left(\frac{\overline{\mathrm{d}f}}{\mathrm{d}p}\right)^{-0.733}}$$

(iii)

$$a' = 1 - \left(\frac{\overline{\mathrm{d}f}}{\mathrm{d}p}\right)^{-0.733}$$

$$= 1 - \left(\frac{42.9}{3.2}\right)^{-0.733}$$

$$= \underline{0.851}$$

PROBLEM 3.15

The following table pertains to chemostat culture data of *E. coli* (aerobic, glucose as the growth-limiting substrate with its concentration c_{s_0} in the fresh medium = 0.968 kg m^{-3}

Table 3.5

Dilution rate, $D(h^{-1})$	Limiting substrate concentration, c_s (kg m^{-3})	Cell concentration, x (kg m^{-3})
0.06	0.006	0.427
0.12	0.013	0.434
0.24	0.033	0.417
0.31	0.04	0.438
0.43	0.064	0.422
0.53	0.102	0.427
0.60	0.122	0.434
0.66	0.153	0.422
0.69	0.170	0.430
0.71	0.221	0.390
0.73	0.210	0.352

(a) What are μ_m and K_s values?
(b) What are m and Y_G values?
(c) What are q_{O_2} and Y_{G_0} values?
(d) Taking the chemical formula of cells as $C_5H_7NO_2$, discuss the stoichiometry of glucose metabolized into cell material at $\mu = 0.5\ h^{-1}$, provided $RQ = 1$. No metabolite other than the cell, CO_2 and H_2O need be considered. See Fig. 3.9.

Solution

(a) $$\mu = \frac{\mu_m c_s}{K_s + c_s}; \quad \frac{1}{\mu} = \frac{K_s}{\mu_m c_s} + \frac{1}{\mu_m}$$

$\mu = D$, and

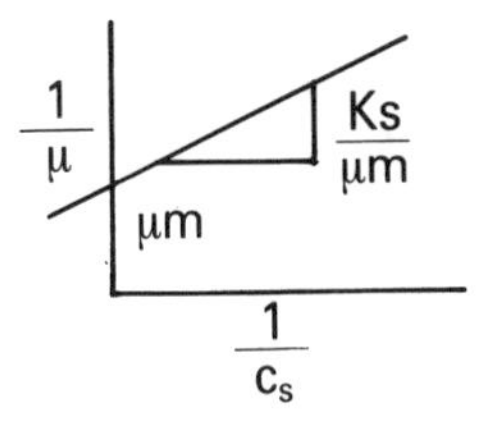

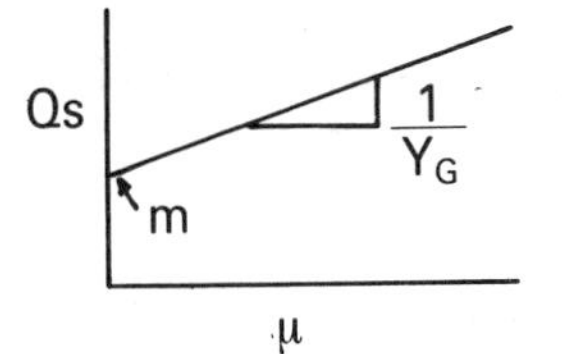

$\mu_m = 1.9\ h^{-1}$

$Ks = 77\ mg\ L^{-1}$

Metablic products;
CO_2 & H_2O mainly
ATP coupling, no waste

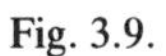

Fig. 3.9.

$$\mu = \frac{D(c_{s_0} - c_s)\ Y_{x/s}}{x}$$

and

$$Q_s = \frac{1}{x}\left(\frac{-dc_s}{dt}\right)$$

$$= m + \frac{1}{Y_{x/s}}\mu$$

Since,

$$\frac{dr}{dt} = Y_{x/s}\left(\frac{-dc_s}{dt}\right) - mY_{x/s}x$$

$$Q_s = \frac{q_{O_2}}{A_{O_2}} + \frac{1}{A_{O_2}}\left(B + \frac{1}{Y_{G_{O2}}}\right)\mu$$

since $Q_{O_2} = q_{O_2} + \frac{1}{Y_{G_{O2}}}\ \mu$, and $A\upsilon = B\mu + Q_{O_2}$

$$A_{O_2} = 6\ \frac{\text{mol } O_2}{\text{mol glucose}}$$

$$C_5H_7NO_2 + 5O_2 \rightarrow 5CO_2 + 2H_2O + NH_3(\text{cell})$$

$$\Rightarrow B = \left(5\ \frac{\text{mol } O_2}{\text{mol cell}}\right) \times \frac{1 \text{ mol cell}}{113 \text{ gm-mol cell}}$$

$$= \underline{0.00442 \text{ mol } O_2\ (\text{g mol})^{-1} \text{ cell}}$$

(b) In some cases the Q_s and μ relationship may exhibit pattern shown in Fig. 3.10 which is abnormal; something seems to be wrong (carbon balance also a factor to be checked)

$m = 0.05$ kg glucose kg^{-1} cell h
$Y_G = 0.53$ kg cell kg^{-1} glucose

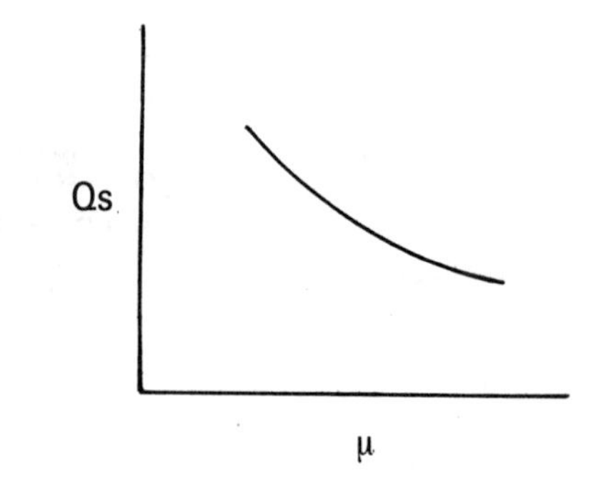

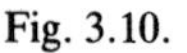
Fig. 3.10.

(c) Based on O_2 balance

$$q_{O_2} = A_m$$

$$Y_{G_{O2}} = \frac{1}{\frac{A}{Y_g} - B}$$

When complete combustion use the following data:

$q_{O_2} = 0.0018$ kg mol O_2 kg^{-1}h
$Y_{G_{O2}} = 45$ g cell mol^{-1} O_2

(d) The stoichiometric equation is given by

$$C_6H_{12}O_6 + aNH_3 + bO_2 \rightarrow cC_5H_7O_2 + dCO_2 + fH_2O$$

c = fraction of biomass carbon per g glucose

$$= \frac{\sigma_b}{\sigma_s} Y_{x/s}$$

where:

σ_b = weight fraction of carbon in cell biomass
σ_s = weight fraction of carbon in glucose
$Y_{x/s}$ = growth yield.

$$= 0.53 \frac{\text{kg cell produced}}{\text{kg glucose consumed}}$$

Therefore,

$$c = \frac{60/113}{72/180} \times 0.53$$

$$= 0.7$$

From stoichiometry,

Carbon balance:

$$6 = 5c + d$$

or,

$$d = 2.5 \qquad (3.15.1)$$

Hydrogen balance:

$$12 + 3a = 7c + 2f$$

or,

$$2f = 7.1 + 3a \qquad (3.15.2)$$

Oxygen balance:

$$6 + 2b = 2c + 2d + f$$

or,

$$2d + f = 4.6 + 2b \qquad (3.15.3)$$

$(\because c = 0.7)$

Nitrogen balance:

$$a = c = 0.7 \qquad (3.15.4)$$

From equations (3.15.1), (3.15.2), (3.15.3) and (3.15.4), we get,

$a = 0.7$
$b = d = 2.5$ (which proves $RQ = 1$)
$f = 4.6$

Therefore the balanced stoichiometry will be

$$C_6H_{12}O_6 + 0.7NH_3 + 2.5O_2 \rightarrow 0.7\ C_5H_7NO_2 + 2.5CO_2 + 4.6\ H_2O$$

Glucose carbon converted to cell material = 70%.

PROBLEM 3.16

The following correlation between the respiration rate R (kg O_2 m^{-3} h^{-1}) and the specific growth rate μ_G l h^{-1} of micro-organisms is suggested,

$$R = K\mu_G(c_{s_0} - c_s) \qquad (3.16.1)$$

where c_s is substrate concentration (kg m^{-3}), c_{s_0} is the initial substrate concentration (kg m^{-3}), and K is a constant, e.g. 0.4 kg O_2 kg^{-1} substrate, for bacteria.

Discuss the physiological meanings of the correlation.

Solution

The equation implies that the respiration by microbes might depend on the growth rate; the product of the specific growth rate, μ_G, and the consumption of substrate, $c_{s_0} - c_s$, may be accepted as representative of growth rate, by assuming that the cells are produced in proportion to the consumption of substrate. The equation could be used to assess an approximate value of oxygen demand by the microorganism concerned, and the estimated value may be used for the design of a fermenter and decisions about the operational condition of agitation and aeration, when growth data is available, but not about oxygen consumption. The assessment is not of course complete — there is no consideration of respiration for maintenance.

PROBLEM 3.17

The yields of cell mass based on consumed substrate and oxygen, and heat generation for cell reproduction are important factors to assess the cost for production of microbial protein. Table 3.6 shows three yield coefficinets of *Pseudomonas fluorescens* grown on various carbon sources (Abbot & Claman, 1973).

Table 3.6 — Cell yields of *Pseudomonas fluorescens* grown on various carbon sources and their cost

Substrate	Cost [¢ / subst.–kg]	Y_s [cell-kg / subst.–kg]	Y_{O_2} [cell-kg / O_2–kg]	Y_{heat} [cell-kg / kJ]
Glucose	4.0	0.51	1.47	0.19
Acetic acid	13.0	0.36	0.70	0.095
Methanol	4.0	0.40	0.44	0.054
Ethanol	13.0	0.68	0.61	0.082
Isopropanol	11.0	0.43	0.23	0.034
n-Paraffin	9.0	1.03	0.50	0.073
Methane	2.0	0.62	0.20	0.028

Discuss the advantages and disadvantages of each carbon source for the production of microbial cell (SCP). The production cost for a raw material (carbon source), oxygen supply and cooling may be assessed with nominal values of costs of substrates shown in the table, and conversion factors for estimation of costs of oxygen supply and cooling, 0.75 [¢/O_2–kg] and 0.05425[¢/kJ], respectively.

Reference

Abbot, B. J. & Claman, A. (1973) *Biotechnol. Bioeng.*, **15**, 117.

Solution

The cost for a raw material, power consumption for oxygen supply and cooling to remove heat generated during cultivation can be roughly estimated dividing the unit cost of substrate by the substrate yield, oxygen supply by the oxygen yield, and cooling by heat generation yield, respectively. Table 3.7 shows a calculated result.

Table 3.7 — Effect of yield coefficients on the cost for cell production

Substrate	Breakdown of cost for cell production [¢/cell–kg]			
	Raw material	Oxygen supply	Cooling	Total
Glucose	7.84	0.51	1.21	9.56
Acetic acid	36.10	1.07	2.42	39.59
Methanol	10.00	1.70	4.26	15.96
Ethanol	19.10	1.23	2.80	23.13
Isopropanol	25.60	3.26	6.76	35.62
n-Paraffin	8.74	1.50	3.15	13.39
Methane	3.22	3.75	8.21	15.18

A couple of comments arise from the table even though the calculated values themselves are not accurate for practical use and the unit costs of raw material, oxygen supply and cooling, strongly depend on the economic situation.

(1) *n*-Paraffin, methanol and methane are more promising raw materials than ethanol, isopropanol and acetic acid because of their low unit costs. The total costs of using them, however, are still greater than of using glucose.

(2) In spite of a much lower unit cost for methane, the total cost is not lower compared with that for the use of glucose because of the high cost of oxygen supply and cooling. Consequently, the above-mentioned yield coefficients are very important factors for selection of raw materials for cultivation of a microorganism, reminding that the cost for raw materials, power consumption for oxygen supply and cooling have big parts in the total cost.

PROBLEM 3.18

(a) Derive and discuss the relationship between cell productivity, product productivity, product yield, cell yield and operation conditions such as operation time and initial conditions in the batch culture where it is assumed that $\mu = \text{constant}$ and that the production rate, p, is given as Leudeking and Piret (1959) model. Total operation time consists of batch culture time t_f and processing time which is needed for washing t_{p_b} sterilization, feeding and so on. Productivity will be defined as the amount of cell or product produced per unit operation time per unit volume.

(b) Discuss the effect of t_f and t_{p_b} on the cell productivity in Problem 3.17 where μ is subject to Monod-type equation (μ = constant) by the numerical calculation. Use the same parameters as Problem 3.17.

Reference

Leudeking, R. & Piret, E. L. (1959) *J. Biochem. Microbiol. Tech. & Eng.*, **1**, 431.

Solution

(a) In this case, the following equation should be considered,

$$\frac{dx}{dt} = \mu x \tag{3.18.1}$$

$$\frac{d(c_s)}{dt} = -\mu x - \gamma p x \tag{3.18.2}$$

$$\frac{d(c_p)}{dt} = px \tag{3.18.3}$$

where,

$$p = \alpha\mu + \beta \tag{3.18.4}$$

α and β are constant parameters. Here μ is assumed to be constant. Initial conditions are; $x(0) = 1$, $c_p(0) = 0$, $c_s(0)$ = given. The performance criteria, cell productivity Q_x and product productivity Q_p can be written as

$$Q_x = \frac{x(t_f) - 1}{t_f + t_b} \tag{3.18.5}$$

$$Q_p = \frac{c_p(t_f)}{t_f + t_b} \tag{3.18.6}$$

From equation (3.18.1)

$$x(t) = \exp(t) \tag{3.18.7}$$

Substituting equation (3.18.7) into equation (3.18.3), the following solution

$$c_p = (\alpha + \beta/\mu)\ \{\exp(t) - 1\} \tag{3.18.8}$$

can be obtained. On the other hand, by integration of equation (3.18.2) using equation (3.18.7),

$$c_s = (1 + \gamma(\alpha + \beta/\mu))\ (1 - \exp(t)) + c_s(0) \tag{3.18.9}$$

is derived. Then production yield Y_p with respect to substrate consumed becomes as

$$Y_p = \frac{c_p(t_f)}{c_s(0) - c_s} = \frac{(\alpha + \beta/\mu)}{1 + \gamma(\alpha + \beta/\mu)} \tag{3.18.10}$$

and cell yield Y_x becomes as

$$Y_x = \frac{x(t_f) - x(0)}{c_s(0) - c_s} = \frac{1}{1 + \gamma(\alpha + \beta/\mu)} \tag{3.18.11}$$

Of course Y_p and Y_x can be derived from direct manipulation of equations (3.18.1), (3.18.2) and (3.18.3). On the other hand, product productivity Q_p and cell productivity Q_x are given as

$$Q_p = \frac{(\alpha + \beta/\mu)\,(\exp(t_f) - 1)}{t_f + t_b} \tag{3.18.12}$$

$$Q_x = \frac{\exp(t_f) - 1}{t_f + t_b} \tag{3.18.13}$$

(b) For various t_f and t_b, Q_x is calculated as shown in Table 3.8 where parameters are taken as the same as in Problem 3.17.

Table 3.8

t_b	t_f	x_f	Q_x
0.18	4.68	100.0	20.40
0.18	5.41	200.0	35.60
0.18	5.53	217.0	37.84
0.18	5.54	217.5	37.85
0.18	5.56	217.9	37.76
1.0	5.54	217.5	33.11

From the above result, optimal operational time t_f is found to exist and it is 5.54 ($x_f = 217.5$). The optimal value of Q_x is 37.85. Of course, processing t_b gives a negative effect on Q_x as is shown in the above numerical example.

Fed batch culture is obviously neither batch nor continuous, but the kinetics of the culture lies somewhere in between. Fed batch culture is practised for various purposes, namely: (a) to reduce the concentration of toxic metabolites, (b) to dilute the product concentration, (c) to improve kinetics, (d) to maximize the activity of participating cells which are in dense culture with normally higher doubling time, and (e) to reduce the size of reactor relative to product concentration obtainable either in batch or continuous culture.

PROBLEM 3.19

In a fed-batch culture, the following problems require solution: (a) computation of the feed rate of substrate in order to keep the value of μ constant (μ can be represented by a function of substrate concentration), and (b) state the conditions under which 'switching' from batch to fed-batch can be achieved without any lag time. 'Switching' can be attributed to either concentration of substrate or product including cells (if growth is the objective function).

Solution

(a) In a fed-batch culture, the following equations should be considered.

$$\frac{d(Vx)}{dt} = \mu Vx \tag{3.19.1}$$

$$\frac{d(Vc_s)}{dt} = -\mu Vx + Dc_{s_f} \tag{3.19.2}$$

$$\frac{dV}{dt} = D \tag{3.19.3}$$

where D is used instead of D_i in general equation. Concentration c_s should be constant because μ is a function of c_s and should be kept constant. Then dc_s/dt must be zero. From equation (3.19.2),

$$\frac{dc_s}{dt} = \mu Vx + D(c_{s_f} - c_s) = 0 \tag{3.19.4}$$

Then

$$D = \frac{-\mu Vx}{c_{s_f} - c_{s_0}} \tag{3.19.5}$$

must be satisfied where c_{s_0} is a value of c_s which gives the desired value of μ. On the other hand, using constant μ in equation (3.19.1).

$$Vx = (Vx)_0 \exp(\mu t) \tag{3.19.6}$$

where $(Vx)_0$ means the total cell mass in the medium when the fed-batch operation starts. Then

$$D = \frac{(Vx)_0}{c_{s_f} - c_{s_0}} \mu \exp(\mu t) \tag{3.19.7}$$

Finally, feed rate, D, becomes an exponentially increasing function as shown in equation (3.19.7). Then, this type of operation is called an exponential fed-batch culture. However, for calculation of D and F, the concentration, c_{s_0} and x_0 which are the desired substrate concentration and cell concentration at the start of the fed-batch operation, respectively, should be given. If these values are not correct, the exponential feed rate shown in equation (3.19.7) cannot regulate μ constant from the start of this fed-batch culture.

(b) In the batch culture, as mentioned in Problem 3.17

$$\frac{dx}{dt} + \frac{dc_s}{dt} = 0 \tag{3.19.8}$$

Integrating equation (3.19.8),

$$x + c_s = x(0) + c_{s_0} = 1 + c_s(0) \tag{3.19.9}$$

Then,

$$x(0) = c_{s_0} - c_0 + 1 \tag{3.19.10}$$

If

$$c_s(0) = c_{s_f} \text{ and } c_s(0) - c_{s_0} \gg 1, \tag{3.19.11}$$

then

$$\frac{(Vx)_0}{c_{s_f} - c_{s_0}} = 1 \tag{3.19.12}$$

that is,

$$D = \mu \exp(\mu t) \tag{3.19.13}$$

or

$$\underline{F = V(0)\mu_m \mu \exp(\mu t)} \tag{3.19.14}$$

In equations (3.19.13) and (3.19.14), they do not include c_{s0} or x_0 except initial liquid volume. It is very easy to give the flow rate according to equation (3.19.13) or (3.19.14) and the quasi-steady state $\dot{x} = \dot{c}_s = 0$, $\dot{V} \neq 0$ will be obtained immediately after changing from the batch to the fed-batch operation. For other conditions of the switching from a batch to the fed-batch operation except for one satisfying equation (3.19.11), it takes lag-time, more or else, to attain a quasi-steady state if the initial cell and substrate concentrations are not given correctly.

PROBLEM 3.20

Analyse the dynamics of the fed-batch culture with respect to x and c_s when the feed rate is kept constant.

Solution

The fed-batch culture where the feed rate is kept constant is called here a constant fed-batch culture. The constant fed-batch culture will be the most easy operation among the various type of fed-batch operations.

In a fed-batch culture, the following equations should be considered.

$$\frac{d(Vx)}{dt} = \mu V x \tag{3.20.1}$$

$$\frac{d(Vc_s)}{dt} = -\mu V x + D c_{s_f} \tag{3.20.2}$$

$$\frac{dV}{dt} = D \tag{3.20.3}$$

From equations (3.20.1) and (3.20.2), equation (3.30.4) is derived,

$$\frac{d(Vx)}{dt} + \frac{d(Vc_s)}{dt} = D c_{s_f} \tag{3.20.4}$$

and integrating this equation,

$$Vx + Vc_s = c_{s_f}Dt + x_0 + c_{s_0} \quad (3.20.5)$$

Because $Dt = V - 1$ from equation (3.20.3),

$$\underline{Vx = V(c_{s_f} - c_s) + (x_0 + c_{s_0} - c_{s_f})} \quad (3.20.6)$$

On the other hand, equation (3.20.1) shows that Vx is an increasing function along time. And equation (3.20.2) shows that Vc_s is the decreasing function with respect to the time. For the right-hand side in equation (3.20.2), $-\mu Vx$ is decreasing and Dc_{s_f} is constant. Then, from a certain time, right-hand side of equation (3.20.2) becomes negative. After all, Vx approached Vc_{s_f} or $c_{s_f}Dt$ as seen in equations (3.20.5) or (3.20.6) when the operation time has been advanced. And Vx increases linearly along operation time t and cell concentration approaches to c_{s_f}. It means that all substrate ideally converted to the cell and that the substrate concentration becomes 0.

After all, Vx increases linearly with respect to time t in the constant fed-batch culture.

PROBLEM 3.21

Find the feed rate of substrate in order to keep μ constant in a fed-batch culture where substrate is also utilized for the production of a bio-product where both specific growth rate μ and production rate ρ are functions depending on only substrate concentration.

Solution

In this case, system equation becomes,

$$\frac{d(Vx)}{dt} = \mu Vx \quad (3.21.1)$$

$$\frac{d(Vc_s)}{dt} = -\mu Vx - \gamma\rho Vx + Dc_{s_f} \quad (3.21.2)$$

$$\frac{d(Vc_p)}{dt} = \rho Vx \quad (3.21.3)$$

$$\frac{dV}{dt} = D \quad (3.21.4)$$

From equation (3.21.2) using the condition that $dc_s/dt = 0$,

$$\begin{aligned} D &= \frac{Vx}{c_{s_f} - c_s}(\mu + \gamma\rho) \\ &= \frac{(Vx)_0}{c_{s_f} - c_s}(\mu + \gamma\rho)\exp(\mu t) \\ &\equiv \alpha \exp(\mu t) \end{aligned} \quad (3.21.5)$$

where,

$$\alpha \triangleq \frac{(Vx)_0}{c_{s_f} - c_s}(\mu + \gamma\rho)$$

The feed rate D also becomes an exponentially increasing function. Integrating equation (3.21.1),

$$Vx = (Vx)_0 \exp(\mu t) \tag{3.21.6}$$

Also substituting equation (3.21.5) into equation (3.21.4) and integrating it.

$$V = (\alpha/\mu)(\exp(\mu t) - 1) + 1 \tag{3.21.7}$$

Then from equations (3.21.6) and (3.21.7),

$$x = \frac{\mu(Vx)_0 \exp(\mu t)}{\alpha(\exp(\mu t) - 1) + \mu} \tag{3.21.8}$$

From this equation, it is shown that x is not constant but approaches to

$$\lim_{t \to \infty} x = \frac{\mu(Vx)_0}{\alpha} \tag{3.21.9}$$

when the operation time has been advanced.

On the other hand, integrating equation (3.21.3), we get:

$$Vc_p = \frac{\rho}{\mu}(Vx)_0 (\exp(\mu t) - 1) + (Vc_p)_0 \tag{3.21.10}$$

$$c_p = \frac{\rho(Vx)_0 (\exp(\mu t) - 1) + \mu(Vc_p)_0}{\alpha(\exp(\mu t) - 1) + \mu} \tag{3.21.11}$$

Then c_p is not also constant but approaches to

$$\lim_{t \to \infty} c_p = \frac{\rho(Vx)_0}{\alpha} \tag{3.21.12}$$

when operation time increases. And the ratio (c_p/x) approaches to ρ/μ. If ρ can be represented by $\rho = \beta\mu$, (β = constant parameter), it can be shown that the situation become the same as that shown in Problem 3.18.

PROBLEM 3.22

During a batch culture of baker's yeast on glucose medium under aerobic conditions, the analytical data of Table 3.9 were recorded. (Initial substrate concentration = 8.96 kg m^{-3}.)

From the data of the first phase of a diauxic growth experiment, the following parameters should be calculated: (experimental conditions for gases: 20°C, 95,990.4 N m^{-2}).

(a) yield coefficient with respect to glucose, $Y_{x/s}$;
(b) the respective specific oxygen uptake rate and carbon dioxide release rate, q_{O_2} and q_{CO_2};
(c) the respiratory quotient, RQ; and
(d) the carbon balance, at time $t = 9.0$ h. (See Fig. 3.11.)

Table 3.9 — Experimental data of aerobic cultivation of baker's yeast on glucose

Time (h)	Dry weight ($kg\ m^{-3}$)	Ethanol ($kg\ m^{-3}$)	Substrate ($kg\ m^{-3}$)	CO_2 content (vol%)	O_2 content (vol%)	Air flow ($m^3\ h^{-1}$)	Reactor volume (m^3)
1.0	0.05	0.15	8.65	0.26	0.06	100.6×10^{-3}	5.95×10^{-3}
3.0	0.14	0.30	8.44	0.54	0.07	100.6×10^{-3}	5.43×10^{-3}
5.0	0.35	0.66	6.38	1.13	0.16	103.2×10^{-3}	4.92×10^{-3}
7.0	0.82	1.55	2.58	1.27	0.15	198.0×10^{-3}	4.41×10^{-3}
9.0	1.68	2.61	0.00	0.16	0.31	200.5×10^{-3}	3.89×10^{-3}

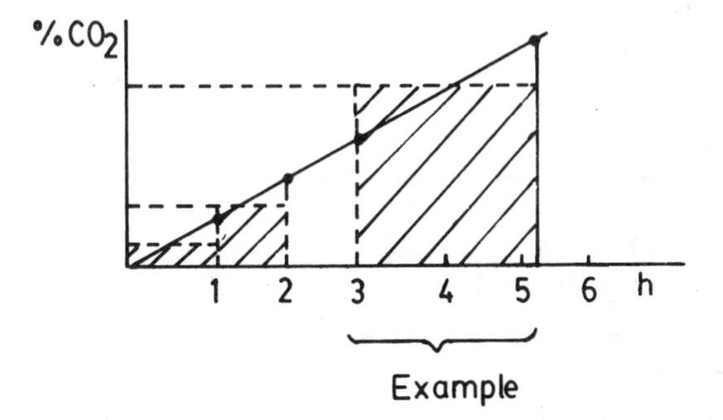

Fig. 3.11 — Example for the interval between 3.0 and 5.0 h.

Solution

(a) $Y_{x/s} = \frac{x}{s} = \frac{1.68}{8.96} = 0.19 \frac{\text{g cells}}{\text{g glucose}}$

(b) $q_{CO_2} = \frac{\text{aeration rate } K_{SC}\ \%CO_2\text{-content}}{\text{volume dry weight} \times 22.4} \times 100\ [\text{mol } CO_2 \text{ g } DW^{-1}\ h^{-1}]$

$$= \frac{200.5 \times 0.8864 \times 0.16 \times 100}{3.89 \times 1.68 \times 22.4}$$

$$= 1.94 \times 10^{-3} \frac{\text{kg mol } CO_2}{\text{kg } DW \text{ h}}$$

$$q_{O_2} = \frac{\text{aeration rate } K_{SC}\ \%O_2 \text{ content}}{3.89 \times 1.68 \times 22.4} \times 100$$

[K_{SC} = conversion factor to SC]

$$= \frac{200.5 \times 0.8864 \times 0.31 \times 100}{3.89 \times 1.68 \times 22.4}$$

$$= 3.76 \times 10^{-3} \frac{\text{kg mol O}_2}{\text{kg } DW \text{ h}}$$

(c) $RQ = \dfrac{\Delta\%\text{CO}_2}{\Delta\%\text{O}_2} = \dfrac{q_{\text{CO}_2}}{q_{\text{O}_2}} = \dfrac{1.94 \times 10^{-3}}{3.76 \times 10^{-3}}$

$= 0.516$

(d) Carbon balance:

$C_{\text{substrate}} = 8.96 \times \dfrac{72}{180}$

$= 3.58 \text{ kg m}^{-3}$

$C_{\text{dry weight}} = 1.68 \times 0.5$

$= 0.84 \text{ kg m}^{-3}$

$C_{\text{ethanol}} = 2.61 \times \dfrac{24}{46}$

$= 1.36 \text{ kg m}^{-3}$

C_{CO_2} = integrated amount over the entire batch

Average of CO_2 production between the third and the fifth hour after inoculation:

$$\frac{(5-3)[\Delta\%\text{CO}_2(\text{at 5h}) \text{ aeration rate} + \Delta\%\text{CO}_2(\text{at 3 h}) \text{ air rate}]}{2 \times 100}$$

$$= \frac{(5-3)[1.13\% \times 103.2 \times 10^{-3} + 0.54\% \times 100.6 \times 10^{-3}]}{V \times 22.4 \times 2 \times 100}$$

$= \text{kg mole CO}_2 \text{ m}^{-3}$ medium

t (0 h) until t_1 (1 h): 0.043 kg CO_2 m^{-3}
t_1 (1 h) until t_2 (3 h): 0.2778 kg CO_2 m^{-3}
t_2 (3 h) until t_3 (5 h): 0.6488 kg CO_2 m^{-3}
t_3 (5 h) until t_4 (7 h): 0.5498 kg CO_2 m^{-3}
t_4 (7 h) until t_5 (9 h): 0.342 kg CO_2 m^{-3}

Total C_{CO_2}

$= 0.043 + 0.2778 + 0.6488 + 1.5498 + 1.342$

$= 3.8614$

$\therefore$ carbon in $CO_2 = 3.8614 \times \dfrac{12}{44} = 1.053$ kg carbon m^{-3}

Thus, the carbon recovery will be:

$$\text{carbon balance } (CB) = \frac{C_{\text{DW}} + C_{\text{ethanol}} + C_{\text{CO}_2}}{C_{\text{substrate}}}$$

$$= \frac{0.84 + 1.36 + 1.053}{3.58}$$

$$= \frac{3.253}{3.58}$$

$$= 0.909$$

or,

$$= \underline{90.9\%}$$

PROBLEM 3.23

A microorganism grows on glucose. The rate of cell production is

$$q_x = \frac{4}{3} \frac{c_s x}{c_s + 4} \frac{g \text{ cells}}{m^3 \, h}$$

The cell yield coefficient is $Y_{s/x} = 0.1$

1. Consider a reactor of volume of one m^3 and a feed with $c_{s_f} = 60 \text{ g m}^{-3}$, $x_f = 0$.
   (a) Calculate the feed flow υ_0 for which maximum cell production is obtained.
   (b) Calculate the largest feed flow υ_{max} which can be handled in the given reactor.
   (c) Assuming that a feed flow $\upsilon_0 = 2.5 \text{ m}^3 \text{h}^{-1}$ has to be handled by the reactor system how would you do this in the best possible way without rebuilding the reactor?
   (d) Assuming that an additional reactor is made available for treatment of the $2.5 \text{ m}^3 \text{h}^{-1}$ feed stream determine its size V_1 for maximum cell production per m^3 in the combined system $(V_1 + 1)$ m^3. Consider both series and parallel coupling of the two reactors.
   (e) A centrifuge is installed. A fraction α of the cells is returned to the reactor. Determine the recirculation rate R as a function of α if (for one reactor of volume $V = 1 \text{ m}^3$) the cell production rate of 1(a) is to be maintained also for $\upsilon_0 = 2.5 \text{ m}^3 \text{h}^{-1}$
2. Consider a general situation where a chemostat is operated at steady state with feed composition $(c_{s_f}, x_f) = (c_{s_f}, 0)$ and outlet composition $(c_s, x) = (c_{s_0}, x_0)$. The holding time $\tau = V/\upsilon$ is changed from τ_0 (which corresponds to (c_{s_0}, x_0)) at time $t = 0$ to $n\,\tau_{crit}$ where

$$\tau_{crit} = \frac{K_m + c_{s_1}}{kc_{s_f}}$$

   (a) Show that the outlet cell concentration for $t > 0$ follows

$$kt = (a+1)\frac{n}{n-1}\left\{\ln\frac{x}{x_0} - \frac{na}{n-1+na}\ln\frac{b - x/x_0}{b-1}\right\}$$

   where

$$b = \frac{c_{s_f}\, Y_{x/s}}{x_0}\frac{(a+1)(n-1)}{n-1+na}$$

   For $n > 1$, $b = (x/x_0)t \rightarrow \infty$

$$q_x = \frac{kc_s x}{K_m + c_s} = \frac{kSx}{a + S}$$

where $S = c_s/c_{s_f}$ and $a = K_m/c_{s_f}$.

(b) Now return to the reactor studied in the first part of the exercise. For $t \leqslant 0$ let $v = 1\ m^3/h$. At $t = 0$ v is changed to $1.5\ m^3/h$.
Determine (and plot) x/x_0 as a function of t (h).

(c) Consider an incident where for 10 minutes the feed stream contains traces of a microorganism with maximum growth rate $k_1 = 8/3\ h^{-1}$. Otherwise the kinetics is the same as for the microorganism in question 1. After 10 minutes the concentration of the foreign microorganism is $10^{-2}\ g\ m^{-3}$. Thereafter the feed stream contains no microorganisms ($x = x_1 = 0$ and $c_{s_f} = 60\ g\ m^{-3}$).
Write a model for $x(t)$ and $x_1(t)$. What happens qualitatively?
Solve the unsteady mass balances on a computer.

Solution

1(a) The cell production rate P_x per m^3 reactor is equal to q_x and since $\frac{v}{V} = D = \frac{q_x}{x} P_x$ is also $= xD$, if $x_f = 0$. Now $D = \frac{q_x}{x} = \frac{q_s}{c_{s_f} - c_s}$ and from the last two expressions $X = \frac{x}{Y_{x/s}\, c_{s_f}} = 1 - S = 1 - \frac{c_s}{c_{s_f}}$ which holds when $x_f = 0$. i.e. for no maintenance there is a simple relation between X and S: $X = 1 - S$ or $x = Y_{x/s}(c_f - c_s)$

$$q_x = \frac{kc_s x}{c_s + K_m} = \frac{kS}{S + a}\,(1 - S)\; Y_{x/s} c_{s_f}$$

q_x is maximum if $\frac{S(1-S)}{S+a}$ is maximum:

$$\mathrm{d}\left(\frac{q_x}{kY_{x/s}c_{s_f}}\right) = 0 \text{ for } (1 - 2S)\,(S + a) - S(1 - S) = 0$$

or,

$$S^2 + 2aS - a = 0$$

$S = -a + \sqrt{a + a^2}$ for which $\frac{S(-1-S)}{S+a} = (\sqrt{1+a} - \sqrt{a})^2$ $\quad a = \frac{1}{15}$; $Y_{s/x} = 0.1$;

$c_{s_f} = 60\ g\ m^{-3}$

$S_{opt} = \frac{1}{5} \rightarrow c_s = 12\ g\ m^{-3}$; $P_x = (\sqrt{16/15} - \sqrt{1/15})^2 \frac{4}{3}(0.1)\,60$

$$= \underline{4.8\ g\ m^{-3}\ h^{-1}}$$

(b) $\tau_{min} = \frac{K_m + c_{s_f}}{kc_{s_f}} = \frac{4 + 60}{\frac{4}{3}60} = 0.8\ h$

$\underline{v_{max} = 1.25\ m^3\ h^{-1} \text{ per } m^3 \text{ reactor}}$

(c) Since maximum productivity is obtained when $\tau = 1$ h, i.e. when 1 $m^3 h^{-1}$ passes the reactor, it is best to bypass 1.5 $m^3 h^{-1}$.

(d) Parallel coupling is easy (and best). Buy a new chemostat of volume 1.5 m^3 and run it in parallel with the 1 m^3 reactor. Split the feed stream and pass 1.5 $m^3 h^{-1}$ through the new reactor. Maximum cell productivity 4.8 g $m^{-3} h^{-1}$ is obtained.

Serial coupling is more difficult.

Obviously the new reactor must be the first in the series, and it must be greater than 2 m^3 to avoid washout ($\tau_{min} = 0.8$ h). Choosing a new reactor of volume 2.5 m^3 would give optimum performance of the new installation (productivity 4.8 g cells $m^{-3} h^{-1}$) but we still have the old reactor of 1 m^3 volume which may be used downstream to give some further cell production (see Fig. 3.12). But should we choose $V_1 = 2.5$ m^3?

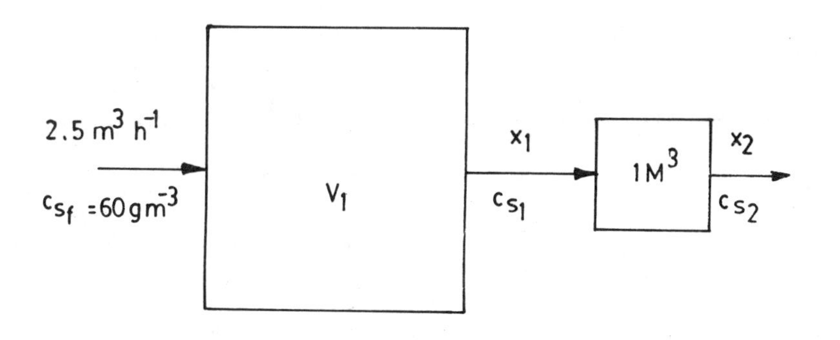

Fig. 3.12.

Surely, V_1 must be less than V_{opt}.

$$2 \text{ m}^3 < V_1 < 2.5 \text{ m}^3$$

We wish to maximize $P = \dfrac{2.5\,x_2}{V_1 + 1} = \dfrac{\text{produced cells (g h}^{-1})}{\text{total volume}}$ subject to the following mass balance restrictions

$$\text{reactor 1: } \frac{2.5}{V_1} = \frac{c_1}{4 + c_{s_f}} \frac{4}{3} \rightarrow V_1 = \frac{1.875\,(4 + c_{s_1})}{c_{s_f}} \tag{3.23.1}$$

$$\text{reactor 2: } \frac{2.5}{1} = 2.5 = \frac{c_{s_2}[(60 - c_{s_1})0.1 + 0.1(c_{s_1} - c_{s_2})]}{(4 + c_{s_2})0.1(c_{s_1} - c_{s_2})} \frac{4}{3}$$

$$= \frac{4}{3}\, c_{s_2} \frac{(60 - c_{s_2})}{(4 + c_{s_2})\,(c_{s_1} - c_{s_2})}$$

or:

$$0.875\, c_{s_2}^2 + (67.5 - 1.875 c_{s_1}) c_{s_2} - 7.5 c_{s_2} = 0 \tag{3.23.2}$$

It is possible — but not worth the effort — to differentiate P with respect to (say) c_{s_1} and equate $\frac{dP}{dc_{s_1}}$ to zero: $\frac{dx_2}{dc_{s_1}} = -Y_{x/s}\frac{dc_{s_2}}{dc_{s_1}}$, $\frac{dV_1}{dc_{s_1}} = \frac{-7.5}{c_{s_1}^2}$ and $\frac{dc_{s_2}}{dc_{s_1}}$ is found by differentiating (3.23.2). It is much easier (and this is a rather general statement) to run a short table:

c_{s_1}	12	12.5	15	17.5	20	18.25	18.75
V_1	2.5	2.475	2.375	2.303571	2.25	2.285959	2.275
						(using (3.23.1))	
c_{s_2}	1.92774	2.04464	2.69566	3.47855	4.4281	3.74412	3.92999
						(using (3.23.2))	
x_2	5.8072	5.7955	5.7304	5.65214	5.5572	5.62558	5.6070
						$= 0.1\,(60 - c_{s2})$	
P	4.1480	4.1695	4.2448	4.2773	4.2748	4.28002	4.28015

The best solution is consequently to let reactor 2 have a little more substrate to work on ($c_{s_1} \simeq 18.55$ g m^{-3} rather than $c_{s_1} = 12$ for optimal performance of V_1). The best 'serial result' 4.28 g m^{-3} h^{-1} is seen to be smaller than 4.8 g m^{-3} h^{-1} for parallel operation (we have shown this to be best!).
Please note: Optimization for production as shown here maybe far from industrial optimization criteria, e.g. very small c_{s_2}.

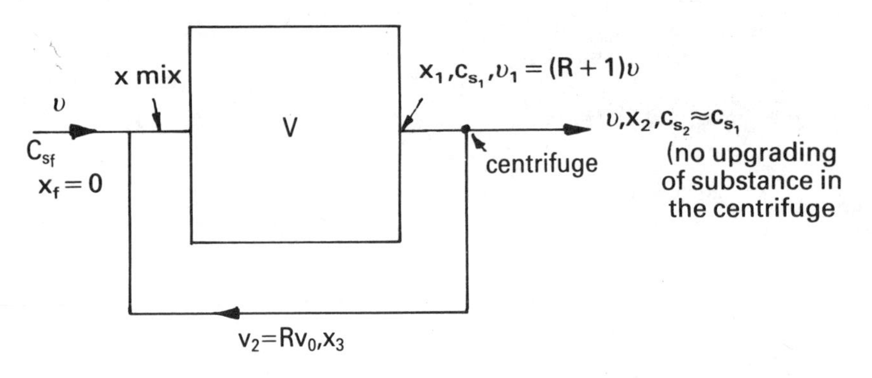

Fig. 3.13.

(e) Let the ratio of cell concentration in the recirculation stream to the cell concentration in the feed stream to the centrifuge be $\beta = \frac{x_3}{x_1} > 1$ (see Fig. 3.13).

A fraction $\alpha = \frac{x_3 v R}{x_1(R+1)v} = \frac{\beta R}{R+1}$ of the cells is returned to the fermenter. A mass balance around the centrifuge yields $x_1(R+1) = Rx_3 + x_2$
or

$$\frac{x_2}{x_1} = 1 - (\beta - 1)R \tag{3.23.3}$$

A mass balance for the total reactor system yields

$$\frac{V}{\upsilon}=\tau=\frac{x_2}{q_x x_1}=\frac{x_1[1-(\beta-1)R]}{x_1\dfrac{kc_{s_1}}{K_m+c_{s_1}}}=\frac{K_m+c_{s_2}}{kc_{s_2}}[1-(\beta-1)R] \tag{3.23.4}$$

Now choose $c_{s_2}=12\text{ g m}^{-3}$ (i.e. $x_2=0.1\,(60-12)=4.8\text{ g m}^{-3}$).

$$\frac{K_m+c_{s_2}}{kc_{s_2}}=1\text{ h}$$

and since

$$\tau=\frac{V}{\upsilon}=\frac{1}{2.5}=0.4\text{ h}$$

we must choose $\underline{1-(\beta-1)R=0.4.}$ This is the desired answer to question (e)

$$\left(\text{or } \alpha=\frac{\beta R}{R+1}=\frac{0.6+R}{R+1} \text{ where } R \text{ is } \frac{0.6}{\beta-1}\right)$$

but let us study the consequences of the design.

From (3.23.3),

$$x_1=\frac{x_2}{0.4}=12\text{ g m}^{-3};\ x_3=\beta x_1=14.4\text{ g m}^{-3};\ x_{mix}=\frac{Rx_3}{R+1}$$

Now let $\beta=1.2$. This means that $R=0.6/0.2=3$ and $\alpha=0.9$ (90% of the cells are returned to the reactor).

$$x_{mix}=\frac{3(14.4)}{4}=10.8\text{ g m}^{-3}$$

A mass balance for the reactor yields

$$Vq_x=\upsilon(R+1)(x_1-x_{mix});\ q_x=\frac{kc_{s_2}x_1}{K_m+c_{s_2}}=x_1 \tag{3.23.5}$$

i.e. $0.4x_1=4(x_1-10.8)$

or, $0.4\times 12=4\times 1.2$ which checks!!

Now let

$$c_{s_2}=3\text{ g m}^{-3},\ x_2=5.7\text{ g m}^{-3},\ \frac{K_m+c_{s_2}}{kc_{s_2}}=\frac{7}{4}=1.75\text{ h}$$

and we obtain

$$1-(\beta-1)R=\frac{0.4}{1.75}=0.22857$$

Again let

$$\beta = 1.2 \text{ and } R = \frac{1 - 0.22857}{1.2 - 1} = 3.857,$$

$\alpha = 0.9529$; $x_1 = \dfrac{5.7}{0.22857} = 24.93\ \text{g m}^{-3}$; $x_3 = 1.2\ x_1 = 29.925\ \text{g m}^{-3}$; $x_{mix} =$ $23.765\ \text{g m}^{-3}$

Check the mass balance for the reactors.

Note that the cell concentration in the recirculation loop and the reactor rapidly increase when $c_{s_2} \to 0$. This is a drawback of the cell recirculation method (which has obvious advantages as seen in this problem). In principle one may obtain $c_s = 0$ (i.e. 2.5 g cells h^{-1} from the reactor but $x_1 \to \infty$. Any maintenance requirement of the cells will kill the project!

2(a) Dynamic mass balances for substrate and for cells are:

$$\upsilon c_{s_f} - \upsilon c_s = q_s V + V \frac{dc_s}{dt} \tag{3.23.6}$$

$$-\upsilon x = -qx\ V + V \frac{dx}{dt} \tag{3.23.7}$$

Let $-q_s = -Y_{s/x} qx$, multiply the cell balance with $Y_{s/x}$ and add:

$$c_{s_f} = (c_s + Y_{s/x} x) + \frac{d(c_s + x)}{d(t/\tau)} \tag{3.23.8}$$

where $\tau = V/\upsilon$ and $t/\tau = Dt = \theta$.

The equation (3.23.8) (inhomogeneous first order differential equation, linear with constant coefficients) is easily integrated to $c_s + Y_{s/x} x = A \exp(-\theta) + c_{s_f}$.

At $t = 0$ the chemostat is working at steady state with $\tau = \tau_0$ and $c_{s_0} + Y_{s/x} x_0 = c_{s_f}$ where (c_{s_0} and x_0) are the concentrations of substrate and cells which correspond to τ_0. Hence $A = 0$ and for $t \geqslant 0$ we must also have $c_s + Y_{s/x}\, x = c_{s_f}$, i.e. the steady state mass balance holds also during the transient. *This is a highly unusual case, winged on the simple proportionality between q_s and q_x.*

We use $c_s + Y_{s/x}\, x = c_{s_f}$ to eliminate c_s from the mass balance for cells. For

$$q_x = \frac{kc_s x}{K_m + c_s}: \quad \frac{dx}{d\theta} = \frac{k\tau(c_{s_f} - Y_{s/x} x)x}{K_m + c_{s_f} - Y_{s/x} x} - x \tag{3.23.9}$$

Inserting $\tau = n\tau_{opt} = n\dfrac{K_m + c_{s_f}}{kc_{s_f}}$ and separation of variables

$$d\theta = \frac{(K_m + c_{s_f} - Y_{s/x} x)dx}{(n-1)\ (K_m + c_{s_f})x - Y_{s/x} x^2 \left\{ (n-1) + \dfrac{nK_m}{c_{s_f}} \right\}} \tag{3.23.10}$$

$$= \frac{(K_m + c_{s_f} - Y_{s/x} x)}{Bx(C-x)}\,dx$$

$$B = Y_{s/x}\left\{(n-1) + \frac{nK_m}{c_{s_f}}\right\} \text{ and } C = \frac{(K_m + c_{s_f})\,(n-1)}{B}$$

$$d\theta = \frac{1}{B}\left\{\frac{K_m + c_{s_f}}{C}\,\frac{1}{x} + \frac{-Y_{s/x} + (K_m + c_{s_f})/C}{C-x}\right\}dx \qquad (3.23.11)$$

Or, integrating

$$\theta = \frac{c_{s_f}}{n(K_m + c_{s_f})}\,kt = \frac{K_m + c_{s_f}}{c_{s_f}}\,\frac{n}{n-1}\left\{\ln\frac{x}{x_0} - \frac{na}{n-1+na}\ln\frac{b - x/x_0}{b-1}\right\}$$

(3.23.12)

or

$$kt = (a+1)\frac{n}{n-1}\left\{\ln\frac{x}{x_0} - \frac{na}{n-1+na}\ln\frac{b - x/x_0}{b-1}\right\} \qquad (n \neq 1) \qquad (3.23.13)$$

$$a = \frac{K_m}{c_{s_f}} \text{ and } b = \frac{c_{s_f} Y_{x/s}}{x_0}\,\frac{(a+1)\,(n-1)}{n-1+na}$$

From the relation

$$D = \mu = \frac{1}{n}\,\frac{c_{s_f}}{K_m + c_{s_f}} = \frac{c_{s_f}}{K_m + c_s}$$

It is easy to prove that $b = \left(\frac{x}{x_0}\right)_{t \to \infty}$ in the case $n > 1$.

$$\text{For } n = 1\text{: } kt = \frac{c_{s_f} Y_{x/s}}{x_0}\,\frac{(1+a)^2}{a}\left(\frac{x_0}{x} - 1\right) + \left(1 + \frac{1}{a}\right)\ln\frac{x}{x_0} \qquad (3.23.14)$$

(b) $\tau = \frac{1}{1.5} = \frac{2}{3}\text{ h} = \frac{5}{6}\tau_{crit}$ since $\tau_{crit} = \frac{4}{5}$ h (question 1(b)).

$$a = \frac{K_m}{c_{s_f}} = \frac{4}{60} = \frac{1}{15}\,; \qquad b = \frac{c_{s_f}\,Y_{x/s}}{x_0}\,\frac{(a+1)(n-1)}{n-1+na} = \frac{60(0.1)}{4.8}\,\frac{\left(1 + \frac{1}{15}\right)}{\frac{-1}{6} + \frac{1}{18}}\,\frac{(-1)}{6} = 2$$

$$\text{Hence, } kt = \frac{4}{3}t = \frac{16}{15}(-5)\left\{\ln\frac{x}{x_0} + \frac{1}{2}\ln\left(2 - \frac{x}{x_0}\right)\right\}$$

$$\text{or, } t = -4\left\{\ln\frac{x}{x_0} + \frac{1}{2}\ln\left(2 - \frac{x}{x_0}\right)\right\} \qquad (3.23.15)$$

The results are shown graphically in Figs. 3.14 and 3.15.

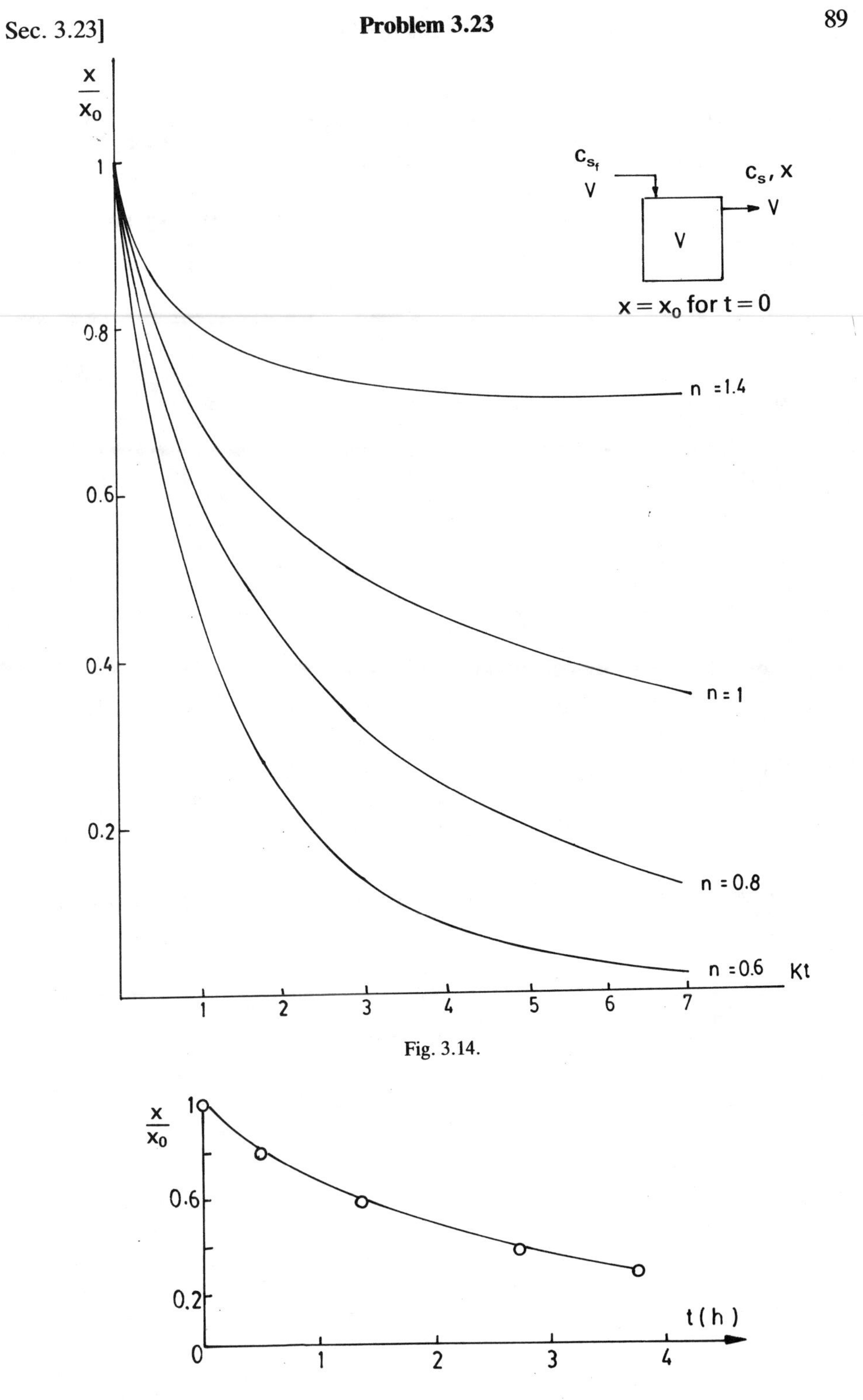

Fig. 3.14.

Fig. 3.15.

(c) $$\frac{dx}{d\theta} = -x + \tau\mu x \qquad \tau = 1\text{ h}, \quad \mu = \frac{4}{3}\frac{c_s}{4 + c_s}$$

$$\frac{dx_1}{d\theta} = -x_1 + 2\tau\mu x_1$$

$$x(t=0) = x_0 = 4.8\text{ g m}^{-3}$$

$$\frac{dc_s}{d\theta} = c_{s_f} - c_s - (\mu x + 2\mu x_1)Y_{s/x}\tau$$

$$x_1(t=0) = x_{10} = 10^{-2}\text{ g m}^{-3}$$

$$c_s(t=0) = 12\text{ g m}^{-3}$$

$$c_{s_f} = 60\text{ g m}^{-3}$$

or, in dimensionless form:

$$\frac{dy_1}{d\theta} = -y_1 + k\tau\frac{y_3}{a + y_3}y_1$$

$$\frac{dy_2}{d\theta} = -y_2 + 2k\tau\frac{y_3}{a + y_3}y_2$$

$$\frac{dy_3}{d\theta} = 1 - y_3 - k\tau\frac{y_3}{a + y_3}\left(\frac{Y_{s/x}x_0}{c_{s_f}}y_1 + 2\frac{Y_{s/x}x_{10}}{c_{s_f}}y_2\right)$$

$$y_1 = \frac{x}{x_0},\ y_2 = \frac{x_1}{x_{10}},\ y_3 = \frac{c_s}{c_{s_f}}$$

$y_1(0) = y_2(0) = 1$ and $y_3(0) = 0.2$

$$k\tau = \frac{4}{3}.1,\ a = \frac{1}{15},\ \frac{Y_{s/x}x_0}{c_{s_f}} = 0.8,\ \frac{Y_{s/x}x_{10}}{c_{s_f}} = \frac{5}{3}\ 10^{-3}$$

We are able to calculate the final values of Y_2 and Y_3. See Fig. 3.16 and Table 3.10.

Since $\tau = 1$ h we obtain the steady state values of c_s and x_1 from

$$\tau = 1\text{ h} = \frac{4 + c_s}{\frac{4}{3}2 + c_s} \rightarrow c_s = 2.4\text{ g m}^{-3}\ \left(Y_3 = \frac{2.4}{60} = 0.04\right)$$

$$x_1 = 0.1(60 - 2.4) = 5.76\text{ g m}^{-3}$$

$$Y_2 = \frac{x_1}{x_{10}} = 576$$

Table 3.10

Time (h)	y_1	y_2	y_3	Time (h)	y_1	y_2	y_3
0.02	1.0000	1.0202	0.1999	5.31	0.8249	137.7714	0.1105
0.08	1.0000	1.0833	0.1997	5.70	0.7650	175.5450	0.0954
0.26	0.9999	1.2966	0.1992	6.12	0.6928	217.9276	0.0825
0.62	0.9993	1.8629	0.1983	6.57	0.6094	264.2517	0.0721
0.96	0.9984	2.6081	0.1975	7.07	0.5169	313.6800	0.0637
1.29	0.9972	3.6269	0.1966	7.64	0.4187	364.8383	0.0569
1.62	0.9956	5.0126	0.1955	8.32	0.3185	416.2100	0.0516
1.94	0.9935	6.8957	0.1940	9.18	0.2194	466.3240	0.0472
2.27	0.9906	9.4519	0.1919	10.44	0.1229	514.7777	0.0437
2.59	0.9868	12.9189	0.1892	11.99	0.0581	547.0859	0.0417
2.91	0.9816	17.6159	0.1855	13.60	0.0264	562.8724	0.0407
3.23	0.9745	23.9687	0.1806	15.44	0.0106	570.7251	0.0403
3.55	0.9647	32.5396	0.1740	17.70	0.0034	574.2982	0.0401
3.89	0.9513	44.0541	0.1656	20.71	0.0008	575.6226	0.0400
4.22	0.9327	59.4104	0.1548	25.14	0.0001	575.9588	0.0400
4.57	0.9070	79.6235	0.1417	32.85	0.0000	575.9990	0.0400
4.93	0.8718	105.6156	0.1265	50.00	0.0000	576.0000	0.0400

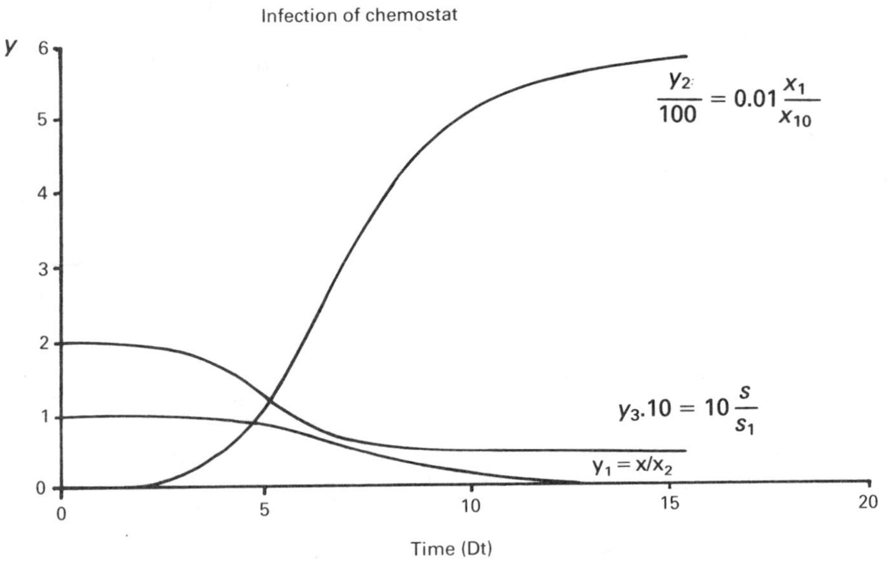

Fig. 3.16.

The original cell population seems to survive quite well for the first couple of hours ($Y_1 > 0.99$), but between 5 and 10 hours after the infection the original cell population looses out to the faster-growing species. Calculations such as those shown here may help the operator to do something useful with the plant during the 'grace period' before the inevitable catastrophe occurs. Eventually the whole tank has to be emptied and cleaned.

4

Death and containment

Contributors: Tarun Ghose, S. N. Mukhopadhyay, K. B. Ramachandran

PROBLEM 4.1

It was desired in a batch penicillin production plant of 40 m^3 capacity to supply sterile air through the bioreactor at the rate of 1 vvm. Average contaminant load in the incoming air was 3000 microorganisms per cubic metre of air. The operating period of the bioreactor was 100 hours. Considering chances of penetration 1 in 1 million, determine the depth of the filter required for the process. (Assume cross-sectional area of the air sterilizer vessel $= 0.42\ m^2$).

Reference

Richard, J. W. (1968) *Introduction to Industrial Sterilization*, Academic Press, New York.

Solution

Volumetric air flow rate $= 1$ vvm

So,

$$F = 40 \times 60\ m^3/h$$
$$= 2{,}400\ m^3\ h^{-1}$$

Incoming microbial load in air

$$= 3{,}000\ m^{-3}$$

$\therefore$Total number of organisms entering in 100 hours (1 batch)

$$N_1 = 2{,}400 \times 100 \times 3{,}000$$
$$= 72 \times 10^7 \text{ microorganisms}$$

Number of microorganisms acceptable in outlet sterile air,

$$N_2 = 1 \text{ in } 10^6 = 10^{-6}$$

Since cross-sectional area of the filter bed $= 0.42\ m^2$,

$$\text{Linear velocity of air} = \frac{\text{volumetric flow rate}}{\text{cross-sectional area}}$$

$$= \frac{2{,}400\ \text{m}^3\ \text{h}^{-1}}{0.42\ \text{m}^2}$$
$$= 5714.2857\ \text{m h}^{-1}$$

Now,

$$\ln \frac{N_1}{N_2} = 2.3 \log_{10} \frac{72 \times 10^7}{10^{-6}}$$
$$= 2.3 \log_{10} 72 \times 10^{13}$$

Now, as per log penetration law of sterilization,

$$\log \frac{N_1}{N_2} = kL$$

where, L = length of filter bed, and k = rate constant.

The value of k was ascertained from the plot of k versus linear velocity to be 40 m^{-1}. So,

$$L = \frac{1}{k} \log \frac{N_1}{N_2}$$
$$= \frac{1}{40} \times 2.3 \log_{10} 72 \times 10^{13}$$
$$= \frac{1}{40} \times 2.3(13 + \log_{10} 72)$$
$$= \underline{0.854\ \text{m}}$$

PROBLEM 4.2

A 40-m^3 working capacity bioreactor will be used to carry out a fermentation (28°C) lasting 4 days. It has been decided that a volumetric flow rate equivalent to 0.1 vvm be used. The incoming air contains on an average 4,000 microorganisms per cubic metre at an average bacterial diameter of 1 micrometre. Fibrous filter material having average fibre diameter of 19 micrometres and a void fraction of 0.95 is available for the construction of an air filter. An allowable risk of 0.001 has been designated in terms of bacterium penetration. Estimate the length of filter required. [Assume air velocity = 0.05 m s^{-1}.]

Solution

The volumetric flow rate of air, F

$$= 40 \times 0.1 \times 60 \times 24\ \text{m}^3\ \text{d}^{-1}$$
$$= 5{,}760\ \text{m}^3\ \text{d}^{-1}$$

Total number of organisms entering the filter, $N_1 = 4 \times 5{,}760 \times 4000 = 9216 \times 10^4$.
Now,

$$\ln \frac{N_1}{N_2} = \ln \left(\frac{9{,}216 \times 10^4}{10^{-3}} \right)$$
$$= 2.3 \log_{10} 9{,}216 \times 10^7$$
$$= 2.3\ (7 + \log_{10} 9{,}216)$$
$$= 25.218$$

$$\eta_0 = \frac{\pi\ d_{fi}(1 - f_f)}{4L\ f_f\ (1 \times 4.5\ f_f)} \ln \frac{N_1}{N_2} = \frac{15}{N_R\ N_{Pe}} \tag{4.2.1}$$

Now given that,

$$d_{fi} = 19\ \mu m = 19 \times 10^{-6}\ m$$

$$1 - f_f = 0.95 = \text{void fraction}$$

$$f_f = 1 - 0.95 = 0.05$$

$$N_R = \frac{d_p}{d_{fi}} = \frac{1.0 \times 10^{-4}}{19 \times 10^{-4}} = 5.27 \times 10^{-2}$$

$$\rho_{air}(25°C) = 1.2\ kg\ m^{-3}$$

$$\mu_{air}(25°C) = 1.8 \times 10^{-5}\ N\ S\ m^{-2}$$

So,

$$N_{Re} = \frac{d_{fi} v \rho}{(-f_f)\mu_{air}} = \frac{19 \times 10^{-6} \times 0.05 \times 1.2}{0.95 \times 1.8 \times 10^{-5}}$$
$$= 7 \times 10^{-2}$$

$$\mathcal{D}_{BM} = \frac{c\ k\ T'}{3\pi\mu_{air} d_p} = \frac{1.16 \times 1.38 \times 10^{-16} \times (273 + 25)}{3 \times 3.14 \times 1.8 \times 10^{-4} \times 1.0 \times 10^{-4}}$$
$$= 2.78 \times 10^{-11}\ m^2\ s^{-1}$$

$$N_{Sc} = \frac{\mu_{air}}{\rho\ \mathcal{D}_{BM}} = \frac{1.8 \times 10^{-5}}{1.2 \times 2.78 \times 10^{-11}}$$
$$= 5.4 \times 10^5$$

so,

$$N_{Pe} = N_{Sc}\ N_{Re} = 5.4 \times 10^5 \times 7.0 \times 10^{-2}$$
$$= 3.78 \times 10^4$$

$$\therefore N_R = N_{Pe}^{1/3}\ N_{Re}^{1/18} = (5.27 \times 10^{-2}) \times (3.78 \times 10^4)^{1/3} \times (7.0 \times 10^{-2})^{1/18}$$
$$= 1.53$$

From Graph 9.6 in *Biochemical Engineering* (2nd edn., 1973) by Aiba *et al.* (p. 228), we have for,

$$N_R\ N_{Pe}^{1/3}\ N_{Re}^{1/18} = 1.53$$

the value of

$$\eta_0 \, N_R \, N_{Re} = 15$$
$$\therefore \eta_0 = \frac{15}{N_R \, N_{Pe}}$$
$$= \frac{15}{(5.27 \times 10^{-2})\,(3.78 \times 10^4)}$$
$$= 7.53 \times 10^{-3}$$

Putting this value in equation (4.3.1), we have,

$$7.53 \times 10^{-3} = \frac{\pi \times 19 \times 10^{-6} \times 0.95}{4 \times L \times 0.05(1 + 4.5 \times 0.05)} \times \ln \frac{9216 \times 10^4}{10^{-3}}$$

or,

$$L = \frac{3.14 \times 19 \times 10^{-6} \times 0.95}{4 \times 0.05 \times 1.225 \times 7.53 \times 10^{-3}} \times 25.218$$
$$= \underline{0.775 \text{ m}}$$

PROBLEM 4.3

In an experiment of sterilization of medium in a 5,678.1 m^3 vessel the results of heating and cooling cycles obtained by Deindoerfer have been shown in Fig. 4.1. It is

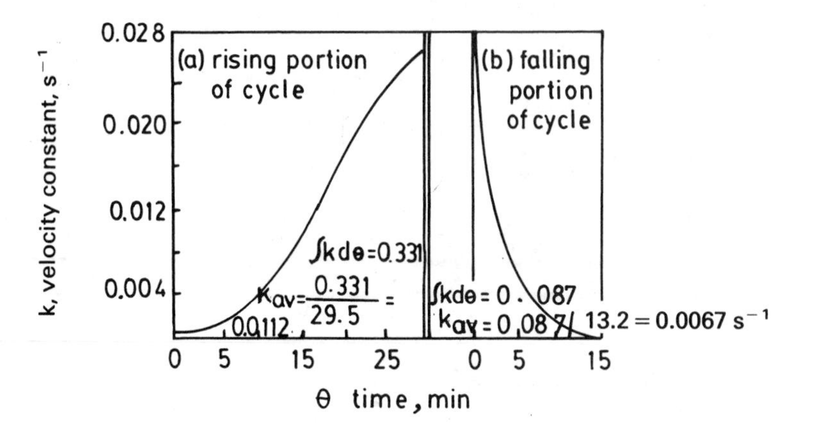

Fig. 4.1.

required to sterilize 45 m^3 of medium. The initial count of microorganisms was 2×10^{13} spore cells m^{-3}. During sterilization (121°C) assume 1 in 1000 chance of a single spore surviving. Determine the time required for the sterilization of the medium. [Assume no sterilization below 121°C.]

Reference

Aiba, S., Humphrey, A. E. and Millis, Nancy F. (1973) *Biochemical Engineering*, 2nd edn., Univ. of Tokyo Press, p. 256.

Solution

From the given data the total number of organisms in the vessel contents

$$N_1 = 2 \times 10^{13} \times 45$$
$$= 9 \times 10^{14} \text{ spores}$$

From the heating–cooling profile,

$$\begin{aligned} \text{Time of heating cycle} &= 29.5 \times 60 \text{ s} \\ \text{gives } k_{av} &= 0.0112 \text{ s}^{-1} \\ \text{Time of cooling cycle} &= 13.2 \times 60 \text{ s} \\ \text{gives } k_{av} &= 0.0066 \text{ s}^{-1} \end{aligned}$$

so total time of sterilization cycling $= 42.7 \times 60$ s,
gives

$$k_{av} = 0.0089 \text{ s}^{-1}$$

Considering first order kinetics of sterilization we have,

$$\frac{dN}{dt} = -kN$$

On integrating between limits 0 to t and N_1 to N_2 or,

$$t = \frac{2.3}{k} \log \frac{N_1}{N_2}$$

or,

$$t = \frac{2.3}{0.0089} \log \frac{9 \times 10^{14}}{N_2}$$

or,

$$\log \frac{9.0 \times 10^{14}}{N_2} = \frac{0.0089 \times 60 \times 42.7}{2.3}$$

solving,

$$\log N_2 = \log 9.0 \times 10^{14} - \frac{0.0089 \times 60 \times 42.7}{2.3}$$
$$= (14 + \log 9) - 9.91$$
$$\therefore N_2 = 1.1 \times 10^5 \text{ spores}$$

Hence heating and cooling cycle reduce microbial count from 9.0×10^{14} to 1.14×10^3 spores. This N_2 now becomes the new N_1 which would be held at 121°C in order to reduce N_1 from 1.14×10^3 spores to $N_2 = 0.001$ spores.

[From Fig. 4.1 with

$$\frac{1}{T} = \frac{1}{121 + 273}$$

the corresponding $k_{av} = 0.0265 \text{ s}^{-1}$.] So, the time required is,

$$t = \frac{2.3}{0.0265} \times \log \frac{1.14 \times 10^3}{1 \times 10^{-3}}$$

$$t = \underline{528 \text{ s}}$$

PROBLEM 4.4

During sterilization of a fermentation medium in a given fermenter, the heat-up from 100°C to 121°C took 25 min, while cool-down period from 121°C to 100°C took 15 min. The characteristic values of rate constant (k) and the degree of sterilization (∇) at different temperatures are $\nabla^{100-121°C}_{1°C/min} = 12.549$ and $k_{121°C} = 2.538 \text{ min}^{-1}$. If the total value of ∇ required for the whole sterilization process is 45.0, what should be the holding period at 121°C?

References
1. Richards, J. W. (1968) *Industrial Sterilization*; 2. Aiba, S., Humphrey, A. E. & Millis, Nancy F. (1973) *Biochemical Engineering* 2nd edn., Univ. of Tokyo Press, p. 253.

Solution
From the given data,

$$\nabla^{100-121°C}_{\text{heating}\,(1°C/min)} = 12.549$$

so, heat-up at the rate of 21°C in 25 min gives,

$$\nabla_{\text{heating}} = \frac{12.549 \times 25}{21} = 14.95$$

Since the cooling cycle to cool the medium from 121°C to 100°C takes 15 min,

$$\nabla_{\text{cooling}} = \frac{12.549 \times 15}{21} = 8.970$$

Now, the required value of

$$\begin{aligned} \nabla_{\text{holding}} &= \nabla_{\text{total}} - \nabla_{\text{heating}} - \nabla_{\text{cooling}} \\ &= 45 - 14.950 - 8.970 \\ &= 21.08 \end{aligned}$$

At holding temperature 121°C the given value of k is 2.538 min^{-1}. So holding period is

$$\begin{aligned} t_{\text{holding}} &= \frac{\nabla_{\text{holding}}}{k} \\ &= \frac{21.08}{2.538} \\ &= \underline{8.3 \text{ min}} \end{aligned}$$

PROBLEM 4.5

The survival of contaminants at the exit of a continuous sterilization system of a fermentation medium is given by:

$$(\bar{N}_c)_{x=1} = \left(\frac{N_c}{N_{c0}}\right)_{x=L}$$

$$= \frac{4\zeta e^{N_{Pe}B/2}}{(1+\zeta)^2 \exp\left(\frac{N_{Pe}B}{2}\zeta\right) - (1-\zeta)^2 \exp\left(\frac{-N_{Pe}B}{2}\zeta\right)} \tag{4.5.1}$$

where,

$$\zeta = \sqrt{1 + \frac{4N_r}{N_{Pe}B}}; \text{ and } N_r = kL\sqrt{u}$$

show that

$$(\bar{N}_c)_{x=1} = \left(\frac{N_c}{N_{c0}}\right)_{x=L} = \exp(-N_r) \text{ for piston flow}$$

and

$$(\bar{N}_c)_{x=1} = \left(\frac{N_c}{N_{c0}}\right)_{x=L} = \frac{1}{1+Nr} \text{ for completely mixed flow}$$

Solution

$$(\bar{N}_c)_{x=1} = \left(\frac{N_c}{N_{c0}}\right)_{x=L}$$

$$= \frac{4\zeta \exp\left(\frac{N_{Pe}B}{2}\right)}{(1+\zeta)^2 \exp\left(\frac{N_{Pe}B}{2}\xi\right) - (1-\zeta)^2 \exp\left(\frac{-N_{Pe}B}{2}\zeta\right)} \tag{4.5.2}$$

again,

$$\zeta = \sqrt{1 + \frac{4N_r}{N_{Pe}B}}$$

Squaring both sides,

$$\zeta^2 = 1 + \frac{4N_r}{N_{Pe}B}$$

or,

$$1 - \zeta^2 = -\frac{4N_r}{N_{Pe}B}$$

or,

$$\frac{N_{Pe}B}{2} = -\frac{2N_r}{1-\zeta^2} \tag{4.5.3}$$

For piston flow, we have from equation (4.5.2),

$$(\bar{N}_c)_{x=1} = \frac{4\zeta \exp\left(\frac{N_{Pe}B}{2}\right)}{(1+\zeta)^2 \exp\left(\frac{N_{Pe}B}{2}\zeta\right) - (1-\zeta)^2 \exp\left(\frac{-N_{Pe}B}{2}\zeta\right)} \tag{4.5.3}$$

also for piston flow, as

$$N_{Pe}B \rightarrow \infty$$

therefore from (4.5.3) we have,

$$\zeta \rightarrow 1$$

the equation (4.5.2) becomes,

$$(\bar{N}_c)_{x=1} = \frac{4 \exp\left(\frac{N_{Pe}B}{2}\right)}{(1+1)^2 \exp\left(\frac{N_{Pe}B}{2}\zeta\right) - 0} \tag{4.5.4}$$

$$= \frac{4 \exp\left(\frac{N_{Pe}B}{2}\right)}{4 \exp\left(\frac{N_{Pe}B}{2}\zeta\right)}$$

$$= \exp\left[\frac{N_{Pe}B}{2}(1-\zeta)\right]$$

$$= \exp\left[-\frac{2N_r}{1-\zeta^2}(1-\zeta)\right]$$

$$= \exp\left[-\frac{2N_r}{1+\zeta}\right]$$

as $\zeta \rightarrow 1$, therefore

$$(\bar{N}_c)_{x=1} = \exp\left[-\frac{2N_r}{1+1}\right]$$

$$= \exp[-N_r]$$

For complete mixed flow, $N_{Pe}B \rightarrow 0$. By dividing the numerator and denominator of equation (4.5.2) by $\exp\left(\frac{N_{Pe}B}{2}\right)$ we get,

$$(\bar{N}_c)_{x=1} = \frac{4\zeta}{(1+\zeta)^2 \exp\left\{\frac{N_{Pe}B}{2}(\zeta-1)\right\} - (1-\zeta)^2 \exp\left\{\frac{-N_{Pe}B}{2}(1+\zeta)\right\}}$$

$$= \frac{4\zeta}{(1+\zeta)^2 \exp\left\{\frac{-2N_r}{1-\zeta^2}(\zeta-1)\right\} - (1-\zeta)^2 \exp\left\{\frac{-2N_r}{1-\zeta^2}(1+\zeta)\right\}}$$

$$= \frac{4\zeta}{(1+\zeta)^2 \exp\left\{\frac{2N_r}{1+\zeta}\right\} - (1-\zeta)^2 \exp\left\{\frac{2N_r}{1-\zeta}\right\}} \quad (4.5.5)$$

On expansion of the exponent of equation (4.5.5) up to the second term, we have,

$$(\bar{N}_c)_{x=1} = \frac{4\zeta}{(1+\zeta)^2\left(1+\frac{2N_r}{1+\zeta}\right) - (1-\zeta)^2\left(1+\frac{2N_r}{1-\zeta}\right)} \quad (4.5.6)$$

$$= \frac{4\zeta}{(1+\zeta)^2 - (1-\zeta)^2 + \{(1+\zeta)^2 N_r - (1-\zeta)^2 N_r\}}$$

$$(\bar{N}_c)_{x=1} = \frac{4\zeta}{4\zeta + 4N_r\zeta}$$

$$= \frac{1}{1+N_r}$$

PROBLEM 4.6

Derive that the denaturation rate constant, k', of a spore suspension at a given temperature is the reciprocal of mean life span, $\bar{t}$, of spores at the given temperature, provided the most probable life span distribution of spores is logarithmic as shown below:

$$\frac{-\mathrm{d}N}{\mathrm{d}t} = a\mathrm{e}^{-bt} \quad (4.6.1)$$

where, N = number of viable spores, and a,b = empirical constants.

Solution

$$\frac{-\mathrm{d}N}{\mathrm{d}t} = a\mathrm{e}^{-bt}$$

$$\therefore N = \int_0^\infty -\frac{\mathrm{d}N}{\mathrm{d}t}\,\mathrm{d}t \quad (4.6.2)$$

so,

$$\bar{t} = \frac{1}{N_0} \int_0^\infty -\frac{\mathrm{d}N}{\mathrm{d}t}\, t\, \mathrm{d}t \tag{4.6.3}$$

$$N = \int_0^\infty a\mathrm{e}^{-bt}\, \mathrm{d}t = \frac{a}{b}$$

or,

$$\bar{t} = \frac{1}{N_0} \int_0^\infty a\mathrm{e}^{-bt}\, t\, \mathrm{d}t$$

$$= \frac{1}{N_0} \frac{a}{b^2} = \frac{1}{b}$$

$$\therefore a = N_0\, \bar{t} \tag{4.6.4}$$

and

$$b = 1/\bar{t} \tag{4.6.5}$$

$$\frac{-\,\mathrm{d}N}{\mathrm{d}t} = N_0 \bar{t}\mathrm{e}^{-t/\bar{t}}$$

$$\therefore \frac{\mathrm{d}N}{N_0} = \bar{t}\mathrm{e}^{-t/\bar{t}}\, \mathrm{d}t$$

or,

$$\frac{-N}{N_0} = -\mathrm{e}^{-t/\bar{t}}$$

Since, $N = N_0 \mathrm{e}^{-kt}$

$$k = -\frac{1}{t}$$

PROBLEM 4.7

If the mechanism due to diffusion is controlling in the collection of air-borne bacteria by a fibrous filter, how could the single fibre collection efficiency, η_0 due to diffusion be increased when the particle size is reduced to two-thirds of the original one? It is assumed that the operation conditions, other than the particle size, remain unchanged.

Solution

The single fibre collection efficiency

$$\eta_0 \propto N_{\mathrm{Sc}}^{-2/3}\, N_{\mathrm{Re}}^{-11/18} \tag{4.7.1}$$

Now,

$$N_{Sc} = \frac{\mu}{\rho \mathcal{D}_{BM}} = \frac{\mu}{\rho} \frac{3\pi\mu d_P}{ckT} \tag{4.7.2}$$

and

$$N_{Re} = \frac{v d_{fi} \rho}{\mu} \tag{4.7.3}$$

$$\therefore \eta_0 \propto \left(\frac{3\pi\mu^2}{\rho ckT}\right)^{-2/3} \left(\frac{v d_{fi}\rho}{\mu}\right)^{-11/18} (d_p)^{-2/3} \tag{4.7.4}$$

$$\frac{\eta_{0_2}}{\eta_{0_1}} = \left(\frac{d_{P_2}}{d_{P_1}}\right)^{-2/3}$$

$$\frac{d_{P_2}}{d_{P_1}} = \frac{2}{3}$$

$$\frac{\eta_{0_2}}{\eta_{0_1}} = \left(\frac{2}{3}\right)^{-2/3} = 1.31$$

Therefore, the increase of efficiency will be 1.31 times.

PROBLEM 4.8

For air sterilization using fibrous filters, the following equation is given:

$$\eta_0 = \frac{\pi d_{fi}(1-f_f)}{4L\, f_f} \ln \frac{N_0}{N_0 - N}$$
$$= \frac{\pi\, d_{fi}(1-f_f)}{4L\, f_f} \ln \frac{1}{1-\bar{\eta}} \tag{4.8.1}$$

where,

N_0 = number of organisms in original aerosol; N = number of organisms retained by fibrous filter; d_{fi} = fibre diameter; L = thickness of filter bed; f_f = volume fraction of fibrous materials in the bed; $\bar{\eta}$ = overall collection efficiency of filter; η_0 = collection efficiency of single fibres whose volume fraction is f_f.

Assuming that the collection eficiency η_0, does not depend on the filter bed thickness, L, derive the above equation.

Solution

$$\dot{V}N = \dot{V}(N + dN) - d\dot{a} \tag{4.8.1}$$
$$\dot{V}dN = -d\dot{a}$$

$\dot{a}$ = amount collected

$= d_{fi} l u N \eta_0$

l = total length of the fibre
u = gas velocity

f_f = volume fraction of fibre

$$= \frac{\pi/4\ d_f^2 l}{AL} \tag{4.8.2}$$

$$\dot{a} = d_{fi}\left(\frac{AL}{\pi/4\ d_{fi}^2}\right) uN\eta_0$$

$$= \frac{f_f AL}{\pi/4\ d_{fi}} uN\eta_0$$

$$d\dot{a} = \frac{f_f A dL}{\pi/4\ d_{fi}} Nu\eta_0$$

$$\dot{V} dN = \frac{-f_f A dl}{\pi/4\ d_{fi}} Nu\eta_0$$

$$\dot{V} = AV_0 = AV\varepsilon$$

$$AV\varepsilon\ dN = \frac{-f_f A\ dL}{\pi/4\ d_{fi}} Nu\eta_0$$

$$\varepsilon dN = \frac{-f_f dL}{\pi/4\ d_{fi}} N\eta_0$$

$$\frac{dN}{N} = \frac{-f_f\ dL}{\pi/4\ d_{fi}\ \varepsilon} \eta_0$$

$$\frac{dN}{N} = \frac{-4f_f\ \eta_0}{\pi\ d_{fi}\ \varepsilon} dL$$

For small value of L and for small ΔL $\Big\}$ $\eta_0 \neq f(L)$

$$\int_{N_0}^{N_0-N} \frac{dN}{N} = \frac{-4f_f\eta_0}{\pi d_{fi}\varepsilon} \int_0^L dL$$

$$\ln N \Big|_{N_0}^{N_0-N} = \frac{-4f_f\eta_0}{\pi d_{fi}\varepsilon} L$$

$$\ln (N_0 - N) - \ln N_0 = \frac{-4f_f\eta_0}{\pi d_{fi}\varepsilon} L$$

$$\ln N_0 - \ln(N_0 - N) = \frac{4f_f\eta_0}{\pi d_{fi}\varepsilon} L$$

$$\ln \frac{N_0}{N_0 - N} = \frac{4\ f_f\ L}{\pi\ d_{fi}\varepsilon} \eta_0$$

$$\eta_0 = \frac{\pi\ d_{fi}\ \varepsilon}{4\ f_f\ L} \ln \frac{N_0}{N_0 - N}$$

$$\eta_0 = \frac{\pi\, d_{fi}(1-f_f)}{4\, f_f\, L} \ln \frac{N_0}{N_0 - N}$$
$$= \frac{\pi\, d_{fi}(1-f_f)}{4\, f_f\, L} \ln \frac{1}{1-\bar{\eta}}$$

PROBLEM 4.9

Show that a probability of unsuccessful sterilization,

$$1 - P = 1 - (1 - e^{-kt})^{N_0} \tag{4.9.1}$$

can be reduced approximately to

$$1 - P = N_0\, e^{-kt} \tag{4.9.2}$$

for large values of N_0 and kt provided:

N_0 = number of contaminants at time, $t = 0$,

and, k = reaction rate constant.

Solution

$$1 - P = 1 - (1 - e^{-kt})^{N_0}$$
$$= 1 - \{1 + N_0(-e^{-kt}) + \frac{N_0(N_0-1)}{2!} \times (e^{-kt})^2 + \ldots\} \tag{4.9.3}$$
$$= 1 - 1 + N_0(e^{-kt}) + \ldots$$
$$= \underline{N_0\, e^{-kt}}$$

for large values of N_0 and kt other terms are neglected.

PROBLEM 4.10

Air at room temperature (20°C) is passed through a fibrous bed of glass fibre (fibre diameter, $d_{fi} = 19$ micrometres and volume fraction, $\alpha = 3.3\%$) with superficial air velocity, $u_s = 0.5\ \mathrm{m\ s^{-1}}$. Calculate the pressure drop of air per unit length (m) of filter. Values of viscosity, μ, and density, ρ, of air at 20°C are as follows:

$$\mu = 1.8 \times 10^{-5}\ \mathrm{N\ s\ m^{-2}}$$
$$\rho = 1.20\ \mathrm{kg\ m^{-3}}$$

Solution

$$N_{Re} = \frac{d_{fi} u \rho}{\mu} = \frac{d_{fi} u_s\, \rho}{\mu(1-f_f)} \quad \left(\because\ u = \frac{u_s}{1-f_f}\right) \tag{4.10.1}$$
$$= \frac{19 \times 10^{-6} \times 0.5 \times 1.2}{1.8 \times 10^{-5} \times (1 - 0.0333)}$$
$$= 0.654$$

From p. 291 of *Biochemical Engineering*, Aiba *et al*. (2nd edn., 1973),

$$C_{Dm} = 75$$

$$1 - \varepsilon = 0.033, \; m = 1.35$$

$$C_{Dm} = \frac{\pi g_c \, d_{fi} \, \Delta P}{2\rho \, L u^2 \, (1-\varepsilon)^m}$$

$$\frac{\Delta P}{L} = \frac{2\rho \, C_{Dm} \, u^2 \, (1-\varepsilon)^m}{\pi \, g_c \, d_{fi}}$$

$$= \frac{2\rho \, C_{Dm} \, u_s^2 \, (1-\varepsilon)^m}{\pi \, g_c \, d_{fi} \, (1-u_s)^2}$$

$$= \frac{2 \times 1.2 \times 75 \times (0.5)^2 \times (0.0333)^{1.35}}{3.14 \times 9.81 \times (19 \times 10^{-6})(1-0.0333)^2}$$

$$= \underline{822 \text{ kg m}^{-2} \text{ m}^{-1}}$$

PROBLEM 4.11

The data of Table 4.1 were obtained in thermal inactivation of spores at 121°C. Does it follow the logarithmic or non-logarithmic death rate? Propose a model which will fit the data and estimate the parameters of the model with the data.

Table 4.1

Time, t (s)	Number of spores (N)
0	100
1	99
2	98
3	95
4	91
5	88
6	83
7	79
8	75
9	70
10	66
15	46
20	31
30	13
40	5
50	2

Solution

The data of Table 4.1 are normalized and presented in Table 4.2.

Table 4.2

Time, t (s)	$\ln (N/N_0)$
0	0
1	−0.01
2	−0.02
3	−0.05
4	−0.094
5	−0.128
6	−0.186
7	−0.236
8	−0.288
9	−0.357
10	−0.416
15	−0.776
20	−0.171
30	−2.041
40	−2.996
50	−3.912

It can be seen from the value of $\ln (N/N_0)$ against t, that it does not follow a logarithmic death profile. So if we propose non-logarithmic death profile in the following way,

$$N_R \xrightarrow{k_R} N_s \xrightarrow{k_s} N_D \quad (4.11.1)$$

where, N_R stands for the active spores, N_s represents the sensitive intermediate spores and, N_D means the inactive or dead spores,

Thus, $$\frac{dN_R}{dt} = -k_R N_R \quad (4.11.2)$$

and,

$$\frac{dN_s}{dt} = k_R N_R - k_s N_s \quad (4.11.3)$$

Solving the two differential equations we get,

$$\frac{N}{N_0} = \frac{k_R}{k_R - k_s}\left[\exp(-k_s t) - \frac{k_s}{k_R}\exp(-k_R t)\right] \quad (4.11.4)$$

where $N = N_R + N_s$ and N_0 is the initial concentration of viable cells. Fitting the data to the above model by non-linear optimization (Kuester & Mize, 1973) gives,

$\underline{k_s = 4.0 \text{ spores s}^{-1}}$

$\underline{k_R = 6.0 \text{ spores s}^{-1}}$

Reference
Kuester, J. L. and Mize, J. H. (1973) *Optimization Techniques with Fortran*, McGraw-Hill, New York.

5

Enzyme reactions and technology

Contributors: John Villadsen, Subhash Chand, V. S. Bisaria, K. B. Ramachandran

PROBLEM 5.1

The reaction ATP⇌ADP+Pi is catalysed by the enzyme ATPase. The molecular weight of the enzyme is 5×10^4 daltons and the Michaelis–Menten constant $(K_m) = 1 \times 10^{-4}$ kg moles m^{-3} and turnover number $k_2 = 166\ s^{-1}$. The enzyme gets deactivated according to first order kinetics with half life $(t_{1/2}) = 414$ s. If 10 mg of an enzyme preparation are added to a batch reactor of volume 0.001 m^3 and the product concentration after 12 hours (measured as Pi) is 0.002 kg moles m^{-3} determine the purity of the preparation given that the initial substrate (ATP) concentration is 0.02 kg moles m^{-3}.

Solution

A first order enzyme deactivation kinetics can be described as

$$\frac{dc_e}{dt} = k_d c_e \tag{5.1.1}$$

Integrating and rearranging, we get

$$= c_{e_0} e^{-k_d t} \tag{5.1.2}$$

and
$$k_d = \frac{\ln 2}{t_{1/2}} \tag{5.1.3}$$

where k_d = decay rate constant (s^{-1}). Here,

$$k_d = \frac{\ln 2}{414} = 1.66 \times 10^{-3}\ s^{-1}$$

Since the initial substrate concentration (0.02 kg moles m^{-3}) is very large compared to K_m (1×10^{-4} kg moles m^{-3}); a zero order reaction kinetics can be reasonably assumed. Therefore, the Michaelis–Menten rate expression simplifies to

$$\upsilon = k_2 c_{e_0} e^{-0.0016t} = \frac{-dc_s}{dt}$$

where k_2 = turnover number, and c_{e_0} = initial enzyme concentration.

For a batch reactor,

$$-\int_{c_{s0}}^{c_s} dc_s = k_2 c_{e_0} \int_0^t e^{-0.0016t}$$

$$= \frac{-k_2 c_{e_0}}{0.0016} [e^{-0.0016t}]_0^t$$

$$= \frac{k_2 c_{e_0}}{0.0016} (1 - e^{-0.0016t})$$

$$[c_s]_{12h} = 0.02 - 0.002 = 0.018 \text{ kg moles m}^{-3}$$

$$0.002 = \frac{k_2 c_{e_0}}{0.0016} \quad (\text{at } t = 12 \times 60 \times 60 \text{s})$$

$$= \frac{166 c_{e_0}}{0.0016}$$

$$\therefore c_{e_0} = \frac{0.0016 \times 0.002}{166} = \underline{2 \times 10^{-8} \text{ kg moles m}^{-3}}.$$

$$= 2 \times 10^{-8} \times 0.001 = \underline{2 \times 10^{-11} \text{ kg moles}}$$

$$= 2 \times 10^{-11} \times 5 \times 10^4 = \underline{1 \times 10^{-6} \text{ kg}}$$

$$\therefore \text{purity} = \left(\frac{c_{e \text{ actual}}}{c_{e \text{ crude}}}\right) = \frac{1 \times 10^{-6}}{10 \times 10^{-6}} = \underline{0.1 \ (10\%)}$$

PROBLEM 5.2

Inversion of sucrose by invertase follows a substrate inhibited enzyme kinetics. An immobilized invertase preparation is to be used in a CSTR with a 100 mol m^{-3} feed concentration. If the reaction velocity passes through a maximum at sucrose concentration of 20 mol m^{-3} and the Michaelis–Menten constant (K_m) and maximum reaction velocity (V_m) for the preparation is 8 mol m^{-3} and 4.45 × 10^{-3} mol m^{-3} s^{-1}. Calculate the feed flow rate to get the maximum productivity from the reactor. Reactor volume = 0.001 m^3.

Solution

Haldane, for the enzyme reaction inhibited by high substrate concentration, proposed the following reaction mechanism. It involves the formation of a second complex which contains two molecules of substrate/mole of enzyme.

$$\begin{array}{c} \mathrm{S} \\ + \\ \mathrm{E+S} \underset{k_{-1}}{\overset{k_1}{\rightleftharpoons}} \mathrm{ES} \overset{k_2}{\rightarrow} \mathrm{E+P} \\ k_s \downarrow\uparrow k_{-s} \\ \mathrm{ESS} \end{array}$$

where $k_1, k_{-1}, k_s, k_{-s}, k_2$ are rate constants for individual steps.

For the above mechanism, the rate expression can be derived as

$$\upsilon = \frac{k_2 c_{e_0}}{1 + \dfrac{K_m}{c_s} + \dfrac{c_s}{K_s}}; \qquad K_s = \frac{k_{-s}}{k_s}$$

Now, given that at $c_s = 20 \text{ mol m}^{-3}$, $\dfrac{d\upsilon}{dc_s} = 0$.

$$\therefore \frac{d\upsilon}{dc_s} = k_2 c_{e_0}\left[-1\mathrm{x}\left(\frac{-K_m}{c_s^2} + \frac{1}{K_s}\right) \Big/ \left(1 + \frac{K_m}{c_s} + \frac{c_s}{K_s}\right)^2\right] = 0$$

or, $$\frac{K_m}{c_s^2} = \frac{1}{K_s} \quad \therefore c_s = \sqrt{K_m K_s} \quad \text{(for maximum reaction velocity)}$$

Since $K_m = 8 \text{ mol m}^{-3}$,

$$K_s = \frac{20^2}{8} = 50 \text{ mol m}^{-3}$$

For CSTR:

$$\tau = \frac{Xc_{s_0}}{-\upsilon} = \frac{\dfrac{Xc_{s_0}}{k_2 c_s}}{1 + \dfrac{K_m}{c_s} + \dfrac{c_s}{K_s}}$$

But $Xc_{s_0} = (c_{s_0} - c_s)$ and $k_2 c_{s_0} = V_m$

$$\tau = \frac{\left(100 - 20\right)\left(1 + \dfrac{8}{20} + \dfrac{20}{50}\right)}{4.4 \times 10^{-3}} = 32{,}727.3 \text{ s} = 9.091 \text{ h}$$

$$F = \frac{V}{\tau} = \frac{0.001}{9.091} = 1.0999 \times 10^{-4} \text{ m}^3 \text{ h}^{-1} = \underline{3.0555 \times 10^{-8} \text{ m}^3 \text{ s}^{-1}}$$

Inference: For a substrate inhibited enzyme catalysed reaction, the velocity and therefore productivity in a CSTR passes through a maximum with respect to $[c_s]$.

Reference

Roberts, D. V. (1977) *Enzyme kinetics*, Cambridge University Press, Cambridge.

PROBLEM 5.3

Enzymes can reversibly bind certain ligands (e.g. substrates, cofactors, etc.). The binding of NAD^+ to yeast glyceraldehyde-3-phosphate dehydrogenase was studied by equilibrium dialysis. A known concentration of the enzyme is taken in a dialysis sac and dialysed against a buffer containing varying concentrations of NAD^+. The concentration of free NAD^+ at equilibrium is monitored and is noted below:

$[NAD^+]_{total}$ μ mol.l^{-1}	41	78	132	187	230	285	374	474
$[NAD^+]_{free}$ μ mol.l^{-1}	13	21	30	39	48	68	125	211

If the enzyme concentration used is 71 μmol/l, deduce the nature of binding and number of NAD^+ binding sites on the enzyme molecule.

Solution

The binding of ligands to biological macromolecules can be expressed by the Scatchard equation:

$$\frac{r}{[NAD^+]} = \frac{n}{k_d} - \frac{r}{k_d}$$

where r = concentration of ligand (NAD^+) bound per unit concentration of the macromolecule (enzyme), n = number of binding sites on the enzyme molecule, k_d = dissociation constant and $[NAD^+]$ = concentration of free ligand.

At each value of $[NAD^+]_{total}$ (see Table 5.1), we can evaluate $[NAD^+]_{bound}$ by

Table 5.1

$[NAD]_{total}$ (μmol/l)	$[NAD]_{free}$ (μmol/l)	$[NAD]_{bound}$ (μmol/l)	r	$\frac{r}{[NAD]_{free}}$ (μmol/l)
41	13	28	0.394	0.030
78	21	57	0.803	0.038
132	30	102	1.436	0.048
187	39	148	2.084	0.053
230	48	182	2.563	0.053
285	68	217	3.056	0.045
374	125	249	3.507	0.028
474	211	263	3.704	0.017

subtraction; r is obtained by dividing $[NAD^+]_{bound}$ by concentration of enzyme (71 μmol/l).

A plot of the data in Table 5.1 according to the Scatchard equation is shown in Fig. 5.1. From the nature of the curve, it can be deduced that the binding of NAD^+ to yeast glyceraldehyde-3-phosphate dehydrogenase exhibits non-equivalent binding sites on the enzyme with positive cooperativity (allosteric behaviour).

Further, plotting the same data (Fig. 5.2) in a double reciprocal plot $\left(\frac{1}{r}\right.$ versus

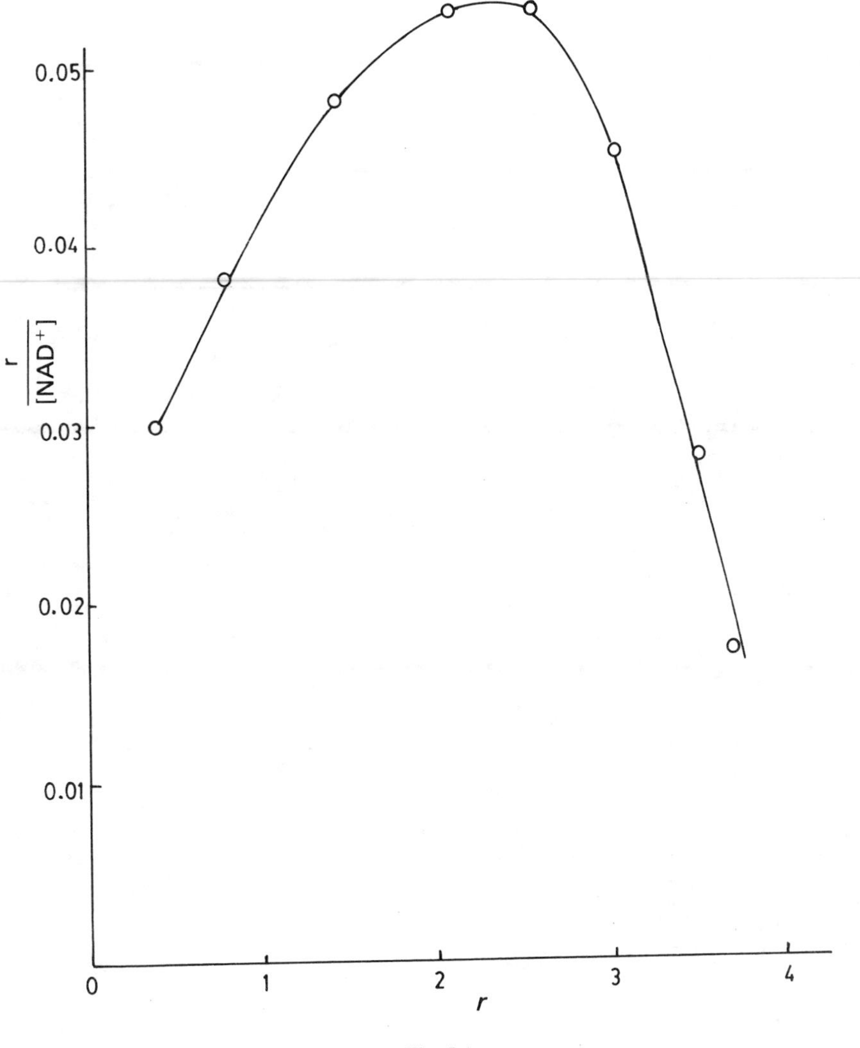

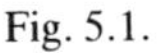
Fig. 5.1.

$\left.\frac{1}{[NAD^+]}\right)$ also confirms a positive cooperativity with an intercept on the $\frac{1}{r}$ axis 0.25, indicating the number of binding sites as four. The enzyme is also known to be a tetramer.

Reference
Hannies, C. G. & Wu, C. W. (1974) *Ann. Rev. Biophys. Bioeng.* **3**, 1.

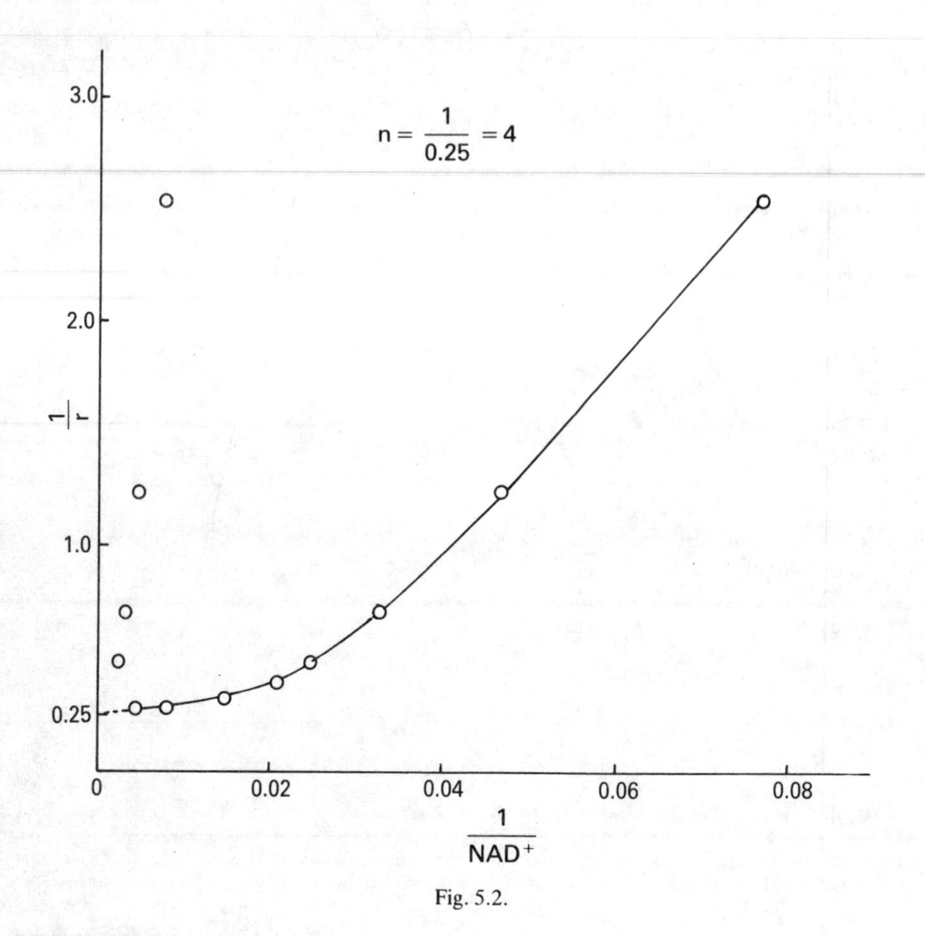

Fig. 5.2.

PROBLEM 5.4

Enzymatic isomerization of glucose to fructose can be kinetically expressed by a reaction mechanism of the type:

$$S + E \underset{k_{-1}}{\overset{k_1}{\rightleftharpoons}} SE \underset{k_{-2}}{\overset{k_2}{\rightarrow}} P + E$$

which on kinetic analysis by steady state approximation leads to the rate expression

$$-\frac{dc_s}{dt} = \frac{(V_{ms}/K_{ms})(c_s) - (V_{mp}/K_{mp})(c_p)}{1 + \dfrac{c_s}{K_{ms}} + \dfrac{c_p}{K_{mp}}}$$

where k_{ms} and K_{mp} are Michaelis–Menten constants for forward and reverse reactions and V_{ms} and V_{mp} are respective maximum reaction velocities. Saini and Veith evaluated the kinetic parameters and the resultant rate expression as:

$$\upsilon = \frac{0.128[c_s] - 0.098[c_p]}{0.096 + 0.383[c_s] + 0.25[c_p]}.$$

If the feed substrate (glucose) concentration is 1 kg mol m^{-3} and the desired conversion is 40%, compare the productivities of the above reaction, if carried out in a CSTR/PFR of similar volume and enzyme content. It has been confirmed experimentally that the reactor performance is controlled by the intrinsic reaction kinetics only.

Reference
Saini, R. & Veith, W. R. (1975) *J. Appl. Chem. Biotechnol.* **25**, 115.

Solution

Let the initial substrate concentration be $c_{s_0} = 1.0$ kg mole m^{-3}.

$$\text{Conversion,} \quad X = \frac{[c_{s_0}] - [c_s]}{[c_{s_0}]} = 0.4$$

$$\therefore \; [c_s] = 0.6 \text{ kg moles m}^{-3}; \; [c_p] = 0.4 \text{ kg moles m}^{-3}$$

For CSTR

$$\tau_{\text{CSTR}} = \frac{Xc_{s_0}}{-v} = 0.4 \times 1 \left[\frac{0.096 + (0.383 \times 0.6) + (0.25 \times 0.4)}{(0.128 \times 0.6) - (0.098 \times 0.4)} \right]$$

$$= 4.53 \text{ time units}$$

[The time units depend upon the units of enzyme activity used in the rate expression].

For PFR

$$\tau_{\text{PFR}} = c_{s_0} \int_0^X \frac{dX}{-v}$$

$$= \int_0^X \left\{ \frac{0.096 + 0.383(1-X) + 0.25X}{0.128(1-X) - 0.098X} \right\} dX$$

$$= \int_0^X \left\{ \frac{0.479 - 0.133X}{0.128 - 0.226X} \right\} dX$$

Put $0.128 - 0.226X = u$

$$\therefore X = \frac{0.128 - u}{0.226} \quad \text{and} \quad dX = \frac{-du}{0.226}.$$

when $X = 0; \quad u = 0.128$

$X = 0.4; \; u = 0.0376$

$$\therefore \tau_{\text{PFR}} = \int_{0.128}^{0.0376} \frac{1}{u} \left\{ \left[0.479 - 0.133 \left(\frac{0.128 - u}{0.226} \right) \right] \frac{du}{-0.226} \right\}$$

$$= \left\{ 1.785 \ln \frac{0.128}{0.0376} + 2.604(0.128\text{–}0.0376) \right\}$$

$$= 2.42 \text{ time units}$$

Productivity for a given fractional conversion (X) and initial substrate concentration is inversely proportional to space time.

Hence,

$$\frac{(Pr)_{PFR}}{(Pr)_{CSTR}} = \frac{4.53}{2.42} = \underline{1.87}$$

Inference: The productivity of a product inhibited enzyme catalysed reaction (or with reversible reaction mechanism) is higher in a PFR compared to a CSTR.

PROBLEM 5.5

The hydrogen ion concentration (measured as pH) in the micro-environment of an enzyme catalysed reaction influences the ionization state of the amino acid functional groups constituting the active site on the enzyme molecule. Since the enzyme catalyses a reaction only in a specific ionization state at its active site, pH significantly influences the rate of an enzyme catalysed reaction.

The effect of pH on the maximum reaction velocity of an enzyme catalysed reaction (L-Malate$\rightleftharpoons$Fumarate + H_2O) by fumarase was studied at 25°C and the following data obtained.

pH	4.5	5.0	5.5	6.0	6.5	7.0	7.5	8.0	8.5
V_m (Arbitrary units)	18	45	108	181	217	200	140	47.5	20.0

(a) Determine the appropriate pK values for the ionizing groups involved in the catalytic activity of the above enzyme.

(b) Suggest a suitable reaction scheme to represent the above pH dependence data and develop an expression that relates V_m to hydrogen ion concentration.

(c) If it is found that at 35°C the lower pK differs by 0.1 units from that at 25°C, what is the enthalpy of ionization of this group?

Solution

(a) Change of pH of the enzyme catalysed reaction leads to the change in the ionization state of amino acid side chains in the enzyme. Variation of V_m with pH reflects the variation of the product forming rate constant of the generalized scheme (Michaelis–Menten). A pH versus log V_m plot therefore represents a sequence of straight lines, which are recorded usually as a smooth bell-shaped curve. Extrapolation of these straight lines gives the pK values of the ionizable groups involved in the catalysis.

The data on the effect of pH on V_m for the fumarase catalysed reaction are plotted in Fig. 5.3 as ln V_m (y-axis) versus pH (x-axis). The plot suggests that two ionizing groups are involved in the catalytic activity of fumarase. From the points of intersection of the linear portions, the pK_as can be determined as 5.8 and 7.5 for pK_a and pK_b respectively.

(b) A scheme of reactions involving two ionizing groups can be represented as below:

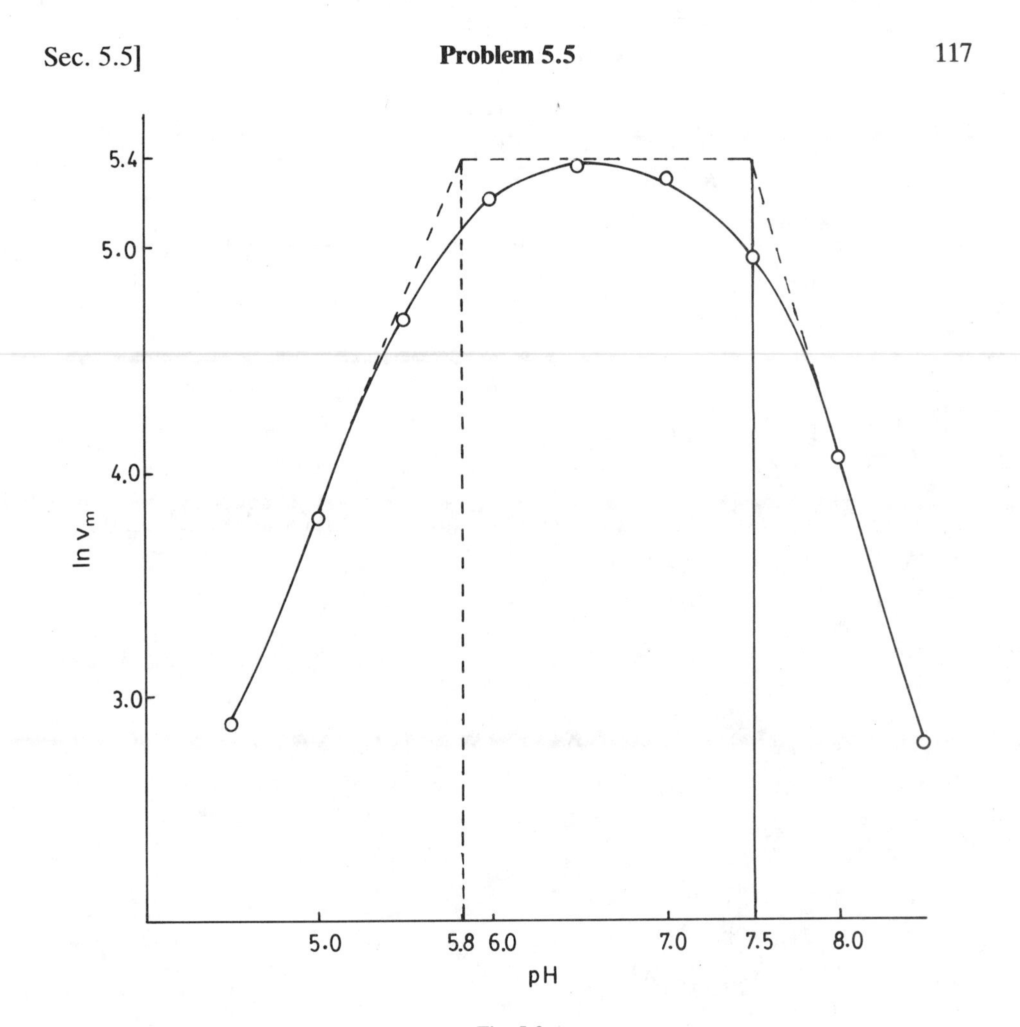

Fig. 5.3.

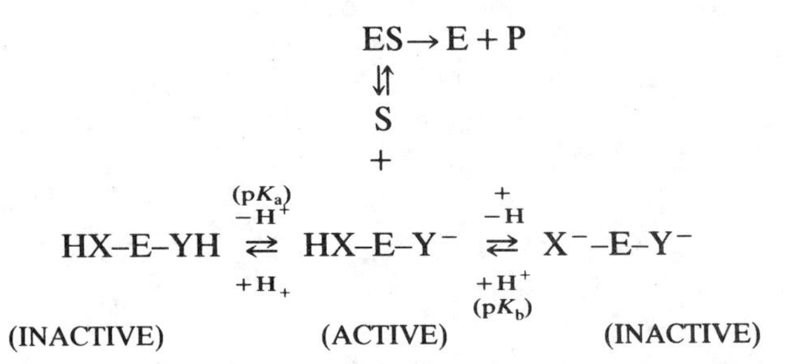

Now, considering that the HX–E–Y$^-$ form of the enzyme is active, and therefore is involved in binding with the substrate, the desired rate expression can be derived as

$$V_m = \frac{\overline{V}_m}{1 + [H^+]/K_a + K_b/[H^+]}$$

where $\overline{V}_m$ is the maximum velocity if all the enzyme is present in the active form (XH–E–Y$^-$) and K_a and K_b are the two ionization constants.

(c) Since the pK_a at 35°C and 25°C differ by 0.1 unit

$$\log\left[\frac{K_a^{308}}{K_a^{298}}\right] = 0.1$$

or

$$\ln\left[\frac{K_a^{308}}{K_a^{298}}\right] = 0.2303$$

Now, by effect of temperature on equilibrium constant (Van't Hoff equation),

$$\frac{\mathrm{d}(\ln K_a)}{\mathrm{d}T} = \frac{\Delta H°}{RT^2}$$

or

$$\int_{K_a^{298}}^{K_a^{308}} \mathrm{d}(\ln K_a) = \int_{298}^{308} \frac{\Delta H°}{RT^2}\mathrm{d}T$$

$$\ln\frac{K_a^{308}}{K_a^{298}} = -\frac{\Delta H°}{R}\left(\frac{1}{308} - \frac{1}{298}\right)$$

Therefore,

$$\frac{\Delta H°}{R}\left(\frac{1}{298} - \frac{1}{308}\right) = 0.2303$$

or,

$$\underline{\Delta H° = 17.5 \text{ KJ mol}^{-1}}$$

Inference: The enzyme under study has only two ionizable groups at the active site with pK values 5.8 and 7.5 units respectively. Also, the effect of temperature on pK values can be used to determine $\Delta H°$ of ionization.

Reference:

Price, N. C. & Stevens, L. (1987) *Fundamentals of Enzymology*, Oxford University Press, Oxford, pp 131–133.

PROBLEM 5.6

An enzyme ATPase, isolated from a recombinant *Bacillus* species, was purified to homogeneity and found to have a molecular weight of 4.5×10^4 daltons, a K_m for ATP of 6×10^{-5} kg mol m^{-3} and a turnover number of 1.33×10^3 molecules of ATP hydrolysed to ADP and inorganic phosphate, (ATP$\rightarrow$ADP + Pi) per molecule of enzyme per second at 37°C. The enzyme's half-life at 37°C was 288 seconds and was found to get inactivated by first order deactivation kinetics. In an experiment with a partially purified enzyme, 20×10^{-3} kg of total protein were added to 1 m^3 reaction mixture containing 0.03 kg mol m^{-3} ATP and incubated at 37°C. After 12 hours no further reaction could be detected, and the concentration of inorganic phosphate increased from initial zero to 0.006 kg mol m^{-3}. Calculate the fraction of the protein which was represented by the enzyme, i.e. how pure the enzyme was.

Solution

The deactivation rate constant k_d of the enzyme ATPase is given by

$$c_{e_t} = c_{e_0}\,e^{-k_d t}$$

or,

$$\tfrac{1}{2}c_{e_0} = c_{e_0}\,e^{-k_d t_{1/2}}$$

$$0.5 = e^{-k_d \times 288}$$

$$-288\,k_d = \ln 0.5 = -0.69$$

$$\therefore k_d = \frac{0.69}{288} = 2.396 \times 10^{-3}\ s^{-1}$$

The rate of appearance of product (inorganic phosphate) is given by

$$\frac{dc_p}{dt} = \frac{V_m \cdot c_s}{K_m + c_s}$$

Since the substrate (ATP) is present in large excess, ($c_s \gg K_m$) and the rate is given by

$$\frac{dc_p}{dt} = V_m = k_2 c_{e_t}$$

k_2 = turnover number of the enzyme

c_{e_t} = concentration of enzyme = $c_{e_0}\,e^{-k_d t}$

$$\frac{dc_p}{dt} = k_2 c_{e_0}\,e^{-k_d t}$$

$$\int_0^{c_p} dc_p = k_2 c_{e_0}\int_0^t e^{-k_d t}\,dt$$

$$c_p = k_2 c_{e_0}\left[\frac{-(e^{-k_d t} - 1)}{k_d}\right] = \frac{k_2 c_{e_0}}{k_d}(1 - e^{-k_d t})$$

As $t = 12$ h ($= 43{,}200$ s) is very large, $e^{-k_d t} \ll 1$

$$\therefore\ c_p = \frac{k_2 c_{e_0}}{k_d}$$

$$c_{e_0} = \frac{k_d c_p}{k_2} = \frac{2.396 \times 10^{-3} \times 0.006}{1.33 \times 10^3} = 1.08 \times 10^{-8}\ \text{kg mol m}^{-3}$$

$$= 1.08 \times 10^{-8} \times 4.5 \times 10^4\ \text{kg mol}^{-3}$$

or,

$$\underline{4.86 \times 10^{-4} \text{kg mol}^{-3}}$$

Since 20×10^{-3} kg of total protein were added, the fraction of the protein represented by the enzyme

$$= \frac{4.86 \times 10^{-4}}{20 \times 10^{-3}} \times 100$$

$$= \underline{2.43\%}$$

PROBLEM 5.7

For hydrolysis of cellulose, the enzyme cellulase needs to be adsorbed. For simplicity, assume that the cellulase system consists of two components, endoglucanase (EG) and cellobiohydrolase (CBH) (the number of isoenzymes in each component notwithstanding). Since the adsorption characteristics of cellulase enzyme onto cellulose are intimately related to the structure and property of both the enzyme and the solid substrate, the different (and sometimes conflicting) results reported in the literature are due to the difference in the properties of both the enzyme and the substrate. Nevertheless, some published studies are useful for understanding the mechanism of cellulose hydrolysis on the basis of the adsorption phenomenon.

Ghose and Bisaria (1979) reported on the measurement of rates and amount of adsorption of the activity of individual enzyme components when adsorbed alone and in combination (Figs. 5.4 and 5.5) on sugarcane bagasse preceding hydrolysis. The xylanase component was included to hydrolyse the hemicellulose in the bagasse. The maximum adsorption values for this system were achieved in about 10 minutes at 5°C, the temperature at which no hydrolysis takes place. Their results indicated that there was not much competition between the three adsorbing components for the surface area (assuming that the adsorption was not controlled by mass transfer) because the rates of adsorption were almost identical for both individual and combined adsorption. However, the presence of a few common sites was implicated in the study, as the adsorption of the components were less when they were present in combination than when they were present individually.

In a recent study conducted by Kyriacou *et al.* (1989), the competition between the cellulase components, EG and CBH, was investigated using labeled purified enzyme components. The equilibrium adsorption values at 5°C, as measured using ^{3}H and ^{14}C labeled tracers for two enzyme components were as follows:

EG — 3.96×10^{-3} kg kg^{-1} cellulose

CBH — 2.74×10^{-3} kg kg^{-1} cellulose.

To elucidate the presence of common or uncommon sites on cellulose, the components were adsorbed simultaneously and sequentially. The results shown in Table 5.2 were obtained.

(a) Comment on the presence of common or uncommon sites for adsorption of EG and CBH components based on these data.

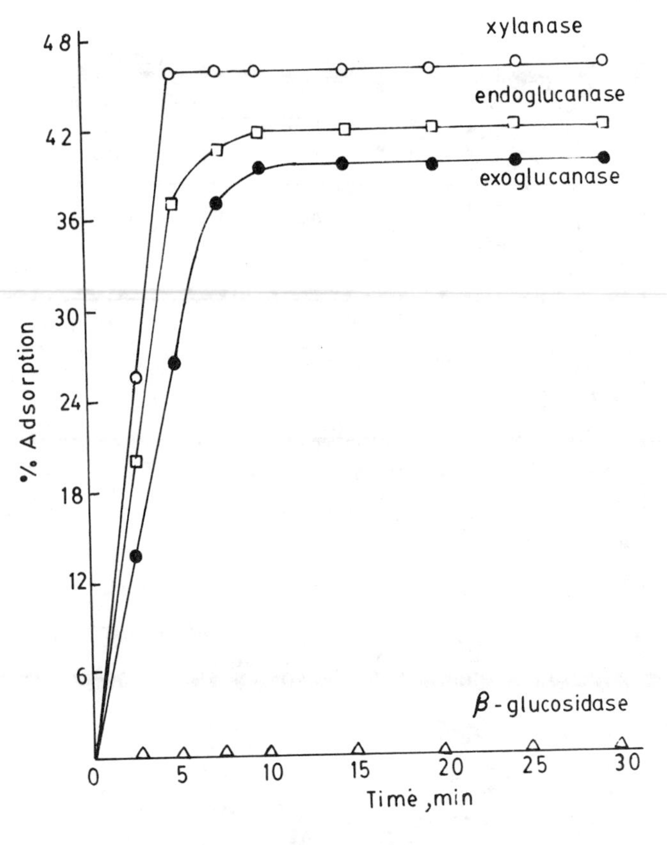

Fig. 5.4.

(b) which of the two components shows preferential adsorption?

(c) which of the two components shows higher binding affinity for the cellulose?

(d) Do these data support or contradict the results reported by Ghose and Bisaria (1979)?

References:
Ghose, T. K. & Bisaria, V. S. (1979) Studies on the mechanism of enzymatic hydrolysis of cellulosic substances, *Biotech. Bioengg*. **21**, 131.
Kyriacou, A., Neufeld, R. J. & MacKenzie, C. R. (1989) Reversibility and competition in the adsorption of *Trichoderma reesei* cellulase components, *Biotech. Bioengg*. **33**, 631.

Solution

(a) The equilibrium adsorption values (see Fig. 5.4):

$$EG = 3.96 \times 10^{-3} \text{ kg kg}^{-1} \text{ cellulose}$$

$$CBH = 2.74 \times 19^{-3} \text{ kg kg}^{-1} \text{ cellulose}$$

During simultaneous adsoprtion of EG and CBH, the percentage adsoption of individual components in mixture as compared to their individual adsorption are:

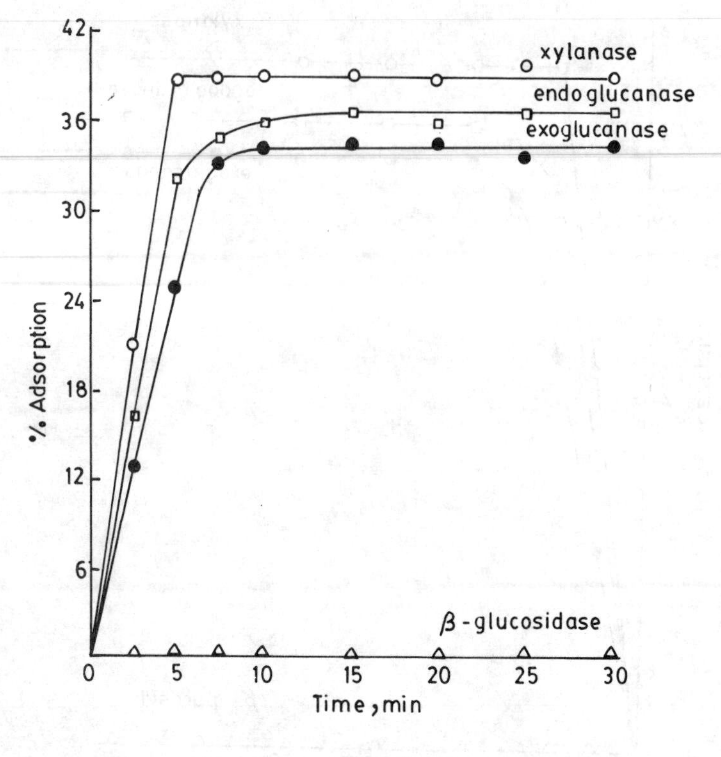

Fig. 5.5.

Table 5.2 — Adsorption of competing EG and CBH fractions at 5°C on cellulose

Cellulase component		Adsorption in kg^{-1} cellulose Fraction 1 adsorbed	Fraction2 adsorbed
(a) Simultaneous adsorption			
Fraction 1	Fraction 2		
CBH	EG	2.6×10^{-3}	1.53×10^{-3}
(b) Sequential adsorption			
Fraction 1	Fraction 2		
EG	CBH	3.45×10^{-3}	2.65×10^{-3}
CBH	EG	2.74×10^{-3}	1.06×10^{-3}

$$\text{CBH} = 2.6 \times 10^{-3} \text{ kg kg}^{-1} \text{ cellulose}$$

and, $$\text{EG} = 1.53 \times 10^{-3} \text{ kg kg}^{-1} \text{ cellulose}$$

that is, 94.8% and 38.6% of their respective adsorption when present alone.

This points to the presence of a few common sites where both EG and CBH could adsorb. Further, the adsorption of CBH seems to block the sites where EG could adsorb.

(b) It points to the preferential adsorption of CBH on cellulose when both CBH and EG are present at the same time.

(c) Based on data on simultaneous adsorption, CBH shows higher binding affinity for the cellulose. One would expect the same results during sequential adsorption also, for this is to be true.

The percentage adsorption of CBH and EG during sequential adsorption is

(i) *For case (i)* [EG followed by CBH]

$$EG = \frac{3.45}{3.96} \times 100 = 87\%$$

$$CBH = \frac{2.65}{2.74} \times 100 = 96.7\%$$

of their respective adsorption when present alone

(ii) *For case (ii)* [CBH followed by EG]

$$EG = \frac{1.06}{3.96} \times 100 = 27\%$$

$$CBH = \frac{2.74}{2.74} \times 100 = 100\%.$$

It is seen that when EG adsorption was followed by CBH, the CBH could be adsorbed to almost its equilibrium level (96.7%) and the displacement of EG was very little. In case (ii), however, EG could not displace any of the CBH component already adsorbed or could not be adsorbed due to blockage of many of its sites by CBH. This finding on sequential adsorption, therefore, also points to the higher binding affinity of CBH.

(d) These results seem to support the results reported by Ghose and Bisaria, which points to the presence of both common and uncommon sites. The results of EG adsorption are, however, not consistent as these varied from 27% (in case (ii) of sequential adsorption) to 87% (in case (i) of sequential adsorption).

PROBLEM 5.8

Semi-synthetic penicillins are produced from 6-amino penicillanic acid as the starting material. The acid is made by enzymatic hydrolysis in repeated batch contacting of fermentation derived penicillin G. The enzyme is immobilized on an ion exchange resin which contains 0.005 kg of enzyme per kg of carrier. The enzyme reaction is carried out in a 0.1-m^3 working volume reactor. The initial penicillin concentration is 10% and the imobilized enzyme loading is 100 kg m^{-3}. The temperature and pH of the reaction is 40°C and 7.0 respectively.

The rate of conversion of penicillin G to 6APA can be expressed as

$$\frac{dc_s}{dt} = -kc_s = \frac{dc_p}{dt} \quad (5.8.1)$$

where c_s is the penicillin concentration and c_p is the 6-APA produced. The rate constant is found to be proportional to the enzyme concentration which includes both free and immobilized enzyme. Its value is $0.7 \times 10^{-3}\,s^{-1}$ when the enzyme concentration is $0.5\,kg\,m^{-3}$ at the above pH and temperature. The reaction proceeds to 99% completion. Then 80% of the solution is decanted after the reaction is over and again $0.8\,m^3$ of 10% penicillin solution is added to the reactor. This process is repeated for 10 consecutive batches. If the free enzyme (X, kg enzyme m^{-3} solution) is in equilibrium with absorbed enzyme (Y, kg enzyme kg^{-1} carrier) and if the equilibrium relationship can be expressed as

$$Y/X = 200 \times 10^{-4} \tag{5.8.2}$$

calculate the time required for 99% conversion, total enzyme concentration for each batch.

Solution

Let

$$\alpha = \frac{\text{kg of enzyme m}^{-3}\text{ of solution}}{\text{kg of enzyme kg}^{-1}\text{ of carrier}} \tag{5.8.3}$$

$$= \frac{X}{Y} = \frac{1}{200 \times 10^{-4}} = 50$$

Let $V = m^3$ of solution; M = kg of carrier; $c_{eT,b}$ = kg of total enzyme in a batch; f_b = fraction of solution withdrawn in each batch; b = batch number. The system is shown in Fig. 5.6.

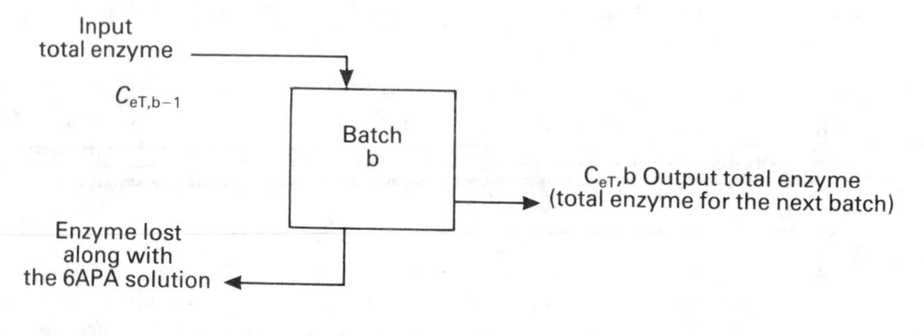

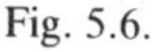
Fig. 5.6.

By material balance:

Input total enzyme to batch b – Output total enzyme from batch b
(coming from batch $b-1$) (total enzyme for the next batch)

= Enzyme lost along with 6APA solution (5.8.4)

Fraction of enzyme lost in each batch = (Enzyme in solution × W) (5.8.5)

$$= \frac{\alpha V}{(\alpha V + M)} f_b \tag{5.8.6}$$

$$\therefore \text{Enzyme lost per batch} = \frac{\alpha V/M}{\left(\dfrac{\alpha V}{M}+1\right)} f_b c_{eT,\,b-1} \tag{5.8.7}$$

Substituting equation (5.8.7) in equation (5.8.4) we get

$$c_{eT,\,b-1} - c_{eT,\,b} = \frac{\left(\dfrac{\alpha V}{M}\right)}{\left(\dfrac{\alpha V}{M}+1\right)} f_b c_{eT,\,b-1} \tag{5.8.8}$$

and $\quad c_{eT,\,b-1} = 0.05 \text{ kg} \quad (5.8.9)$

Solving the difference equation (5.8.8) with the initial condition (5.8.9) gives

$$c_{eT,\,b} = \frac{0.05}{0.733}(0.733)^b$$

$$= \underline{0.0682\,(0.733)^b} \tag{5.8.10}$$

The construction of Table 5.3 is based on equation (5.8.10).

Table 5.3

Batch no.	Total enzyme (kg)	Batch no.	Total enzyme (kg)
1	0.05	6	0.0106
2	0.0367	7	0.00776
3	0.0269	8	0.00568
4	0.0197	9	0.00417
5	0.0144	10	0.00305

Given conversion per batch is 99%

$$\therefore \frac{c_s}{c_{s_0}} = 0.01$$

Conversion time is calculated as follows:

$$\frac{dc_s}{dt} = -kc_s \tag{5.8.11}$$

Also, given the rate constant k being proportional to $c_{eT,\,b}$,

$$\therefore k = Ac_{eT,\,b}$$

But $c_{eT,\,b} = 0.5 \text{ kg m}^{-3} = 0.05$ kg per batch of 0.1 m^3.

$$k = 2.5\,\mathrm{h}^{-1}$$

$$\therefore A = \frac{2.5}{0.05} = 50$$

$$\therefore \ln\frac{c_s}{c_{s_0}} = \frac{-2.5}{0.05} c_{eT,\,b} t \tag{5.8.12}$$

$$\ln\frac{c_s}{c_{s_0}} = \frac{-2.5}{0.05} c_{eT,\,b} t = \frac{-2.5}{0.05} \times 0.0682\,(0.733)^b t$$

$$\ln\frac{1}{100} = \frac{-2.5}{0.05} \times 0.0682\,(0.733)^b t$$

$$t = \frac{\ln 100}{3.41\,(0.733)^b}$$

from the above relation, the time required for ten consecutive batches can be tabulated as in Table 5.4.

Table 5.4

Batch no.	Time for 99% conversion (h)
1	1.84
2	2.51
3	3.42
4	4.68
5	6.40
6	8.69
7	11.87
8	16.22
9	22.09
10	30.20

PROBLEM 5.9

Since immobilized enzyme loses activity on use, it is desirable to maximize the total amount of product formed per unit enzyme or reactor volume. This can be achieved in two ways. In the first method, since conversion is to be kept constant, the flow rate of the substrate through the reactor is varied with time until the enzyme has lost a certain fraction of its original activity after which fresh enzyme is charged into the reactor. In this case total product formed by the reactor will vary with time. In the second method, conversion and flow rate are kept constant and the total product formed is maintained within tolerance limit by using a number of reactors in series and utilizing them with staggered start-up and enzyme loading time. If the enzyme deactivates exponentially with time, for the first method, derive an expression for the flow rate through the reactor to keep conversion constant and also an expression for total product formed. For the second method, derive an expression for tolerance in total product formed in terms of the half life of the enzyme ($t_{1/2}$), time (t_c) for which reactor need to be used before loading with fresh enzyme and the number of reactors (N) required to keep total product formed within the tolerance limit. With the data given below, calculate for the first method the total product formed and for the second method, the number of reactors required in series to achieve a variation of 5% in total product formed.

Data: Half life of the enzyme, $t_{1/2} = 2$ h. Time for which each reactor is to be used, $t_c = 4$ h.

Solution

The total product formed from a reactor can be related to total flow rate F and conversion X by

$$P_t = \int_0^{t_c} FX\mathrm{d}t \tag{5.9.1}$$

For plug flow reactor,

$$\tau = \frac{V}{F} = \int_0^X \frac{\mathrm{d}X}{-\upsilon} \tag{5.9.2}$$

If the enzyme decays exponentially, then

$$V_m = V_{m0}e^{-\lambda t} \tag{5.9.3}$$

$$\therefore \upsilon = \frac{V_{m0}e^{-\lambda t}c_s}{c_s + K_m} = \upsilon_0 e^{-\lambda t}$$

$$\therefore \frac{V}{F} = \int_0^X \frac{\mathrm{d}X}{\upsilon_0 e^{-\lambda t}} \tag{5.9.4}$$

$$\frac{V}{F} = \frac{X}{\upsilon_0 e^{-\lambda t}} \tag{5.9.5}$$

If X is to be constant, then

$$F = F_0 e^{-\lambda t} \tag{5.9.6}$$

So flow rate also has to vary exponentially to keep conversion constant.

$$P_t = \int_0^{t_c} FX\mathrm{d}t$$

$$= \int_0^{t_c} F_0 e^{-\lambda t} X\mathrm{d}t \tag{5.9.7}$$

Integrating,

$$P_t = F_0X\left[\frac{e^{-\lambda t}}{-\lambda}\right]_0^{t_c} \tag{5.9.8}$$

$$= \frac{F_0X}{\lambda}[1 - e^{-\lambda t_c}] \tag{5.9.9}$$

Total product formed if the loss in activity of enzyme is zero

$$= F_0Xt_c \tag{5.9.10}$$

$$\therefore \frac{\text{Total product formed}}{\text{Product formed when no loss of acitivity}} = \frac{1 - e^{-\lambda t_c}}{\lambda t_c} \tag{5.9.11}$$

If enzyme loss activity exponentially,

$$V_m = V_{m0}e^{-\lambda t} \tag{5.9.12}$$

$$\frac{V_{m0}}{V_m} = e^{\lambda t} \tag{5.9.13}$$

$t_{1/2} = 2$ h and $t_c = 4$ h

$$\therefore 2 = e^{2\lambda}$$

$$\therefore \lambda = \frac{0.693}{2}$$

$$\frac{1 - e^{-\lambda t_c}}{\lambda t_c} = \frac{1 - e^{-(0.693/2)\times 4}}{\frac{0.693}{2}\times 4}$$

$$= \underline{0.541}$$

For method 2:

$$\text{product formed} = F_0Xe^{-\lambda t} \tag{5.9.14}$$

$$\text{product formed if enzyme does not deactivate} = F_0X \tag{5.9.15}$$

$$\therefore \text{Tolerance in productivity, } P_r = \frac{F_0Xe^{-\lambda t}}{F_0X} \tag{5.9.16}$$

$$= \underline{e^{-\lambda t}} \tag{5.9.17}$$

The above equation can be used to find time at which the P_r value will drop below the tolerance limit. If each reactor is to be operated above this time till a cycle time of t_c, then

$$t_c = tN \tag{5.9.18}$$

where N = number of reactors in series.
Substituting (5.9.18) in (5.9.17), we get

$$P_r = e^{-\lambda t_c/N} \qquad (5.9.19)$$

For t_c = 4 h and $t_{1/2}$ = 2 h, for P_r = 0.95

$$0.95 = e^{-(0.693/2)\times(4/N)}$$

$$\therefore N = \underline{27}$$

So a minimum of 27 reactors in series will be needed to keep tolerance in product formed between 100% and 95%. Until the time $t = t_c/N$, one reactor is enough to keep the productivity above the tolerance limit. Thereafter, after each $t = t_c/N$ one more reactor has to be added in series to keep the tolerance in productivity between limits. When 27 reactors have been added in series the first reactor will have completed the cycle time of 4 h and will be ready for recharge with fresh enzyme.

PROBLEM 5.10

It is desired to test the activity of an enzyme E in a laboratory reactor (Fig. 5.7). The

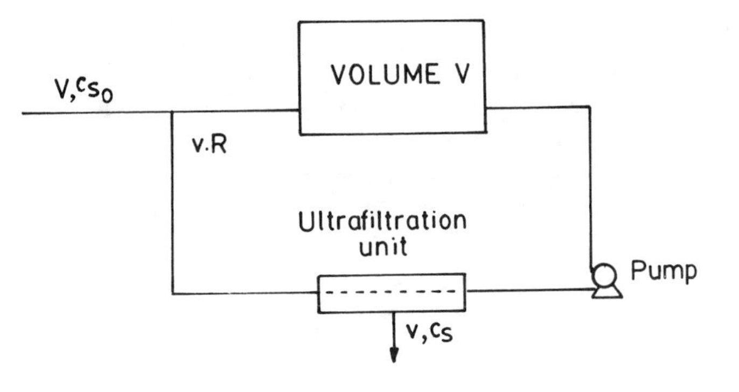

Fig. 5.7.

recirculation loop is fitted with an ultrafiltration unit from which υ ml h^{-1} enzyme-free medium with substrate concentration c_s g l^{-1} is withdrawn. The feed is υ ml h^{-1} with substrate concentration c_{s_0} g l^{-1}. The recirculation rate, R, is high enough to ensure homogeneity in the total reacting volume (tank + tubing) of V *ml*.

(a) The reactor is filled with 490 ml liquid (substrate concentration c_{s_0}) and 10 ml enzyme solution. The recirculation pump is started and inlet + outlet valves opened to give a net flow of υ ml/h through the system. When the outlet substrate concentration has become constant the following measurements are made for c_{s_0} = constant = 10 g l^{-1}:

υ ml h^{-1}	248	68.7	20.8	11.5	5.8
c_s g h^{-1}	18.75	6.10	2.40	1.29	0.63

Determine the kinetic constants in

$$r_s = \frac{k c_{s_0} c_s}{K_m + c_s}$$

(b) Write a transient mass-balance for the reactor.
Use a PC-terminal (or a suitably equipped pocket calculator) to calculate how many hours it takes for the system to get to within 1% of the steady state values for c_s in the above table. An analytical solution can be used to check the results.

(c) The reactor is running at steady state with $\upsilon = 20.8$ ml h^{-1} $c_s = 2.40$ g l^{-1}. At $t = 0$ an experimental error leads to a sudden leakage of 25 ml fresh substrate into the reactor. The leakage is immeiately stopped (in less than 15 s).
Calculate s at t approximately = zero assuming perfect mixing of the pulse into the reactor liquid. With constant $\upsilon = 20.8$ ml h^{-1}, how long does it take for c_s to decrease to 2.42 g l^{-1}? Assume that the deviation from steady state is small. Make a graph of the s time profile from $t = 0$ and until $c_s = 2.42$ g l^{-1}.

(d) The reactor is used to test for the pesticide content of a drinking water sample. With the reactor running at steady state with $\upsilon = 20.8$ ml h^{-1}, a volume nV_p of pesticide containing sample is added. We assume that the molecular size of the pesticide is big enough to prevent it from leaving the system through the ultrafiltration membrane. The following steady state s-values are obtained.

n	1	5	20	50	; $V_p = 1$ ml
c_s g l^{-1}	2.37	2.47	2.80	3.26	; $\upsilon = 20.8$ ml h^{-1}

(i) The inhibition constant K_i is of the order of 10^{-1} mg l^{-1}. Find the pesticide concentration in the water sample.
(ii) List a number of criticisms concerning the proposed assay for pesticide.

Solution

(a) A mass balance for c_s yields $\upsilon(c_{s_0} - c_s) = Vr_s$.

$$r_s = \frac{\upsilon(c_{s_0} - c_s)}{V}$$

υ (ml h^{-1})	248	68.7	20.8	11.5	5.8
r_s (g l^{-1} h^{-1})	0.620	0.536	0.316	0.20	0.109
$\frac{1}{r_s}$ (l h g^{-1})	1.61	1.87	3.16	5.0	9.20
$\frac{1}{c_s}$ (l g^{-1})	0.114	0.164	0.417	0.775	1.59

From the plot of $1/r_s$ against $1/c_s$ (Fig. 5.8)

$$-\frac{1}{K_m} = -0.2$$

$$\therefore K_m = 5 \text{ g l}^{-1}$$

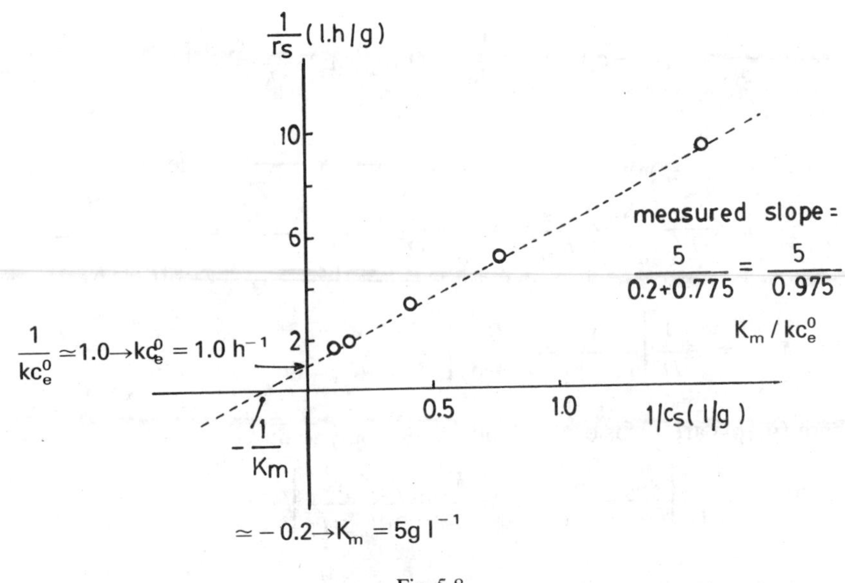

Fig. 5.8.

$$\therefore \text{the kinetics is } r_s = \frac{1.0 \times c_s}{5 + c_s} \text{g l}^{-1}\text{h}^{-1}$$

(b)

$$V\frac{dc_s}{dt} = \upsilon(c_{s_0} - c_s) - Vr_s$$

(accumulation) (in–out) (consumed by reaction)

$$c_s = c_0 = 10 \text{ g l}^{-1} \text{ for } t = 0$$

$$\frac{dc_s}{dt} = D(c_{s_0} - c_s) - r_s$$

where $D = \frac{\upsilon}{V}$; $r_s = \frac{c_s}{5 + c_s}$ $\text{g l}^{-1}\text{ h}^{-1}$

which is solved from $t = 0$, $c_s = c_{s_0} = 10 \text{ gl}^{-1}$ and until $t = t_f$

where $\frac{c_s - c_{s_f}}{c_{s_f}} = 0.01$.

The equation may be solved analytically as shown in the following, but any major effort to solve non-linear differential equations can, in general, not be recommended since excellent differential equation solvers are available, even for pocket calculators. A check on the numerical (or analytical) solution can be made by neglecting the first term $D(c_{s_0} - c_s)$ which is obviously small for $c_s \sim c_{s_0}$. The solution is then (for small t) $t \simeq K_m \ln\frac{c_s}{c_{s_0}} + (c_s - c_{s_0})$. For $c_s = 9 \text{ g l}^{-1}$, $t =$ 1.53 h which may be compared with the exact solution in the table below.

$$\frac{dc_s}{dt} = D(c_{s_0} - c_s) - r_s \rightarrow \frac{dc_s}{d\theta} = D'(c_{s_0} - c_s) - \frac{c_s}{K_m + c_s};$$

$$\theta = \frac{t}{r_{max}}; \quad D' = \frac{D}{r_{max}}$$

$$dt = \frac{K_m + c_s}{D'(c_{s_0} - c_s)(K_m + c_s)} = -\frac{1}{D'}\frac{K_m + c_s}{c_s^2 - \left(c_{s_0} - \frac{1}{D'} - K_m\right)c_s - K_m c_{s_0}}$$

$$= -\frac{1}{D'}\left[\frac{A}{c_s - \rho_1} + \frac{B}{c_s - \rho_2}\right]$$

where (ρ_1, ρ_2) are zeros of the denominator polynomial.

$$dt = \frac{1}{D'}\left[\frac{K_m + \rho_1}{\rho_2 - \rho_1}\frac{1}{c_s - \rho_1} - \frac{K_m + \rho_2}{\rho_2 - \rho_1}\frac{1}{c_s - \rho_2}\right]$$

$$t = \frac{1}{D'(\rho_2 - \rho_1)}\left[(K_m + \rho_1)\ln\frac{c_s - \rho_1}{c_{s_0} - \rho_1} - (K_m + \rho_2)\ln\frac{c_s - \rho_2}{c_{s_0} - \rho_2}\right]$$

For $D' = 20.8/500$; $K_m = 5\,g\,l^{-1}$; $c_{s_0} = 10\,g\,l^{-1}$; $\rho_1 = -21.377$; $\rho_2 = 2.339$ (ρ_2 can be shown to be $= c_s$ at steady state).

$$t = 1.01358\left[-16.3773\ln\frac{c_s + 21.3774}{31.3774} - 7.3389\ln\frac{c_s - 2.3389}{7.6611}\right]$$

$c_s\,(g\,l^{-1})$	9	8	7	6	5	4	3	2.5
t (h)	1.587	3.344	5.364	7.756	10.747	14.89	22.9	33.26

for $\frac{c_s - c_{s_f}}{c_{s_f}} = 0.01$; $c_s = 2.424$ and $t = 38.06$ h.

(c) The reactor volume increases to 525 ml and the outlet substrate concentration is perturbed from its steady-state value. The initial ($t = 0$) outlet concentration is obtained from $(25 \times 10) + (500 \times 2.40) = 525\,c_s$

$$\therefore c_s = 2.762\,g\,l^{-1}$$

The new steady state is c_s^* given by $\frac{20.8}{525}(10 - c_s^*) = \frac{c_s^*}{5 + c_s^*} = 2.226\,g\,l^{-1}$ (compared to $2.3389\,g\,l^{-1} = c_s^*$ for $V = 500$ ml — see the expression for t in question (b). This value is slightly different from that used in the problem formulation ($c_s^* = 2.4\,g\,l^{-1}$). But the data of that table were calculated with $K_m = 5\,g\,l^{-1}$ and $r_{max} = 0.975\,g\,l^{-1}\,h^{-1}$ which is of course in practice indistinguishable from $r_{max} = 1\,g\,l^{-1}\,h^{-1}$ found by the graphical method of the solution to question (a).).

Thus with $c_s^* = 2.226\,\mathrm{g\,l^{-1}}$, $D = \dfrac{20.8}{525}\,\mathrm{h^{-1}}$, $c_{s_0} = 10\,\mathrm{g\,l^{-1}}$

$c_s(t=0) = 2.762\,\mathrm{g\,l^{-1}}$

$$\left.\begin{aligned} 0 &= D(c_{s_0} - c_s^*) - \frac{c_s^*}{K_m + c_s^*} \\ \frac{dc_s}{dt} &= D(c_{s_0} - c_s) - \frac{c_s}{K_m + c_s} \end{aligned}\right\} \frac{dc_s}{dt} = \frac{d(c_s - c_s^*)}{dt} = -D(c_s - c_s^*) - (r_s - r_s^*)$$

Now, $$r_s - r_s^* \simeq \left.\frac{dr_s}{dc_s}\right|_{c_s = c^*_s} (c_s - c_s^*) + \frac{1}{2}\left(\left.\frac{dr_s}{dc_s}\right|_{c_s = c^*_s}\right)^2 (c_s - c_s^*)^2 + \ldots$$

and $$\frac{dy}{dt} = -Dy - \frac{K_m}{(K_m + c_s^*)^2} y \text{ with } y\begin{cases} = 2.762 - 2.226 \\ = 0.536 \text{ at } t = 0 \end{cases}$$

This linear differential equation with constant coefficients is easily solved.

$$\frac{dy}{dt} = -\left(\frac{20.8}{525} + 0.09576\right) y \rightarrow \underline{y = 0.536 \exp(-0.1354t)}$$

For $y = 2.42 - 2.226 = 0.194$ $\underline{t = 7.51\ \mathrm{h}}$

If $c_s^* = 2.338$ (corresponding to $V = 500$ ml) $\underline{t = 11.57\ \mathrm{h}}$ which turns out to be a remarkably different result — indicating that the approximation $V \sim$ constant $= 500$ ml is poor. If the original steady state value $c_s^* = 2.40\,\mathrm{g\,l^{-1}}$ is used $t =$ 24.3 h. This shows the remarkable sensitiveness of this type of calculation to the accuracy of the original data. An exact solution of the differential equation from $(t, c_s) = (0, 2.762)$ to $(t, 2.42)$ yields $t = 7.37$ h, close to the result obtained by linear analysis.

Note: The perturbation (linear) analysis of the present problem is applicable in the study of any system which is disturbed slightly (in some sense) from an equilibrium. It has found universal use in chemical engineering model studies — in particular in process control.

(d) (i) The data suggest that the inhibition is of the competitive type: increase of c_i by a factor of 5 from '1 to 5 ml inhibitor solution added' leads to a considerably smaller change in c_s (and thereby in r_s) than a change in c_i by a factor 2.5 from 20 to 50 ml. Thus,

$$r_s = \frac{r_{max} C_s}{c_s + K_m\left(1 + \dfrac{c_i}{K_i}\right)} = D(c_{s_0} - c_s)$$

where $r_{max} = 1\,\mathrm{g\,l^{-1}\,h^{-1}}$; $K_m = 5\,\mathrm{g\,l^{-1}}$.

The calculations shown in Table 5.5 are made. Since $K_i \sim 10^{-1}\mathrm{mg\,l^{-1}}$ we obtain $1\,\mathrm{mg\,l^{-1}}$ in the original water sample.

Table 5.5

$n\ V_p$ (ml)	1	5	20	50	(added to 500 ml medium)
$D = \frac{v}{V_{total}}$	0.04152	0.04119	0.04000	0.03782	(h^{-1})
c_s	2.37	2.47	2.80	3.26	($g\,l^{-1}$)
c_i/K_i	0.0222	0.09872	0.3844	0.9058	(based on total V)
(c_i/K_i) based on original sample	11.1	9.97	9.99	9.96	

(ii) *Comment*: It takes ages to reach equilibrium and we have seen that calculation of somewhat secondary quantities are sensitive to errors in c_s. Also, the enzyme might presumably deactivate irreversibly owing to the long coexistence with the inhibitor. On the other hand: enzymes are very specific 'reagents' for certain inorganic/organic poisons in quantities which may be very difficult to obtain by other chemical analytical methods.

But the system in question is *too large* and the dynamics too slow. The enzyme assay must be done with test-tubes, etc., where the response is faster. Bacterial cultures on agar plates is one possibility if the enzyme is induced in the metabolism.

PROBLEM 5.11

(a) Let substrate S_1 be degraded in two consecutive steps to S_3

$$E + S_1 \overset{K_m}{\leftrightarrow} ES_1 \overset{k}{\rightarrow} E + S_2 \tag{1}$$

$$E + S_2 \overset{K_m}{\leftrightarrow} ES_2 \overset{k}{\rightarrow} E + S_3 \tag{2}$$

$$\sum_1^3 c_{s_i} = c_{s_i}^{\circ}$$

We wish to design a chemostat operation for maximum productivity of S_2 per m^3 feed volume (kg S_2 produced m^{-3} feed h^{-1}) for a given feed with $c_{s_1} = c_{s_1}^{\circ}$ (kg m^{-3}), and $c_{s_2} = c_{s_3} = 0$ in the feed. The dissociation constants K_m for ES_1 and ES_2 are equal, and the rate constants for the two rate determining steps in 1 and 2 are both k.

(i) Calculate the optimal value of

$$y_1 = \frac{c_{s_1}}{c_{s_2}} \quad (\text{or } x = 1 - y_1)$$

(ii) Calculate the corresponding reactor volume per m^3 feed.

(b) Extend the reaction sequence to include

$$E + S_3 \overset{K_m}{\leftrightarrow} ES_3 \overset{k}{\rightarrow} E + S_4 \tag{3}$$

(i) Repeat the calculation of the optimal value of $c_{s_1}/c^\circ_{s_1}$ for maximum productivity of S_3 (the last but one species in the sequence).

(ii) Calculate the product distribution:

$$\{y_1\} = \left\{\frac{c_{S_1}}{c^\circ_{S_1}}, \frac{c_{S_2}}{c^\circ_{S_1}}, \frac{c_{S_3}}{c^\circ_{S_1}}, \frac{c_{S_4}}{c^\circ_{S_1}}\right\}$$

for this particular value of $c_{s_1}/c^\circ_{s_1} = 1 - x$.

(c) Extend to $N-1$ consecutive reactions with N species:

$$S_1 \rightarrow S_2 \rightarrow \ldots S_{N-1} \rightarrow S_N \tag{4}$$

$$\sum_1^N c_{S_i} = c^\circ_{S_i}$$

(i) Show that

$$y_N = \frac{c_{S_N}}{c^\circ_{S_1}} = x^{N-1} (= (1-y_1)^{N-1})$$

(ii) Prove that maximum productivity of S_{N-1} is obtained for

$$x = \frac{N-2}{N-1} = x_{\text{opt}}$$

(iii) Write down the product distribution $\{y_i\}$, i = 1, 2, ..., N for $x = x_{\text{opt}}$.

(iv) Show that $y_N \rightarrow 1/c_e$ for $N \rightarrow \infty (c_e^{-1} = 0.36\ldots)$ when $x = x_{\text{opt}}$.

(d) Assume that only S_N and S_1 can be measured analytically. The existence of S_2, S_3, ..., S_{N-1} is ignored and the reaction is written as $S_1 \overset{k^*}{\rightarrow} S_N$ with rate r_{S_N}. For N large (>50) show that for all practical purposes the rate r_{S_N} given above is $k^* c_{s1}$ unless x is very close to 1.

Solution

(a)

$$\left.\begin{array}{l} E + S_1 \overset{K_m}{\rightleftharpoons} ES_1 \overset{k}{\longrightarrow} E + S_2 \\ E + S_2 \overset{K_m}{\rightleftharpoons} ES_2 \longrightarrow E + S_3 \end{array}\right\} \quad c_{e_0} = c_e + c_{es_1} + c_{es_2} = c_e\left(1 + \frac{c_{S_1} + c_{S_2}}{K_m}\right)$$

$$K_m = \frac{c_e c_{s_1}}{c_{es_1}} = \frac{c_e c_{s_2}}{c_{es_2}}; \quad -r_{s_1} = kc_{es_1}; \; r_{s_2} = kc_{es_1} - kc_{es_2}$$

Consider the chemostat with given volume V. Feed concentration; $c_{s_1} = c^o_{s_1}$, $c_{s_2} = 0$. The production of S_2 is maximized (in the given V) when Dc_{s_2} is as large as possible. The design equation for the chemostat is

$$D = \frac{-r_{s_1}}{c^o_{s_1} - c_{s_1}} = \frac{r_{s_2}}{c_{s_2}} \rightarrow \frac{c_{es_1}}{c^o_{s_1} - c_{s_1}} = \frac{c_{es_1} - c_{es_2}}{c_{s_2}}$$

or

$$\frac{\frac{c^o_e}{K_m} c_{s_1}}{\left(1 + \frac{c_{s_1} + c_{s_2}}{K_m}\right)(c^o_{s_1} - c_{s_1})} = \frac{\frac{c^o_e}{K_m}(c_{s_1} - c_{s_2})}{\left(1 + \frac{c_{s_1} + c_{s_2}}{K_m}\right) c_{s_2}}$$

or

$$\frac{c_{s_1}}{c^o_{s_1} - c_{s_1}} = \frac{c_{s_1} - c_{s_2}}{c_{s_2}} \tag{5.11.1}$$

$$c_{s_2} = \frac{c_{s_1}}{c^o_{s_1}}(c^o_{s_1} - c_{s_1}) \text{ or } y_2 = y_1 x$$

where x is the degree of conversion of S_1, y_2 is maximum for $x = \frac{1}{2}$ and $y_2(\max) = \frac{1}{4}$.

$$D(y_2 = \tfrac{1}{4}) = \tfrac{1}{2}\frac{kc^o_e}{(K_m + \frac{3}{4}c^o_{s_1})\frac{1}{2}} = \frac{kc^o_e}{K_m + \frac{3}{4}c^o_{s_1}}$$

Maximum production per m^3 reactor is

$$(Dc_{s_2})_{x=1/2} = \frac{c^o_e k c^o_{s_1}}{4K_m + 3c^o_{s_1}}.$$

(b) With an extra enzymatic step

$$E + S_3 \underset{}{\overset{K_m}{\rightleftharpoons}} ES_3 \overset{k}{\longrightarrow} E + S_4$$

one obtains quite analogously to equation (5.11.1),

$$\underset{(5.11.2)}{\frac{c_{s_1}}{c^o_{s_1} - c_{s_1}}} = \underset{(5.11.3)}{\frac{c_{s_1} - c_{s_2}}{c_{s_2}}} = \underset{(5.11.4)}{\frac{c_{s_2} - c_{s_3}}{c_{s_3}}}$$

From (5.11.2) and (5.11.3): $y_2 = (1 - x)x = y_1 x$

From (5.11.2) and (5.11.4): $y_3 = (1 - x)x^2 = y_1 x^2 = \frac{y_2^2}{y_1}$

Maximum y_3 is obtained for $2x(1 - x) - x^2 = 0 \rightarrow x = \frac{2}{3}$.

The product distribution is

$$\underline{(y_1, y_2, y_3, y_4) = (\tfrac{1}{3}, \tfrac{2}{9}, \tfrac{4}{27}, \tfrac{8}{27})}$$

(c) In the general case:

$$\frac{1-x}{x} = \frac{y_1}{1-y_1} = \frac{y_1 - y_2}{y_2} = \ldots = \frac{y_{N-2} - y_{N-1}}{y_{N-1}}$$

$$\sum_1^N y_i = 1 \rightarrow y_N = 1 - \sum_1^{N-1} y_i$$

$$\begin{aligned} y_1 &= 1 - x; & y_2 &= y_1 x; & y_3 &= \frac{y_2^2}{y_1} & \ldots\ y_{N-1} &= \frac{y_{N-2}^2}{y_{N-3}} \\ &= y_1 x^0 & &= y_1 x^1 & &= y_1 x^2 & &= y_1 x^{(N-2)} \end{aligned}$$

$$\therefore y_N = 1 - y_1 \sum_1^{N-1} x^{(i-1)} = 1 - y_1 \left(\sum_1^N x^{(i-1)} - x^{N-1} \right)$$

$$= 1 - (1-x)\left(1 \frac{1-x^N}{1-x} - x^{N-1} \right) = x^{N-1}$$

where we have used the summation formula for a geometrical progression. Thus:

$$y_{N-1} = \frac{y_{N-2}^2}{y_{N-3}} = y_1 x^{N-2} = (1-x)x^{N-2}$$

Maximum y_{N-1} is obtained for

$$\left[(N-2)\left(\frac{1-x}{x} \right) - 1 \right] x^{(N-2)} = 0$$

or

$$x = \frac{N-2}{N-1}$$

And now:

$$y_{N-1} = \frac{1}{N-1}\left(\frac{N-2}{N-1} \right)^{(N-2)} = y_1 x_{\text{opt}}^{(N-2)}$$

The product distribution is

$$\{y_i\} = \left\{ \frac{1}{N-1}, \frac{1}{N-1}\frac{N-2}{N-1}, \frac{1}{N-1}\left(\frac{N-2}{N-1} \right)^2 \cdots \frac{1}{N-1}\left(\frac{N-2}{N-1} \right)^{N-2}, \left(\frac{N-2}{N-1} \right)^{(N-1)} \right\}$$

$$= \left\{ \frac{1}{N-1}\left[1, \frac{N-2}{N-1}, \ldots \left(\frac{N-2}{N-1} \right)^i, \ldots \left(\frac{N-2}{N-1} \right)^{N-1}, (N-1)\left(\frac{N-2}{N-1} \right)^N \right] \right\}$$

The last expression immediately shows that

$$\sum_{1}^{N} y_i = 1$$

For $N>3$, $y_i<y_{i-1}$ for all $i<N$.

For $N>4$, y_N is larger than y_1 (and of course $> y_2 \ldots y_{N-1}$).

Conclusion: If the object is to maximize the production of the last but one species S_{N-1} in a sequence of consecutive enzymatic reactions, each with the same K_m and k, the obtainable yield S_{N-1}/S_1° forms a decreasing sequence for increasing N, the ratio $\dfrac{y_i}{y_{i-1}}$ $(i<N)$ being $\dfrac{N-2}{N-1}$. The ratio $\left(\dfrac{y_N}{y_{N-1}}\right)_{\text{opt}} =$ $N-2 \to \infty$ for $N \to \infty$, whereas $\dfrac{y_N}{y_1} = \left(\dfrac{N-2}{N-1}\right)^{N-1} (N-1)$ $(>1$ for $N>4)$.

As a final note:

$$y_N = \left(\frac{1-2/N}{1-1/N}\right)^{N-1} = \left(\frac{1-2/N}{1-1/N}\right)^{N} \frac{N-1}{N-2}.$$

For $N \to \infty$ $(1-a/N)^N \to e^{-a}$ and, consequently

$$y_N \to \frac{e^{-2}}{e^{-1}} = e^{-1} \text{ for } N \to \infty$$

The last species S_N accounts for 36.79% of the total weight while the remainder $S_1 \ldots S_{N-1}$ accounts for 63.21%.

These conclusions are of course *all* based on maximum production of species S_{N-1} which for $N \to \infty$ accounts for only $\sim \dfrac{1}{e(N-2)}$ of the total weight

(d)

$$D = \frac{k c_e^\circ \dfrac{c_{S_{N-1}}}{c_{S_N}}}{K_m + \sum_{1}^{N} c_{S_i} - c_{S_N}} = \frac{k c_e^\circ \dfrac{1-x}{x}}{c_{S_1}^\circ (f+1-x^{N-1})} \approx k^* \frac{1-x}{x} \text{ unless } x \approx 1$$

Note that the apparent 'first order rate constant' depends on $c_{S_1}^\circ$.

6

Mass transfer and scale up

Contributors: Tarun Ghose, Y. Yoshida, Subhash Chand, S. N. Mukhopadhyay, Purnendu Ghosh, K. B. Ramachandran

PROBLEM 6.1

A 50 m^3 bioreactor ($H/D_r = 2.5$; working volume 60%) equipped with two sets of a standard flat blade turbine is used for yeast growth. The bioreactor is operated continuously at a dilution rate of 0.3 h^{-1}. The organism obeys Monod's equation ($\mu_m = 0.4\,h^{-1}$ and $K_s = 2\,kg\,m^{-3}$). The inlet feed sugar (glucose) concentration is 50 kg m^{-3}. The bioreactor is aerated and agitated at 0.5 volume of air per volume of media per minute and 60 rpm. The yield of biomass based on glucose is 0.5 g cell (dry) per gram glucose consumed. The density and viscosity of the broth are 1200 kg m^{-3} and 0.02 N sec m^{-2}.

State whether the system is mass transfer limited or biochemical reaction controlled.

Solution

The maximum rate of oxygen transfer to the broth is $k_L ac^*$ where $k_L a$ is volumetric oxygen transfer coefficient and c^* is oxygen solubility in the broth. For a respiring organism, the oxygen demand (maximum oxygen uptake rate) can be represented as $x\mu_m z_{O_2}$ where x is cell concentration in the bioreactor, μ_m is maximum specific growth rate and z_{O_2} is oxygen required by unit mass of microorganism.

If $k_L ac^* > x\mu_m z_{O_2}$, then the system is biochemical reaction controlled; otherwise it is mass transfer limited.

Ratio of height (H) to diameter (D_r) of bioreactor = 2.5.

$$\frac{\pi}{4} D_r^2 \times 2.5 \times D_r = 50$$

$$D_r = 2.94\,m$$

Taking the ratio of tank diameter to impeller diameter to be 3,

$$D_i = 0.98\,\text{m}$$

For stirred and aerated bioreactor the correlation between $k_L a$ and operating variables can be expressed as (Bailey & Ollis, 1986)

$$k_L a = 2.0 \times 10^{-3} \left(\frac{P_g}{V}\right)^{0.7} \left(U_{gs}\right)^{0.2}$$

where (P_g/V) is gassed power per unit volume (W m^{-3}). U_{gs} is the superficial gas velocity equal to gas feed volumetric flow rate divided by vessel cross-sectional area times the gas holdup (m s^{-1}).

$$\text{Reynolds number } N_{Re} = \frac{nD_i^2 \rho}{\mu} = 5.8 \times 10^4$$

In the turbulent regime, power number $N_P = 6$. Thus ungassed power requirement (P) can be estimated from

$$N_P = \frac{P g_c}{\rho n^3 D_i^5}$$

$$P = \frac{N_P \rho n^3 D_i^5}{g_c}$$

$$= \frac{6 \times 1200 \times 1^3 \times (0.98)^5}{9.81} = 663\,\text{kg m s}^{-1}$$

$$= 6501.82\,\text{W}$$

If power requirement with two sets of impellers can be estimated by multiplying the value of P by 2, the total ungassed power requirement will be

$$P = 2 \times 6501.82 = 13003.64\,\text{W}$$

Hughmark (1980) correlated gassed to ungassed power for flat blade turbine impellers. This correlation covers the widest range of variables and is represented as

$$\frac{P_g}{P} = 0.10 \left(\frac{nV_L}{F}\right)^{0.25} \left(\frac{n^2 D_i^4}{g W_b V_L^{2/3}}\right)^{-0.20}$$

where n is impeller speed (rps), V_L is liquid volume (m^3), F is volumetric gas flow rate ($\text{m}^3\,\text{s}^{-1}$), D_i is impeller diameter (m), g is acceleration (m s^{-2}) and W_b is width of the impeller blade (m).

Taking $W_b/D_i = 0.20$

$$\frac{P_g}{P} = 0.10\left(\frac{1 \times 30}{0.25}\right)^{0.25}\left(\frac{1 \times (0.98)^4}{9.81 \times 0.20 \times 0.98 \times 30^{2/3}}\right)^{-0.20}$$

$$= 0.60$$

$$P_g = 0.60 \times 13003.64 = 7802.184\,\mathrm{W}$$

$$\frac{P_g}{V_L} = \frac{7802.184}{30} = 260\,\mathrm{W\,m^{-3}}$$

Using the correlation between k_La and operating variables,

$$k_La = 2 \times 10^{-3}\left(\frac{P_g}{V}\right)^{0.7}(U_{gs})^{0.2}\,\mathrm{s^{-1}}$$

where U_{gs} is superficial gas velocity and (P_g/V) is in $\mathrm{W\,m^{-3}}$.

Considering 20% gas hold up,

$$k_La = 2 \times 10^{-3} \times (260)^{0.7} \times \left(\frac{30 \times 0.5 \times 0.20}{60 \times (\pi/4) \times (2.94)^2}\right)^{0.2}$$

$$= 0.058\,\mathrm{s^{-1}}$$

$$k_Lac^* = 0.058 \times 3600 \times 6.8 \times 10^{-3}$$

$$= 1.42\,\mathrm{kg\,m^{-3}\,h^{-1}}$$

According to Monod's equation

$$D = \mu = \mu_m \frac{c_s}{K_s + c_s}$$

$$x = Y_s(c_{s0} - c_s)$$

Using $D = 0.3\,\mathrm{h^{-1}}$, $\mu_m = 0.4\,\mathrm{h^{-1}}$, $K_s = 2\,\mathrm{kg\,m^{-3}}$,

$$c_{s0} = 50\,\mathrm{kg\,m^{-3}}, \quad Y_s = 0.5\,\mathrm{g\,g^{-1}}$$

$$x = 22\,\mathrm{kg\,m^{-3}}$$

$$c_s = 6\,\mathrm{kg\,m^{-3}}$$

In general, the oxygen requirement for cell growth is expressed as (Mateles, 1971)

$$z_{O_2} = \frac{32\mathrm{C} + 8\mathrm{H} - 16\mathrm{O}}{Y_sM} + 0.01\mathrm{O}' - 0.0267\mathrm{C}' + 0.01714\mathrm{N}' - 0.08\mathrm{H}'$$

where z_{O_2} is oxygen required for one gram of cell produced; C, H and O represent number of atoms of carbon, hydrogen and oxygen respectively in each molecule of carbon source (glucose); C′, H′, N′ and O′ are the percentages of carbon, hydrogen, nitrogen and oxygen respectively in the cell and M is the molecular weight of the carbon source.

Considering yeast cell to be $CH_{1.8}O_{0.5}N_{0.2}$

$$Z_{O_2} = \frac{(32\times 6)+(8\times 12)-(16\times 6)}{0.50\times 180} + (0.01\times 32.5) - (0.0267\times 48.8) + (0.0174\times 11.4) - (0.08\times 7.3)$$

$$= 0.771\ \text{g}\ O_2/\text{g cell}$$

$$x\mu_m z_{O_2} = 22\times 0.5\times 0.771 = \underline{8.48\ \text{kg m}^{-3}\ \text{h}^{-1}}$$

Thus, $x\mu_m Z_{O_2} > k_L a c^*$

The system is mass transfer limited.

References
Bailey, J. E. & Ollis, D. O. (1986), *Biochemical Engineering Fundamentals*, 2nd edn, McGraw-Hill, New York.
Hughmark, G. (1980), *Ind. Engg. Chem. (Process Des. Dev.)*, **19**, 638.
Mateles, R. I. (1971), *Biotechnol. Bioengg.* **13**, 581.

PROBLEM 6.2

Cultivation of *Torula utilis* was carried out by Phillips *et al.* (1961) in 0.076 m diameter HRF (horizontal rotary fermenter) at 30°C using $2.5\times 10^{-3}\ \text{m}^3\ \text{s}^{-1}$ air flow rate, 500 rpm agitation speed and $0.5\times 10^{-3}\ \text{m}^3$ working liquid volume in HRF. During cultivation dissolved oxygen concentration in the liquid was monitored by a probe and the percentage of oxygen in the fermenters' exit gas was analysed by a paramagnetic oxygen analyser as shown in Table 6.1. From these data determine the values of $k_L a$, rX and c^* by oxygen balance method of Mukhopadhyay and Ghose (1976). Assume air density at 30°C = $1.2\ \text{kg m}^{-3}$.

Table 6.1

Time (h)	Dissolved oxygen concentration (ppm)	Percentage oxygen in exit air of HRF
0	—	21.00
2	6.48	20.00
4	5.76	20.40
6	5.20	20.00
8	2.40	18.80
10	0.60	16.80

References
Phillips, K. L. *et al.* (1961), *Ind. Engg. Chem.*, **53**(9), 749.
Mukhopadhyay, S. N. and Ghose, T. K. (1976), *J. Ferm. Tech.*, **54**, 509.

Solution

Air flow rate, $F = 2.5 \times 10^{-3}\,\mathrm{m^3\,s^{-1}} = 9\,\mathrm{m^3\,h^{-1}}$. Liquid volume in fermenter, $V = 0.5 \times 10^{-3}\,\mathrm{m^3}$

$$\rho_{a_{30°C}} = 1.2\,\mathrm{kg\,m^{-3}}$$

From the data in the table, we have

c_L (ppm)	$f_i - f_o$
6.48	0.2×10^{-2}
5.76	0.6×10^{-2}
5.20	1.0×10^{-2}
2.40	2.2×10^{-2}
0.60	4.2×10^{-2}

where f_i and f_o are fraction of oxygen in the inlet and outlet air streams to the HRF.

A plot of c_L versus $(f_i - f_o)$ as in Fig. 6.1 provides

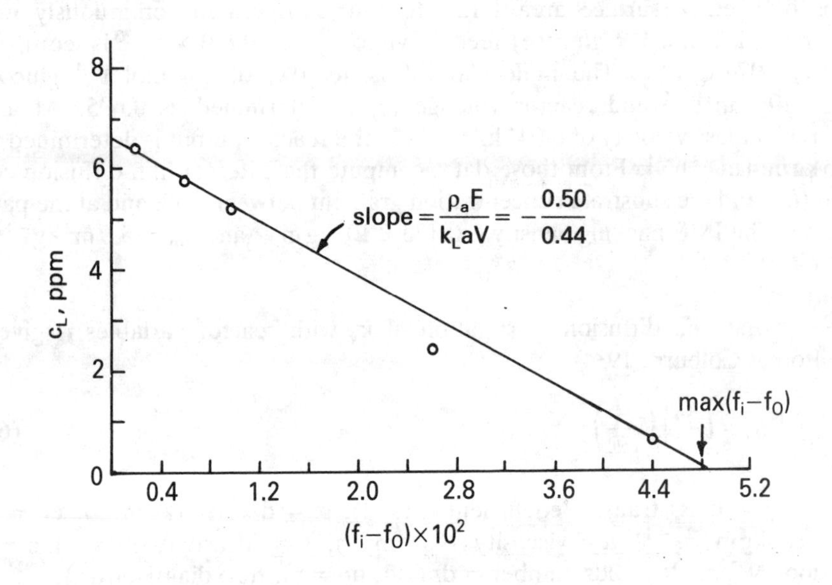

Fig. 6.1.

$$\text{slope} = \frac{-\rho_a F}{k_L a V} = \frac{-0.5}{0.44 \times 10^{-2}}(\text{ppm})^{-1} = \frac{-0.5}{0.44 \times 10^{-2}} \times 10^3 \left(\frac{\text{kg}}{\text{m}^3}\right)^{-1}$$

Putting the values of ρ_a, F and V we have

$$k_L a = \frac{1.2 \times 9.0 \times 0.44 \times 10^{-2}}{0.5 \times 10^{-3} \times 0.5 \times 10^{+3}}$$

$$= 0.19\,\mathrm{h^{-1}}$$

At steady state,

$$rX = \frac{\rho_a F}{V}(f_i - f_o) - \frac{dc_L}{dt}$$

Neglecting dc_L/dt compared to that first term on the right-hand side,

$$rX = \frac{1.2 \times 9}{0.5 \times 10^{-3}} \times 4.88 \times 10^{-5}$$

$$= 1044 \text{ ppm h}^{-1} \quad \text{or} \quad \underline{1.044 \text{ kg m}^{-3} \text{h}^{-1}}$$

PROBLEM 6.3

A packed-bed reactor containing immobilized glucose isomerase in the form of thin chips of average surface area $1.12 \times 10^{-4} \text{m}^2$ is operated continuously with a substrate (1.0 mol l^{-1} glucose) feed of viscosity (μ) $0.601 \times 10^{-3} \text{N sec m}^{-2}$ and density 1074kg m^{-3}. The molecular diffusivity ($\mathcal{D}$) of 1.0 mol l^{-1} glucose is $1.18 \times 10^{-9} \text{m}^2 \text{s}^{-1}$ and reactor voidage (ε) is determined as 0.605. At a feed superficial mass velocity of $0.041 \text{kg m}^{-2} \text{s}^{-1}$ the reactor output is determined to be $0.16 \text{kg mol m}^{-3} \text{h}^{-1}$. From those data, compute the external film diffusion coefficient (k_v) and the substrate concentration gradient between bulk and at the particle surface. The IME packing density is $0.173 \times 10^3 \text{kg m}^{-3}$ and $a_m = 5.9 \text{m}^2 \text{kg}^{-1}$.

Solution

For external film diffusion, correlation of k_v with reactor variables is given by (Chilton & Colburn, 1934)

$$j_D = \left(\frac{k_v}{G}\right)\left(\frac{\mu}{\rho \mathcal{D}}\right)^{2/3} = C(N_{Re})^n \qquad (6.3.1)$$

where k_v = mass transfer coefficient (m s^{-1}), ρ = density (kg m^{-3}), G = mass velocity ($\text{kg m}^{-2} \text{s}^{-1}$), μ = viscosity (N sec m^{-2}), $\mathcal{D}$ = diffusivity ($\text{m}^2 \text{s}^{-1}$), ε = void fraction, N_{Re} = Reynolds number = dpG/μ, dp = particle diameter (m).

For the given reactor geometry, the relationship proposed by Wilson & Geanvopolis (1966) holds good, i.e. for $0.0016 < N_{Re} < 55$, and $0.35 < \varepsilon < 0.75$,

$$j_D = \frac{1.09}{\varepsilon}(N_{Re})^{-2/3} \qquad (6.3.2)$$

Data provided are:

surface area of a chip $= 1.12 \times 10^{-4} \text{m}^2$

diameter of a sphere of 1.12×10^{-4} m² area $= dp = 0.792 \times 10^{-2}$ m .

mass velocity through the reactor $= 0.041 \text{kg m}^{-2} \text{s}^{-1}$

viscosity $\mu = 0.601 \times 10^{-3} \text{N sec m}^{-2}$

$$\therefore N_{Re} = \frac{dpG}{\mu} = \frac{0.792 \times 10^{-2} \times 0.041}{0.601 \times 10^{-3}} = 0.54$$

$$N_{Sc} = \frac{\mu}{\rho \mathcal{D}} = \frac{0.601 \times 10^{-3}}{1074 \times 1.18 \times 10^{-9}} = 474$$

$$\varepsilon = 0.605$$

$$\therefore j_D = \frac{1.09}{0.605}(0.54)^{2/3} = 2.7$$

$$j_D = \left(\frac{k_v \rho}{G}\right)\left(\frac{\mu}{\rho \mathcal{D}}\right)^{2/3}$$

$$\therefore k_v = \frac{2.7 \times 0.041}{(474)^{2/3} \times 1074} = 1.712 \times 10^{-6}\,\mathrm{m\,s^{-1}}$$

Now, reaction rate $= 0.16\,\mathrm{kg\,mol\,m^{-3}\,h^{-1}}$

$$= \frac{0.16 \times 52}{1000 \times 9 \times 3600} = 2.57 \times 10^{-7} \frac{\mathrm{kg\,mol}}{\mathrm{kg\ s}}$$

Surface area of the biocatalyst,

$$a_m = 5.9\,\mathrm{m^2\,kg^{-1}}$$

$$\therefore \text{Observed reaction rate} = k_v a_m (c_{S_b} - c_{S_s})$$

where c_{S_b} and c_{S_s} are substrate concentrations in the bulk and at the surface respectively.

$$c_{S_b} - c_{S_s} = \frac{2.57 \times 10^{-7}}{1.712 \times 10^{-6} \times 5.9} = \underline{0.0254\,\mathrm{kg\,mol\,m^{-3}}}$$

Hence there is negligible external film diffusion resistance.

References
Chilton T. H. and Colburn, A. P. (1934), *Ind. Engg. Chem.*, **26**, 1183.
Wilson, E. J. and Geanvopolis, C. J. (1966), *Ind. Engg. Chem.* (*Fundamentals*), **5**, 9.

PROBLEM 6.4

The dimensionless analysis of the mass transfer coefficient, k_L in agitated and geometrically similar vessels shows:

$$\frac{k_L D_i}{\mathcal{D}} = \alpha \left(\frac{D_i^2 n \rho}{\mu}\right)^{\beta} \left(\frac{\mu}{\rho \mathcal{D}}\right)^{\nu} \tag{6.4.1}$$

where D_i = impeller diameter (m), $\mathcal{D}$ = molecular diffusivity of a species in liquid under consideration ($\mathrm{m^2\,s^{-1}}$), n = rotational speed of impeller ($\mathrm{s^{-1}}$), ρ = liquid density ($\mathrm{kg\,m^{-3}}$), μ = liquid viscosity ($\mathrm{N\,sec\,m^{-2}}$), α = empirical constant, and β, ν= empirical exponents.

In order to achieve the same value of k_L in two geometrically similar vessels in scale up, the power per unit volume of broth, P/V must satisfy the following equation:

$$\frac{P_2/V_2}{P_1/V_1} = \left\{\frac{D_{i2}}{D_{i1}}\right\}^{2-3[(2\beta-1)/\beta]} \tag{6.4.2}$$

where subscript 1 and 2 represent small and large vessels respectively.

Assuming turbulent liquid agitation in each vessel, and also assuming the physical properties of the liquid in the two vessels are identical, derive equation (6.4.2).

Solution

$$k_L \frac{D_i}{\mathscr{D}} = \alpha\left(\frac{D_i^2 n\rho}{\mu}\right)^{\beta}\left(\frac{\mu}{\rho\mathscr{D}}\right)^{\nu} \tag{6.4.1}$$

$$\therefore \frac{k_L}{\mathscr{D}} = \alpha\left(\frac{\rho}{\mu}\right)^{\beta}\left(\frac{\mu}{\rho\mathscr{D}}\right)^{\nu} D_i^{2\beta-1} n^{\beta} \tag{6.4.3}$$

or,

$$(D_{i_1})^{2\beta-1} n_1^{\beta} = (D_{i_2})^{2\beta-1} n_2^{\beta} \tag{6.4.4}$$

$$\therefore \frac{n_2}{n_1} = \left(\frac{D_{i1}}{D_{i2}}\right)^{(2\beta-1)/\beta}$$

Again, since $(P/V) \alpha n^3 D_i^2$, we may rewrite equation (6.4.4) as

$$\frac{P_2/V_2}{P_1/V_1} = \left(\frac{n_2}{n_1}\right)^3 \left(\frac{D_{i_2}}{D_{i_1}}\right)^2$$

$$= \left(\frac{D_{i_1}}{D_{i_2}}\right)^{3\times(2\beta-1)/\beta]} \left(\frac{D_{i_2}}{D_{i_1}}\right)^2$$

$$= \left(\frac{D_{i_2}}{D_{i_1}}\right)^{2-3[(2\beta-1)/\beta]}$$

PROBLEM 6.5

Suppose a spherical particle (diameter = dp) be placed in a quiescent liquid and oxygen (molecular diffusivity = $\mathscr{D}$) diffusion from outside onto the particle surface in steady state. k_2 denotes the liquid film coefficient. Derive then that the Sherwood number = 2. Assuming that the particle ($dp = 2 \times 10^{-6}$ m) simulates a bacterium (*Pseudomonas ovalis*), calculate the maximum rate of oxygen transfer onto the bacterial surface, provided:

dissolved oxygen concentration in the bulk of liquid medium, $c_L = 6 \times 10^{-3}\,\mathrm{kg\,m^{-3}}$

concentration of bacterial cells in the medium, $n = 7.8 \times 10^{13}$ cells $\mathrm{m^{-3}}$

oxygen diffusivity, $\mathcal{D} = 10^{-9}\,\mathrm{m^2\,s^{-1}}$.

The Sherwood number relationship is also assumed in this calculation. The respiration rate measurement for this bacterial suspension, on the other hand, gave the maximum value, V_{max} as of the order of $1.667 \times 10^{-8}\,\mathrm{mol\,m^{-3}\,s^{-1}}$. Comparing the calculation with the experimental data of V_{max}, discuss the point of oxygen uptake in this well-dispersed system.

Solution

$c_L = f(r)$ alone.

Stationary sphere. $\therefore v = 0$.

Steady state condition prevailing and since no oxygen is produced, $R_A = 0$.

$$\therefore \mathcal{D}\frac{1}{r^2}\frac{\partial}{\partial r}\left[r^2\left(\frac{\partial c_L}{\partial r}\right)\right] = 0$$

$$\frac{\partial}{\partial r}\left[r^2\left(\frac{\partial c_L}{\partial r}\right)\right] = 0$$

$$r^2\frac{dc_L}{dr} = c_1$$

$$\frac{dc_L}{dr} = \frac{c_1}{r^2}$$

$$\therefore c_L = \frac{-c_1}{r} + c_2$$

Boundary conditions:

When

$$r = \infty \;; \qquad c_L = c^*$$

$$r = \frac{dp}{2} \;; \qquad c_L = \bar{c}$$

$$\therefore c^* = c_2$$

and

$$\bar{c} = \frac{-c_1}{(dp/2)} + c^*$$

$$\bar{c} - c^* = \frac{-c_1 2}{dp}$$

or

$$c_1 = \frac{dp}{2}(c^* - \bar{c})$$

At $r = dp/2$, rate diffused = rate transferred.

$$\mathscr{D}\frac{dc}{dr}\bigg|_{r=dp/2} = k_L(c^* - \bar{c})$$

$$\mathscr{D}\left[\frac{dp}{2}(c^* - \bar{c})\frac{4}{dp}\right] = k_L(c^* - \bar{c})$$

$$\mathscr{D}(c^* - \bar{c})\frac{2}{dp} = k_L(c^* - \bar{c})$$

$$\left(\frac{2\mathscr{D}}{dp} - k_L\right)(c^* - \bar{c}) = 0$$

$$c^* \neq \bar{c}$$

$$\therefore \frac{2\mathscr{D}}{dp} = k_L$$

$$\text{and, } \frac{k_L dp}{\mathscr{D}} = 2$$

$$n_{O_2'} = k_L a(c^* - \bar{c}) = k_L a c^*$$

$$\text{and, } a = n \times \bar{a}$$

where $\bar{a}$ = surface/cell

$$nO_2' = k_L(\bar{a}n)c^* = \left(\frac{2\mathscr{D}}{dp}\right)(4\pi r^2)nc^*$$

$$nO_2' = \frac{2\times 10^{-9}}{2\times 10^{-6}} \times 4\pi(10^{-6})^2 \times 7.8\times 10^{13} \times 6\times 10^{-13}\,\text{kg}\,\text{m}^{-3}\,\text{s}^{-1}$$

$$= \underline{1.838\times 10^{-4}}\,\text{kg mol m}^{-3}\,\text{s}^{-1}$$

PROBLEM 6.6

The volumetric gas phase transfer coefficient in a fermenter can be estimated by the sulphite-oxidation method. The method is based on the catalytic oxidation of sulphite to sulphate by oxygen in presence of cobalt or copper ions as catalysts at the concentration more than 1.0 mM. The rate of oxidation of sulphite is determined iodimetrically for measurement of oxygen dissolution rate. The method is still recognized as useful because it can be used for bio-reactors of any configuration and scale, flasks, test-tubes, etc., for comparison of their capabilities of oxygen supply without any difficulty in instrumentation and preparation, while it has a limit that the method can be applied only to non-biological system.

Estimate the gas-phase volumetric coefficient for oxygen transfer, k_v, from the data in Table 6.2. Make plots of k_v (kg mol m^{-3}h^{-1} (N m^{-2})$^{-1}$) versus agitation speed N (rpm) and nominal air velocity U_s ($\mathrm{m\,h^{-1}}$) on a logarithmic scale and estimate the values of α and β in the correlation

$$k_v = KN^{\alpha}V_s^{\beta} \tag{6.6.1}$$

Solution

Table 6.2 — Decrease of titration value (ml) in 10 min using 1.0 ml of a sample of sulphite solution from an aerated and agitated fermenter

Agitation rate, N (rpm)	Decrease of titration value (ml per 10 min) — Aeration rate Q (litres $\mathrm{min^{-1}}$)		
	25	50	100
150	1.12	2.15	3.38
175	1.58	2.69	3.93
200	2.45	4.29	7.30
250	4.47	6.71	8.56

The 0.01N iodine solution (factor 1.03) was used for titration of 1.0 ml of the sulphite solution sampled six times with 10 min intervals from a fermenter (working volume: 100 litre; tank diameter: 0.6 m; data were averaged out using 5 experimental values).

Sodium sulphite is oxidized by the following reaction:

$$2Na_2SO_3 + O_2 \rightarrow 2Na_2SO_4 \tag{6.6.2}$$

Then, the volumetric oxygen transfer coefficient can be obtained by the following equation

$$k_v = \frac{c_1 - c_2}{2p_{gm}(t_2 - t_1)} , \tag{6.6.3}$$

where c_1 and c_2 are the sulphite ion concentration [$\mathrm{kg\,mol\,m^{-3}}$] at time t_1 and t_2, respectively, and p_{gm} is the mean value of oxygen partial pressure.

Then, k_v [kg mole O_2 $\mathrm{m^{-3}\,h^{-1}N\,m^{-2}}$] was obtained by the following equation,

$$k_v = \frac{0.01 \times 1.03 \times 10^{-3}\Delta V}{2 \times 2 \times 0.201 \times 10} \times 60 \times 10^3$$

$$= 7.68 \times 10^{-2} \times \Delta V , \tag{6.6.4}$$

where ΔV is the difference in the titration volume at a 10-min interval.

The values of k_v for different agitation and aeration conditions were calculated using the date shown in Table 6.2. The results are listed in Table 6.3.

After converting aeration rate [l min^{-1}] to the nominal air velocity, u_s [m h^{-1}], the values of k_v are plotted against agitation rate N and the nominal air velocity u_s in Figs 6.2 and 6.3 respectively. k_v was well correlated to agitation rate and the nominal air velocity, although estimation of the values of α and β is hardly obtained merely by plotting.

The least squares method was applied to get the values of K, α and β, the equation of the correlation is shown as follows;

$$k_v = 5.57 \times 10^{-7} N^{2.35} u_s^{0.675} \ [\text{kg mol O}_2 \text{ m}^{-3} \text{N m}^{-2}] \tag{6.6.5}$$

The correlation is verified by comparing the experimental and calculated results as shown in Fig. 6.4.

PROBLEM 6.7

Koizumi and Aiba (1984) discussed the delay of the dissolved oxygen (DO) probe. The oxygen molecule diffuses from the medium to the electrode surface through liquid film, membrane, and electrolyte solution (see Fig. 6.5), and the subsequent electrochemical reactions occur on the cathode and anode of a membrane-covered oxygen electrode.

Analyse the oxygen transport in the membrane to discuss delay of the probe.

Reference
Koizumi, J. and Aiba, S. (1984), *Biotechnol. Bioeng.*, **26**, 1131.

Solution

Fick's second law can be used to describe the unsteady-state of oxygen transport in the membrane as follows:

$$\frac{\partial c_m}{\partial t} = \mathcal{D} \frac{\partial^2 c_m}{\partial x^2} \tag{6.7.1}$$

where c_m, $\mathcal{D}$, t and x are the oxygen concentration, the oxygen diffusion coefficient in the membrane, time and the distance from the cathode surface, respectively. The initial and boundary conditions (for a step change) are:

$$c_m = 0 \qquad \text{at } t < 0 \tag{6.7.2}$$

$$c_m = 0 \qquad \text{at } x = 0 \tag{6.7.3}$$

$$c_m = c^* \qquad \text{at } x = b,\ t \geq 0 \tag{6.7.4}$$

where b is the membrane thickness and c^* is the concentration of dissolved oxygen in the bulk flow.

The probe response is proportional to the oxygen flux, N_{O_2} at the membrane-electrode interface;

$$N_{O_2} = \mathcal{D} \left(\frac{\partial c_m}{\partial x} \right)_{x=0} \tag{6.7.5}$$

Table 6.3 — Volumetric oxygen transfer rate k_v at various agitation and aeration rates

Agitation rate (rpm)	k_v (kg mol O_2 m^{-3} h^{-1} atm^{-1})		
	Aeration rate ($l\,min^{-1}$) (nominal velocity of air) ($m\,h^{-1}$)		
	25 (1.5)	50 (3.0)	100 (6.0)
150	0.0861	0.165	0.260
175	0.121	0.207	0.302
200	0.188	0.330	0.561
250	0.344	0.516	0.658

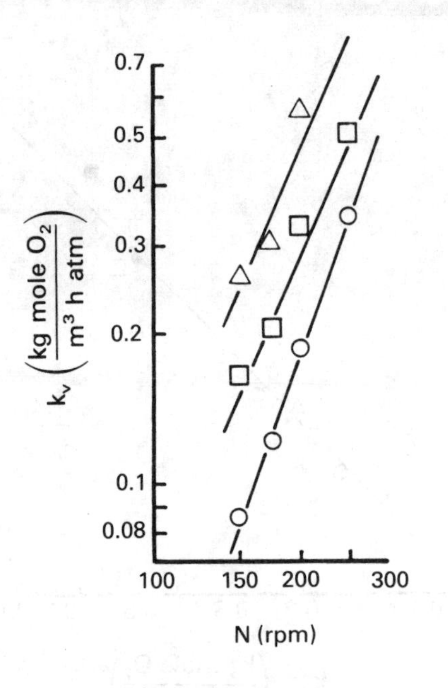

Fig. 6.2 — Correlation between the volumetric oxygen absorption coefficient, k_d and agitation rate, N.

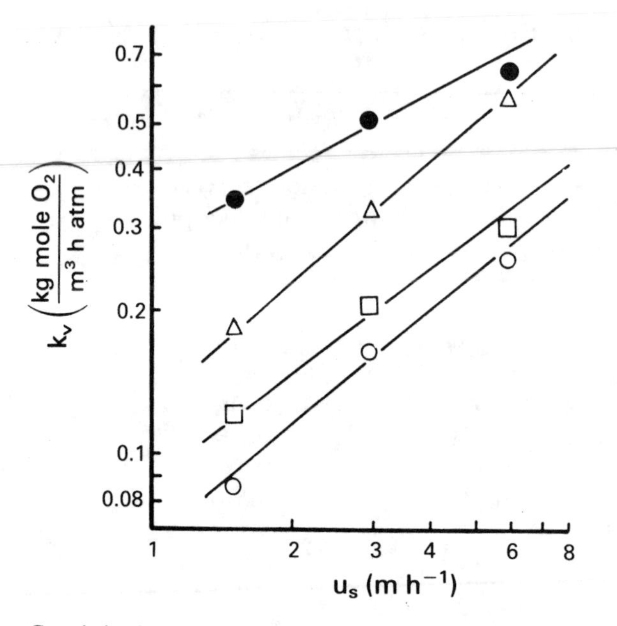

Fig. 6.3 — Correlation between the volumetric oxygen absorption coefficient, k_v and nominal velocity of air, u_s.

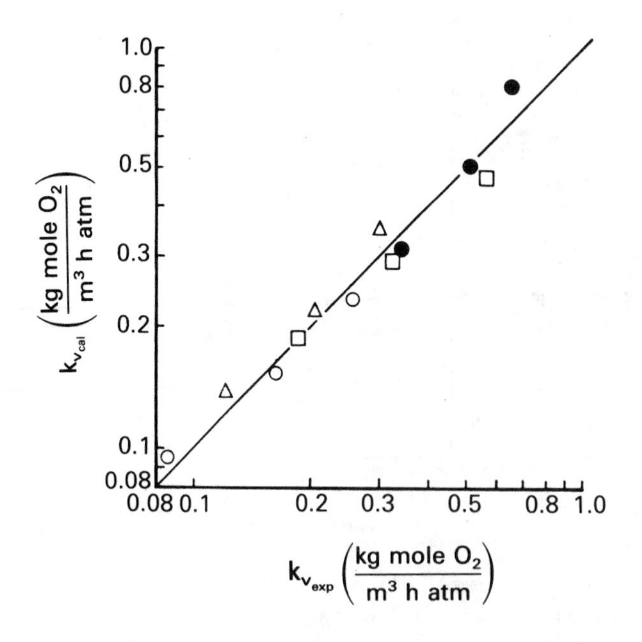

Fig. 6.4 — Comparison of the calculated and experimental value of k_v.

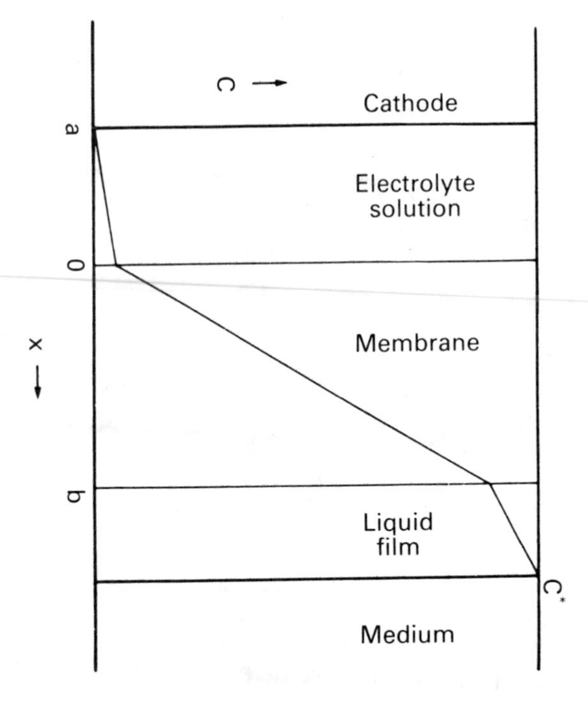

Fig. 6.5 — Schematic diagram of oxygen transport from medium to electrode.

The solution of equation (6.7.1) under the initial and boundary conditions, equations (6.7.2) to (6.7.4) gives:

$$\left(\frac{\partial c_{\mathrm{m}}}{\partial x}\right)_{x=0}=\frac{1}{b}\left[1+2\sum_{n=1}^{\infty}(-1)^n \exp(-n^2\pi^2\mathscr{D}t/b^2)\right] \tag{6.7.6}$$

From equations (6.7.5) and (6.7.6)

$$N=\frac{\mathscr{D}}{b}\left[1+2\sum_{n=1}^{\infty}(-1)^n \exp(-n^2\pi^2\mathscr{D}t/b^2)\right] \tag{6.7.7}$$

Thus, the step response of the probe, $\Gamma(t)$ can be written as,

$$\Gamma(t)=\frac{I_{\mathrm{P(o)}}-I_{\mathrm{P(t)}}}{I_{\mathrm{P(o)}}-I_{\mathrm{P}(\infty)}}$$

$$=1+2\sum_{n=1}^{\infty}(-1)^n \exp(-n^2\pi^2\mathscr{D}t/b^2) \tag{6.7.8}$$

where $I_{\mathrm{P(t)}}$ is the reading of the output current at $t=t$.

$\Gamma(t)$ could be approximated to a broken line as shown in Fig. 6.6. One may approximate the response in the following equation:

$$\text{response of probe signal} = \begin{cases} 0 & t < \tau \\ 1 - \exp[-k(t-\tau)] & t \geq \tau \end{cases} \tag{6.7.9}$$

Consequently, the response of oxygen probe is characterized by two system parameters, τ (dead time) and k (sensitivity). These parameters are strongly affected by the diffusion coefficient of oxygen in the membrane and the thickness of the membrane as shown in equation (6.7.8).

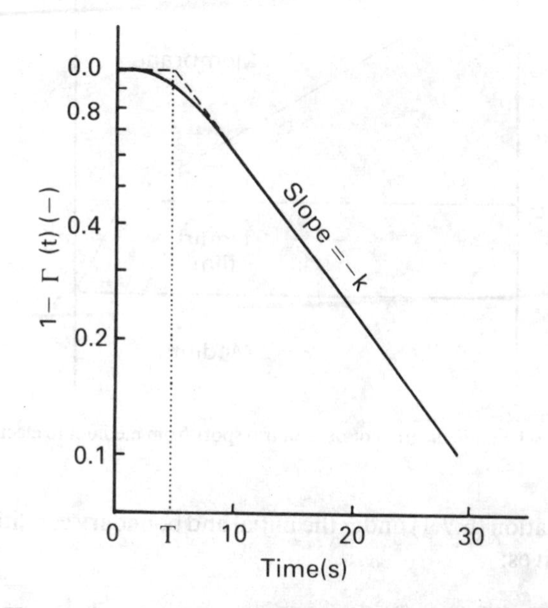

Fig. 6.6 — The response of an dissolved oxygen electrode to a step change of the dissolved oxygen.

PROBLEM 6.8

Discuss the merits and demerits of the following methods for determination of the oxygen transfer coefficient: the sulphite oxidation method, a dynamic method using DO probe, and a mass balance method using a oxygen gas analyser and a DO probe.

Solution

The sulphite oxdation method of the volumetric oxygen transfer coefficient was applied to a fermenter by Hixon *et al*. (1950) and Batholomew *et al*. (1950). It is based on the catalytic oxidation of sulphite to sulphate by oxygen at the presence of cobalt or copper ions as catalysts. The rate of oxidation of sulphite is determined iodimetrically for measurement of oxygen dissolution rate. The method requires only chemicals, not any special equipment, though it is necessary to be careful about the temperature control because heat is generated by the reaction. In this system, the reaction rate is limited by the diffusion in the gaseous film at the interface between the gas and liquid phases while no dissolved oxygen exists in the liquid phase. The mechanism of oxygen transfer in this system is not the same as that in the culture broth, where the oxygen transfer in the liquid phase is the limiting step.

The development of on-line measurement of the dissolved oxygen by a membrane-covered elecrode opened a new method for measurement of the volumetric oxygen transfer coefficient, k_La. Taguchi *et al.* (1966) presented a so-called dynamic method of k_La measurement. It is a method to estimate k_La from the change in the dissolved oxygen concentration: the decrease of the dissolved oxygen concentration after stopping the aeration and the increase of that by re-aeration. This method allows us to estimate k_La in the culture broth at any time during cultivation.

The oxygen consumption rate of the culture, R, is obtained from the slope of the decrease of the dissolved oxygen concentration. The change in the oxygen concentration is expressed by the following equation,

$$\frac{dc}{dt} = k_La(c^* - c) - R \tag{6.8.1}$$

or, assuming the respiration rate is independent of c (which it should be above some critical value),

$$c = \frac{1}{k_La}\left(\frac{dc}{dt} + R\right) + c^* \tag{6.8.2}$$

The value of k_La is obtained from the slope of the straight line of the plot of c versus $(dc/dt + R)$.

Integrating equation (6.8.1)

$$c(\infty) - c(t) = [c(\infty) - c(0)]\exp(k_Lt) \tag{6.8.3}$$

where $c(\infty)$ is the dissolved oxygen concentration before stopping aeration, which might be identical to the steady state concentration after re-aeration. The value of k_La is obtained by plotting $c(\infty) - c$ against t on a semi-logarithmic scale.

A dynamic method is a very useful method for estimation of k_La in an actual cultivation system, though the cessation of aeration must be done so as not to cause any disturbance of metabolism which may happen at dissolved oxygen concentration below the critical value. This method is improved if the gas flow rate is simply varied but not stopped (Howell *et al.* (1984), *J. Wat. Poll. Control*, April, 319–324).

Another method is estimation of k_La by monitoring the dissolved oxygen concentration and the partial pressure of oxygen in the purged gas. The following equation is obtained from the oxygen balance,

$$k_La = R/(c^* - c) \tag{6.8.4}$$

$$R = \frac{Q_N}{V}k_0\left[\frac{(p_O)_{in}}{p_t - (p_O)_{in} - (p_C)_{in}} - \frac{(p_O)_{out}}{p_t - (p_O)_{out} - (p_C)_{out}}\right] \tag{6.8.5}$$

where p_t = total pressure, p_O = partial pressure of oxygen, and p_C = partial pressure of carbon dioxide. Subscript 'in' and 'out' refer to inlet and outlet gas respectively.

Q_N = gassing rate of nitrogen (m^3s^{-1}), V = culture volume (m^3), and k_0 = conversion factor for volume of gas to weight of gas.

This is the recommended method of the three because it does not cause any disturbance to the fermentation and continous monitoring is possible with case on a computer-coupled system, though accurate analysis of gas composition and gas flow measurement are essential, especially in the case of slow oxygen consumption.

References
Hixon, A. W. and Gaden, E. L., Jr. (1950). *Ind. Eng. Chem.*, **42**, 1792.
Bartholomew, W. H., Karow, E. O., Stat, M. R. and Wilhelm, R. H. (1950), *Ind. Eng. Chem.*, **42**, 1801.
Taguchi, H. and Humphrey, A. E. (1966), *J. Ferment. Technol.*, **44**, 881.

PROBLEM 6.9

Discuss oxygen transfer for microbial cultivation in shake flasks considering transport steps of oxygen from the outside of a flask to the medium, and show the methods of estimation of the oxygen transfer coefficient of each step.

Solution

The resistance for the transfer of oxygen from the ambient into a flask through the cotton plug or other materials besides the resistance from gaseous phase to the medium in the flask.

The volumetric rate of oxygen transfer into the medium in a flask is expressed by the following equation,

$$N_{O_2} = \frac{p_O - p}{\frac{V}{k_{v,c}} + \frac{1}{K_v}} \tag{6.9.1}$$

where p_O and p are partial pressure of oxygen in the ambient and in the culture broth, respectively, and V is the volume of culture broth. $k_{v,c}$ and k_v are the oxygen transfer coefficient at the cotton plug and the interface between gas and liquid, respectively. Equation (6.9.1) leads a suggestion for an increase of oxygen transfer rate, namely (1) minimization of $V/k_{v,c}$ by minimization of the volume of culture broth, and (2) maximization of the oxygen transfer coefficient at the cotton plug and maximization of k_v by increasing shaking speed and by having rugged surface of the inner wall of the flask. It may be noted that at low shaking speed diffusion through porous plug is not likely to be affected.

The oxygen transfer coefficient in the cotton plug is estimated by monitoring the change of the oxygen partial pressure inside a flask after purging the air off the flask by nitrogen gas. This has to be done without any liquid from inside the flask.

$$k_{v,c} = \frac{V_G}{22.4}\frac{T_O}{T}\frac{1}{t}\ln\frac{p_O - p^*}{p_O - p_0^*} \tag{6.9.2}$$

where V_G is the volume of gaseous phase inside the flask, T_0 and T are the temperature at the standard condition and the temperature inside the flask, and p_O, p^* and p_0^* are the partial pressure of oxygen in the ambient, in the flask and in the flask at zero time.

The oxygen transfer coefficient between gaseous phase and the liquid phase inside a flask is estimated by the sulphite oxidation method or from a decrease rate of the oxygen partial pressure inside a flask with a rubber stopper containing sulphite solution or a culture with high oxygen consumption. This thus measures $k_g a$ and not $k_L a$.

PROBLEM 6.10

Oxygen transfer in a fermenter with several sets of curved impellers of various scale was studied by Fukuda *et al.* (1968). The data shown in Table 6.4 and 6.5 was obtained.

Compare and discuss the acceptabilities of correlations between oxygen transfer, and dimensions and operational conditions of fermenters, which have been proposed by several investigators, such as Cooper *et al.* (1944), Richards (1961) and Humphrey (1964).

References

Fukuda, H. , Sumino, Y., and Kanzaki, T. (1968) *J. Ferment. Technol.*(Japan) **46**, 823.
Cooper, C. M., Fernstrom, G. A. and Miller, S. A. (1944), *Ind. Eng. Chem.*, **36**, 504.
Humphrey, A. E. (1966) *J. Ferment. Technol.*, **42**, 265.
Richards, J. W. (1961), *Prog. Ind. Microbiol.*, **3**, 141.

Solution

Cooper *et al.* (1944) presented the following equation to correlate the volumetric oxygen absorption coefficient, k_v to the superficial air velocity, u_s and the agitation power per unit volume, P_v,

$$k_v = K P_v^{0.95} u_s^{0.67} \tag{6.10.1}$$

Absorption numbers can be defined as follows:

$$N_{abs} = k_v / u_s^{0.67} \tag{6.10.2}$$

Fig. 6.7 shows the plotting of N_{abs} versus P_v on a log scale. Several different correlations were obtained according to the size of the fermenters, and it was not possible to correlate all the data in a single line.

Another correlation between volumetric oxygen transfer coefficient, $k_L a$ and operational conditions has been proposed by Richards (1961) using the dimensional analysis and an experimental correlation about the gas–liquid interfacial area derived by Calderbank (1958)

$$k_L a = K (P/V)^{0.4} u_s^{0.5} N^{0.5} \tag{6.10.3}$$

According to the correlation, k_v were plotted on a log scale as shown on Fig. 6.8. Parallel lines were obtained with different intercepts by use of different fermenters, resulting the following equation with different values of a constant K for each fermenter:

$$k_v = K[P_g/V)^{0.4} u_s^{0.5} N^{0.5}]^{1.4} \quad [\text{g mol } O_2 \text{ min}^{-1} \text{ m}^{-3} \text{ atm}^{-1}] \tag{6.10.4}$$

The value of K was well correlated to the number of impellers resulting in the following equation,

$$K_v = (2.0 + 2.8 N_i)\, 10^{-3} \tag{6.10.5}$$

Table 6.4 — Data of oxygen transfer rate measured in various scale of fermenters

No.	Fermenter	V (m^3)	A ($kg\ cm^{-2}$)	N (rpm)	Q (%)	Q ($l\ min^{-1}$)	V_s ($cm\ min^{-1}$)	P_g HP	P_0 HP	OTR ($\times 10^{-7}$) ($g\ mol\ ml^{-1}\ min^{-1}$)	k_v ($\times 10^{-6}$) ($g\ mol\ ml^{-1}\ min^{-1}\ atm^{-1}$)
1	F.0.2	0.1	0.1	150	25	25	9.0	0.321	0.342	4.7	1.43
2	F.0.2	0.1	0.1	150	50	50	18.0	0.300	0.342	10.4	2.76
3	F.0.2	0.1	0.1	150	100	100	36.0	0.279	0.342	16.7	4.33
4	F.0.2	0.1	0.1	175	25	25	9.0	0.504	0.515	7.3	2.03
5	F.0.2	0.1	0.1	175	50	50	18.0	0.472	0.515	12.7	3.45
6	F.0.2	0.1	0.1	175	100	100	36.0	0.440	0.515	19.0	5.03
7	F.0.2	0.1	0.1	200	25	25	9.0	0.685	0.707	10.2	3.14
8	F.0.2	0.1	0.1	200	50	50	18.0	0.643	0.707	18.7	5.50
9	F.0.2	0.1	0.1	200	100	100	36.0	0.633	0.707	29.5	0.35
10	F.0.2	0.1	0.1	250	25	25	9.0	1.385	1.445	15.2	5.72
11	F.0.2	0.1	0.1	250	50	50	18.0	1.340	1.445	25.8	8.60
12	F.0.2	0.1	0.1	250	100	100	36.0	1.252	1.445	37.2	10.95
13	F.1	0.5	0.5	160	25	125	19.0	2.95	3.11	12.5	5.40
14	F.1	0.5	0.5	160	50	250	38.0	2.90	3.11	15.7	5.75
15	F.1	0.5	0.5	160	100	500	76.0	2.57	3.11	27.7	9.84
16	F.2.1	1.0	0.5	100	25	250	21.6	2.63	2.63	9.0	3.43
17	F.2.1	1.0	0.5	100	50	500	43.2	2.63	2.63	13.6	4.83
18	F.2.1	1.0	0.5	100	1000	100	86.4	2.63	2.63	19.2	6.51
19	F.2.1	1.0	0.5	120	25	250	21.6	2.68	2.68	5.9	2.06
20	F.2.1	1.0	0.5	120	50	500	43.2	2.63	2.68	10.7	3.68
21	F.2.1	1.0	0.5	120	100	1000	86.4	2.63	2.68	21.4	7.36
22	F.6-1(C)	3.6	0.5	90	20	720	25.0	1.29	1.34	4.1	1.34
23	F.6-1(C)	3.6	0.5	120	20	720	25.0	2.82	3.09	7.7	2.80
24	F.6-1(C)	3.6	0.5	150	20	720	25.0	5.58	5.76	10.2	4.16
25	F.6-1(C)	3.6	0.5	170	20	720	25.0	7.54	9.14	11.8	5.20
26	F.6-1(C)	4.2	0.5	90	10	420	16.0	1.50	1.75	2.6	0.88
27	F.6-1(C)	4.2	0.5	90	40	1680	64.0	1.37	1.75	7.1	2.29
28	F.6-1(C)	4.2	0.5	120	10	420	16.0	3.14	3.67	4.8	1.88
29	F.6-1(C)	4.2	0.5	120	20	840	32.0	3.08	3.67	7.5	2.88
30	F.6-1(C)	4.2	0.5	120	30	1260	48.0	2.90	3.67	10.5	3.75

Table 6.4 (*Continued*)

No.	Fermenter	V (m^3)	A ($kg\ cm^{-2}$)	N (rpm)	Q (%)	Q ($l\ min^{-1}$)	V_s ($cm\ min^{-1}$)	P_g HP	P_0 HP	OTR ($\times 10^{-7}$) ($g\ mol\ ml^{-1}\ min^{-1}$)	k_v ($\times 10^{-6}$) ($g\ mol\ ml^{-1}\ min^{-1}\ atm^{-1}$)
31	F.6-1(C)	4.2	0.5	150	10	420	16.0	5.69	6.92	6.5	3.09
32	F.6-1(C)	4.2	0.5	150	20	840	32.0	5.23	6.92	10.0	3.85
33	F.6-1(C)	4.2	0.5	150	30	1260	48.0	5.10	6.92	14.4	5.76
34	F.6-1(C)	4.2	0.5	150	40	1680	64.0	5.10	6.92	17.9	6.90
35	F.6-1(C)	4.2	0.5	170	10	420	16.0	9.00	9.75	7.0	3.54
36	F.6-1(C)	4.2	0.5	170	20	840	32.0	7.70	9.75	12.0	5.34
37	F.6-1(C)	4.2	0.5	170	30	126	48.0	7.67	9.75	18.4	7.50
38	F.6-1(C)	4.2	0.5	170	40	168	64.0	7.14	9.75	19.6	7.85
39	F.6-1(C)	4.2	0.5	90	20	840	37.0	1.40	1.75	5.4	1.83
40	F.6-1(C)	4.2	0.5	120	20	840	37.0	3.08	3.67	8.6	3.27
41	F.6-1(C)	4.2	0.5	150	20	840	37.0	5.23	6.92	11.9	5.30
42	F.6-1(C)	4.2	0.5	170	20	840	37.0	7.68	9.75	13.0	6.20
43	F.6-1(A)	4.2	0.5	90	20	840	32.0	1.40	1.75	4.6	1.53
44	F.6-1(A)	4.2	0.5	120	20	840	32.0	3.08	3.67	7.0	2.50
45	F.6-1(A)	4.2	0.5	150	20	840	32.0	5.23	6.92	10.3	4.21
46	F.6-1(A)	4.2	0.5	170	20	840	32.0	7.68	9.75	12.4	5.65
47	F.6-1(B)	4.2	0.5	90	20	840	32.0	1.40	1.75	4.3	1.42
48	F.6-1(B)	4.2	0.5	120	20	840	32.0	3.08	3.67	6.4	2.22
49	F.6-1(B)	4.2	0.5	150	20	840	32.0	5.23	6.92	8.6	3.25
50	F.6-1(B)	4.2	0.5	170	20	840	32.0	7.68	9.75	12.1	5.50
51	F.6-0	0	0	112	0	0	—	—	8.0	—	—
52	F.6-0	23.5	0.5	112	0	0	—	—	48.2	—	—
53	F.6-0	36.0	0.5	112	0	0	—	—	69.5	—	—
54	F.6-0	36.0	0.5	112	11	3300	24.8	69.5	69.5	7.0	3.24
55	F.6-0	36.0	0.5	112	20	6000	45.0	69.5	69.5	11.7	5.18
56	F.6-0	36.0	0.5	112	30	9000	67.5	71.0	69.5	18.0	8.00
57	F.6-0	40.0	0.5	112	0	0	—	—	91.2	—	—
58	F.6-0	40.0	0.5	112	10	4000	31.5	83.0	91.2	7.1	3.46
59	F.6-0	40.0	0.5	112	20	8000	63.0	79.0	91.2	12.9	5.73
60	F.6-0	40.0	0.5	112	30	12000	94.5	75.0	91.2	19.3	8.55

Table 6.5 — Dimensions of fermenters

Fermenter	F. 0.2	F. 1	F. 2–1	F. 6–1	F. 60
H_t/D	2.08	2.06	1.87	2.05	2.00
H_v/D	1.64	1.64	1.48	1.64	1.60
H_l/D	0.84	1.21	1.17	1.38	1.32
D_i/D	0.55	0.45	0.50	0.30	0.32
W_t/D	0.127	0.095	0.100	0.070	0.056
R_t/D	0.09	0.10	0.10	0.10	0.10
H_a/D_i				1.00	
H_b/D_i	1.00	2.10	1.11	1.30	1.26
H_c/D_i			0.75	1.30	1.59
H_f/D_i	0.52	0.88	0.48	1.00	1.24
W_i/D_i	0.20	0.45	0.36	0.21	0.21
Working volume (l)	100	700	1,000	4,200	40,000
Impeller	1	1	2	3	2
Type of impeller	Curved turbine	Curved turbine	Curved turbine	Curved disk turbine with guide	Curved disk turbine with guide
Sparger	1	1	1	2	2

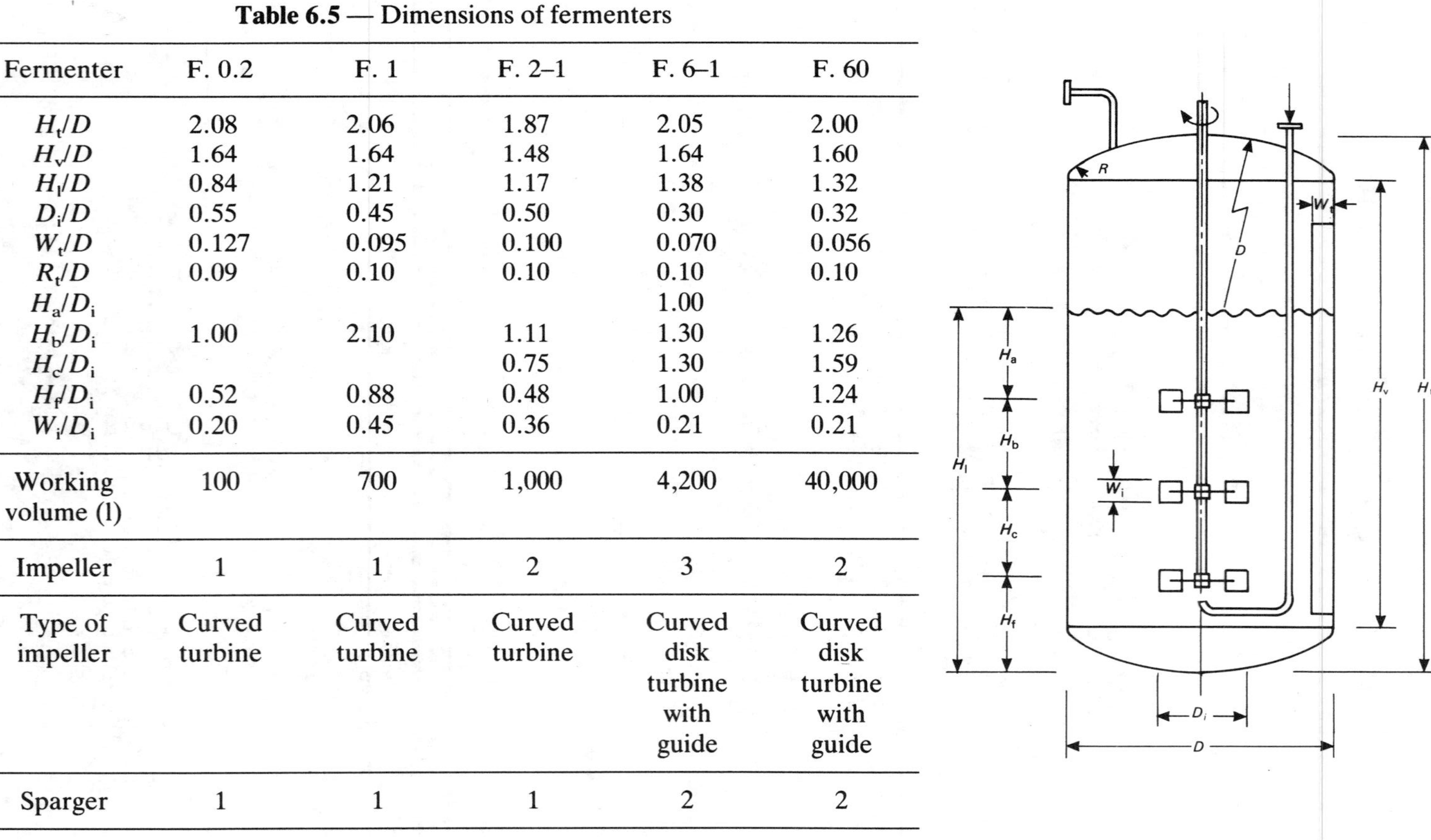

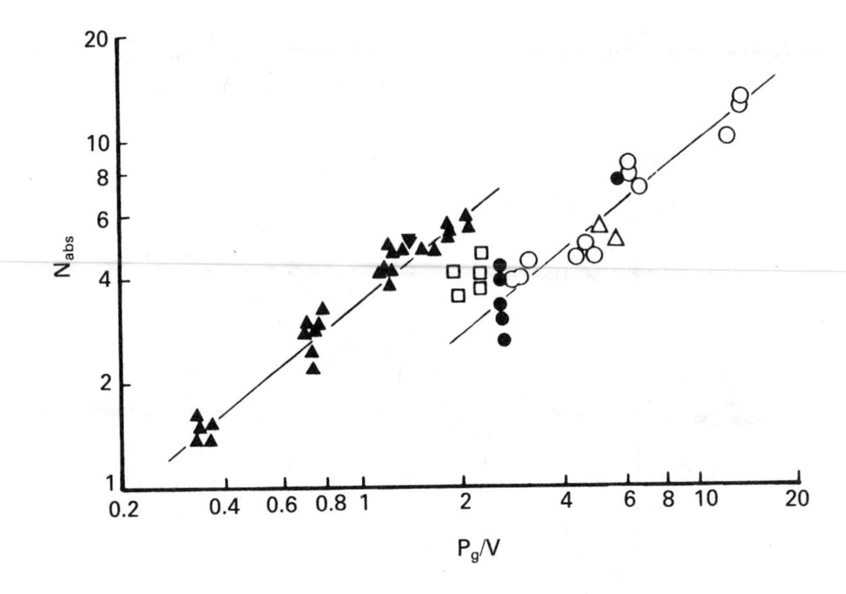

Fig. 6.7 — Plot of the absorption number, N_{abs} against the power requirement per unit volume, $P_v = P_g/V^{-1}$ (Cooper's correlation).

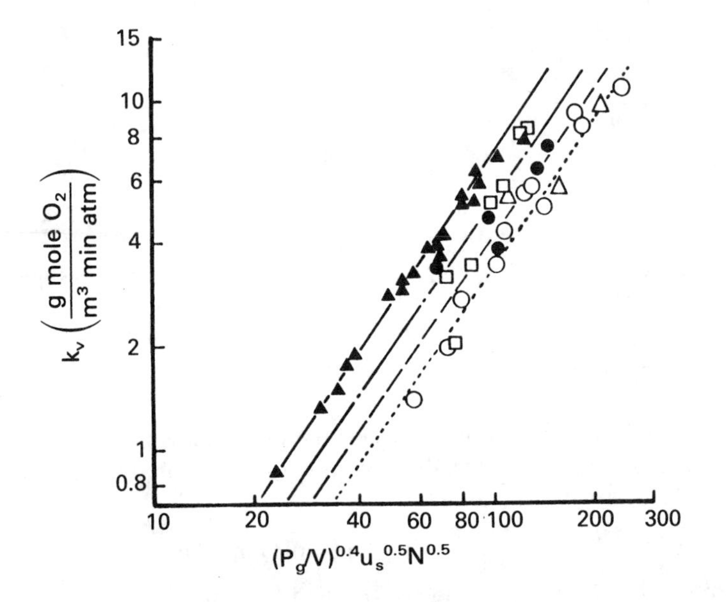

Fig. 6.8 — Plot of the volumetric oxygen absorption coefficient, k_v against $(P_g/V)^{0.4}u_s^{0.5}N^{0.5}$ (Richards' correlation).

Then, the following equation is applicable to correlate k_v to the configuration, dimensions and operational conditions of fermenters as shown in Fig. 6.9.

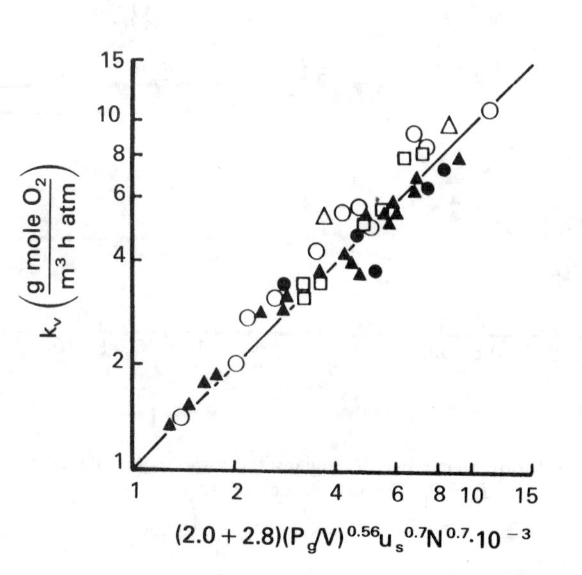

Fig. 6.9 — Plot of k_v versus $(2.0+2.8N_i)(P_g/V)^{0.56}u_s^{0.7}N^{0.7}$.

$$k_v = (2.0+2.8)(P_g/V)^{0.56}u_s^{0.7}N^{0.7}10^{-3} \quad [\text{g mol O}_2\,\text{m}^{-3}\,\text{min}^{-1}\,\text{atm}^{-1}] \qquad (6.10.6)$$

where the units of P_g, V, u_s and N are (HP), (m^3), (cm min^{-1}) and (rpm), respectively.

Next, k_v was plotted against $N^{4/3}D_i^2u_s^{1/2}$ according to the correlation suggested by Humphrey (1964),

$$k_La = \frac{(N^2D_i^3)^{2/3}u_s^{1/2}}{(\sigma\mu_a)^{0.3}} \qquad (6.10.7)$$

Different lines were obtained for different fermenters as shown in Fig. 6.10. Those lines can be extrapolated to go through a common intercept on the y axis at a point $(0, -2)$ to get the following equation,

$$k_v = -2.0 + K_{slope}(N^{4/3}D_i^2u_s^{1/2}) \qquad [\text{g mol O}_2\,\text{m}^{-3}\,\text{min}^{-1}\,\text{atm}^{-1}] \quad (6.10.8)$$

where the units of N, D_i and u_s are (rpm), (m) and (cm min^{-1}).

K_{slope} could be correlated to the diameter of impeller as follows,

$$K_{slope} = 630D_i^{-1.76} \qquad (6.10.9)$$

The above discussion is summarized in the following correlation.

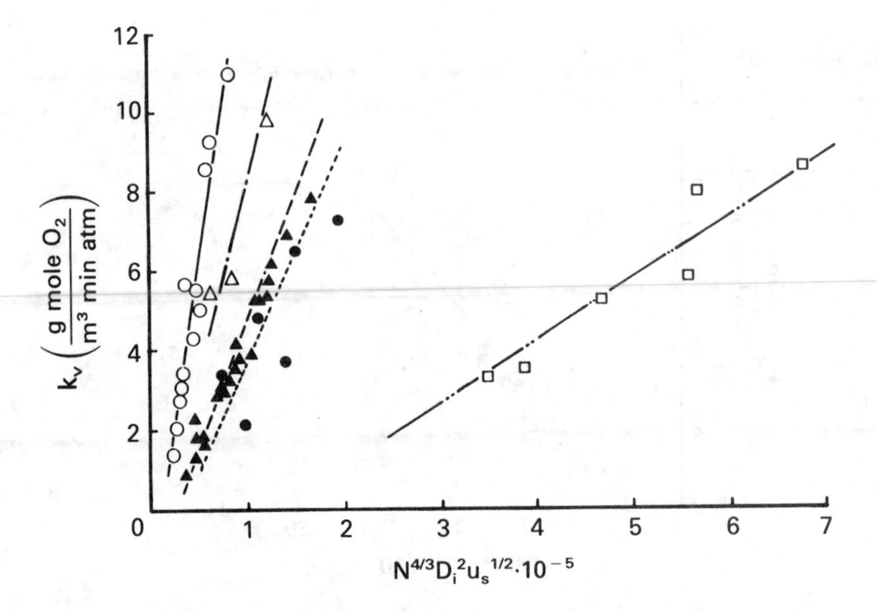

Fig. 6.10 — Plot of k_v versus $N^{4/3}D_i^2u_s$ (Humphrey's correlation).

$$k_v = 0.630 N^{4/3} D_i^{0.24} u_s^{1/2} 10^{-3} - 2.0 \qquad [\text{g mol } O_2\, m^{-3}\, \text{min}^{-1}\, \text{atm}^{-1}] \tag{6.10.10}$$

The correlation was verified by plotting k_v against $N^{4/3}D_i^{0.24}u_s^{1/2}$ as shown in Fig. 6.11.

References
Calderbank, P. H. (1958), *Trans. Instn. Chem. Engrs.*, **36**, 443.
Humphrey, A. H. (1964), *J. Ferment. Technol.*, **42**, 265.

PROBLEM 6.11

The usual procedure of scale-up of a fermenter is to fix one of several criteria involving Reynolds number, power consumption per unit volume of liquid, tip velocity of an impeller, the liquid circulation time and the volumetric oxygen transfer coefficient. The choice of criterion will depend on the fermentation being studied.

Estimate, using two methods, the required speed of an impeller and the power requirements of a production scale fermenter of 60 m^3, to match the volumetric mass transfer coefficient. Following optimum conditions were obtained with a 0.3 m^3 Fermenter:

Density of culture broths = 1200 kg m^{-3}
Liquid volume = 0.18 m^3
Aeration rate = 1.0 v.v.m.

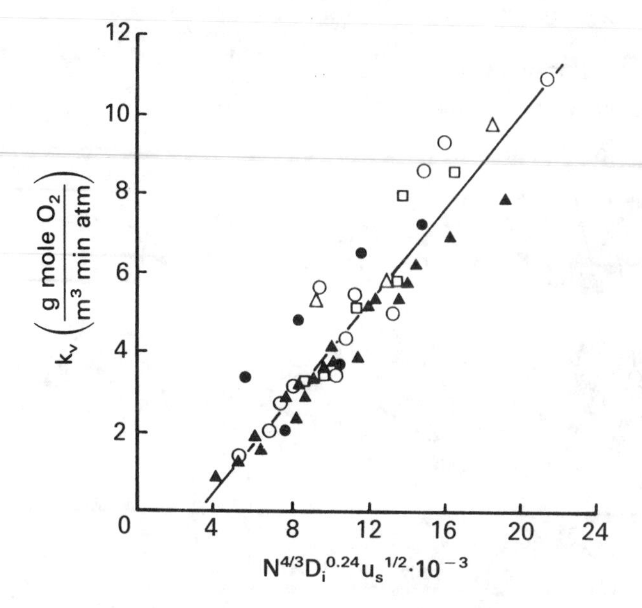

Fig. 6.11 — Plot of k_v versus $N^{4/3}D_i^{0.24}u_s$.

* 60% of Vt = V1

Oxygen transfer rate = $0.25\,\text{kg mol m}^{-3}\,\text{h}^{-1}$
Liquid height = 1.2 × tank diameter.
Two sets of standard flat blade turbine impellers were installed.

Solution
Dimensions of the bioreactors are as follows:

$$V_1 = \frac{\pi}{4} D_{t_1}^2 H_{L_1} = 0.3\pi D_{t_1}^3$$

$$D_{t_1} = \left(\frac{V_1}{0.3\pi}\right)^{1/3} = 0.267\,\text{m}$$ wrong

$$D_{i_1} = \frac{1}{3} \times 0.267 = 0.089\,\text{m}$$

$$H_{L_1} = 1.2 \times 0.267 = 0.32\,\text{m}$$

$$D_{t2} = \left(\frac{V_2}{0.3\pi}\right)^{1/3} = 3.36\text{ m}$$

$$D_{i2} = \frac{1}{3} \times 3.36 = 1.12\text{ m}$$

$$H_{L_2} = 1.2 \times 3.36 = 4.03\,\text{m}$$ 0.18

Air flow rate, $F_1 = 1.0 \times 0.018 = 0.018\,\text{m}^3\,\text{min}^{-1} = 3 \times 10^{-4}\,\text{m}^3\,\text{s}^{-1}$

time

$$\text{Superficial air velocity, } u_{S_1} = \frac{0.018 \times 60}{\frac{\pi}{4}(0.267)^2} = 19.3\,\text{m h}^{-1}$$

$$\text{Partial pressure of oxygen } p_1 = \frac{1 + \left(1 + \frac{H_{L_1}}{10.3}\right)}{2} \times 0.21$$

$$= 0.213\,\text{atm}$$

$$\text{Oxygen transfer rate, } (k_v p)_1 = 0.25\,\text{kg mol m}^{-3}\,\text{h}^{-1}$$

$$\text{Volumetric oxygen transfer coefficent, } (k_v)_1 = \frac{0.25}{0.213}$$

$$= 1.17\,\text{kg mol m}^{-3}\,\text{h}^{-1}\,(\text{atm})^{-1}$$

Cooper *et al.* (1944) correlated k_v, gassed power per unit volume (P_g/V) and superficial air velocity (u_s) for vaned disk impellers as follows:

$$k_v = 0.0635\left(\frac{P_g}{V}\right)^{0.95}(u_s)^{0.67}\,\text{kg mol h}^{-1}\,\text{m}^{-3}\,(\text{atm})^{-1}$$

For standard flat blade turbine, Aiba *et al.* (1965) halved the coefficient of the equation. Thus

$$1.17 = 0.0318\left(\frac{P_g}{V}\right)_1^{0.95}(19.3)^{0.67} \qquad \therefore\ (P_2)_1 = 0.1\ \text{HP}$$

For two sets of impellers, power number $N_P = 2 \times 6 = 12$.
Thus

$$\frac{P_1 g_c}{\rho n_1^3 D_{i_1}^5} = 12$$

$$P_1 = \frac{12 \times 1200 \times (0.089)^5 n_1^3}{9.81} = 8.196 \times 10^{-3} n_1^3\,\text{kg m s}^{-1}$$

$$= 0.108 \times 10^{-3} n_1^3\,\text{HP}$$

Using Michaelis correlation,

$$(P_g)_1 = 0.5\left\{\frac{P_1^2 n_1 D_{i_1}^3}{F_1^{0.56}}\right\}^{0.45}$$

$$0.1 = 0.5\left\{\frac{(0.108 \times 10^{-3} n_1^3)^2 \times (0.089)^3}{(3 \times 10^{-4})^{0.56}}\right\}^{0.45}$$

$$n_1 = 12\,\text{rps} = 720\,\text{rpm}$$

$$P_1 = 0108 \times 10^{-3} n_1^3 = 0.187\,\text{HP}$$

$$\left(\frac{P_g}{P}\right)_1 = \frac{0.10}{0.187} = 0.54$$

For constant power input in geometrically similar vessels, we can write

$$\frac{\rho n_1^3 D_{i_1}^5}{V_1} = \frac{\rho n_2^3 D_{i_2}^5}{V_2}$$

$$n_2 = n_1 \left(\frac{V_2}{V_1}\right)^{1/3} \left(\frac{D_{i_1}}{D_{i_2}}\right)^{5/3}$$

$$= 720 \left(\frac{60}{0.03}\right)^{1/3} \left(\frac{0.089}{1.12}\right)^{5/3} \text{ rpm}$$

$$= \underline{133\,\text{rpm}}$$

For constant impeller tip velocity, we can write

$$n_1 D_{i_1} = n_2 D_{i_2}$$

$$n_2 = n_1 \left(\frac{D_{i_1}}{D_{i_2}}\right) = \underline{57.2\,\text{rpm}}$$

References

Cooper, C. M., Fernstorm, G. A. and Miller, S. AS. (1944) *Ind. Eng. Chem.*, **36**, 504.
Aiba, S., Humphrey, A. E. and Millis, Nancy F. (1965), *Biochemical Engineering*, University of Tokyo Press, p. 164.
Michel, B. J. and Miller, S. A. (1962), *AIChE Journal*, **8**, 262.

PROBLEM 6.12

Maxon and Steel (1966) studied novobiocin fermentation to find several phenomena peculiar to the cultivation of fungi. They measured actual oxygen absorption rate, OAR ($\text{m mol l}^{-1}\text{h}^{-1}$), by monitoring the oxygen partial pressure in the exit gas during fermentation using various scale of fermenters, from $0.02\,\text{m}^{-3}$ to $90\,\text{m}^3$ with different ratios of impeller to tank diameter, D_i/D_t ranging from 0.23 to 0.6. They found:

(a) More power was required for the same performance of fermentation when D_i/D_t was bigger.
(b) OAR was well correlated to the tip speed of an impeller.
(c) Novobiocin production was well correlated to OAR.
(d) Localization of dissolved oxygen in the fermenter was observed and also possibly inside mycelial pellets.

Discuss policies for optimization of operational conditions such as D_i/D_t, impeller rotation speed, and even any suggestion for impeller design.

Reference

Maxon, W. D. and Steel, R. (1966) *Biotechnol Bioeng.*, **8**, 108.

Solution

It is recommended to have a higher impeller speed so long as no serious mechanical damge to the cells of the fungus takes place using a smaller impeller. These conditions will result in good oxygen supply in such a dense culture of fungal cells, owing to a high shear rate maintaining the same power requirement. It is also important to have a well-mixed condition throughout a fermenter to avoid any stagnant regions of poor oxygen supply from the gaseous phase to the liquid phase and also inside the mycelial aggregate. Maxon and Steel suggested a new type of agitating device, an assembly of straight rods which are placed vertically along with a supporting plate fixed on a shaft for rotation. The rods reached almost to both the bottom and the surface of culture broth and the assembly reached to the baffle plate on the wall of the fermenter. This configuration of the agitating device facilitates a good distribution of a strong shear field throughout the fermenter. One of the possible problems is the mechanical strength of the system.

PROBLEM 6.13

Funahashi *et al.* (1987) found that the shear stress is one of the most important physical factors caused by agitation in the production of xanthan gum by *Xanthomonas campestris* ATCC13951. Moreover, it was suggested that the xanthan gum layer around the cells might limit the mass transfer of materials, e.g. oxygen or glucose, from the medium into the cells, although Finn has shown that the liquid film around the cells offers no appreciable resistance to diffusion of the materials.

Interpret oxygen transfer around the cells referring the investigation of Brian, Hales and Sherwood (1969) on the heat and mass transfer between liquids and spherical particles correlating $N_{Sh}/N_{Sc}^{1/3}$ to N_{Re}

The dimensionless numbers are defined as follows:

$$N_{Re} = (P_w/v)^{1/2} d_p^2/v \tag{6.13.1}$$

$$N_{Sh} = k_v d_p/\mathscr{D}_s \tag{6.13.2}$$

$$N_{Sc} = v/\mathscr{D}_s \tag{6.13.3}$$

where $P_{w/v}$ is the agitation power per unit mass of fluid, $v \cdot \dot{\gamma}^2/2$, v is the kinematic viscocity, d_p is the diameter of a cell, k_v is the mass transfer coefficient in the xanthen gum layer around a cell, $\mathscr{D}_s$ is the diffusion coefficient of xanthan gum layer, $\dot{\gamma}$ is the shear rate, and k_v is calculated from the specific oxygen uptake rate (SOUR) as follows,

$$\text{SOUR} = m k_v a_c c_b \ , \tag{6.13.4}$$

where m is the cell number per unit weight of dry cell, a_c is the interfacial area per cell and c_b is the dissolved oxygen concentration out of the xanthan gum layer around the cells.

$$m = 1.4 \times 10^{15}\,[\text{cells (kg-cell)}^{-1}]$$

$$d_p = 9.6 \times 10^{-7}\,[\text{m}]$$

$$c_b = 7.0 \times 10^{-3}\,[\text{kg O}_2\,\text{m}^{-3}]$$

$$\mathscr{D}_s = 2.7 \times 10^{-13}\,[\text{m}^2\,\text{s}^{-1}]$$

$$\rho = 1.0 \times 10^{3}\,[\text{kg}\,\text{m}^{-3}]$$

Data of oxygen uptake rate under various shear conditions are provided in Table 6.6.

Table 6.6 — Data of oxygen up take rate of cells of *Xanthomonas campestris* ATCC13951 during xanthan gum production under various shear conditions

Experiment no.	$\dot{\gamma}$ (s^{-1})	τ (Pa)	Specific oxygen uptake rate $[10^{-5}$ kg O_2 kg^{-1} cells $s^{-1}]$	Experiment no.	$\dot{\gamma}$ (s^{-1})	τ (Pa)	Specific oxygen uptake rate $[10^{-5}$ kg O_2 kg^{-1} cells $s^{-1}]$
1	12300	219	5.73	14	1200	4.3	2.91
2	12300	114	5.52	15	1200	3.9	2.55
3	11000	98.5	5.64	16	1500	173	5.7
4	12400	66.0	5.43	17	3500	76	4.95
5	6400	47.5	5.54	18	1800	55	5.10
6	6300	40.5	5.57	19	1100	22	5.18
7	4100	24.5	4.89	20	2200	22	5.02
8	3100	16.5	4.22	21	3200	20.3	4.98
9	5100	14.5	3.29	22	2300	14.0	4.12
10	3900	13.0	2.93	23	3200	15.0	3.50
11	3100	11.0	3.49	24	2700	12.3	2.76
12	2500	10.5	2.87	25	1500	9.4	3.65
13	1400	7.3	2.79				

References

Funahashi, H., Maehara, M., Yoshida, T. and Taguchi, H. (1987) *J. Chem. Eng. Japan*, **20**, 16.
Brain, P. L. T., Hales, H. B. and Sherwood, T. K. (1969), *AIChE J.*, **15**, 727.

Solution

Substituting $v = \mu_{ap}/10^3 = (\tau/\dot{\gamma})/10^3$ and the value of d_p into equation (6.13.1),

$$N_{Re} = 6.52 \times 10^{-10}\ (\dot{\gamma}^2/\tau)\ . \tag{6.13.5}$$

Substituting k_v = SOUR/$(ma_c c_b)$ into equation (6.13.2),

$$N_{Sh} = 1.25 \times 10^5 \times \text{SOUR}. \tag{6.13.6}$$

For Schmidt number,

$$N_{Sc} = v/D_s = (\tau/\dot{\gamma})/10^3/\mathscr{D}_s = 3.70 \times 10^9\ (\tau/\dot{\gamma}) \tag{6.13.7}$$

Then,

$$\underline{N_{Sh}/N_{Sc}^{1/3} = 81.0 \times \text{SOUR}(\dot{\gamma}/\tau)^{1/3}} \tag{6.13.8}$$

Table 6.7 is the calculated result, and Fig. 6.12 shows plotting of $N_{Sh}/N_{Sc}^{1/3}$ versus N_{Re} together with the correlations obtained by Brian *et al.* (1967). The calculated results of this problem locate on an extension of the correlation of the previous researches on mass and heat transfer.

References

Harriott, P. (1962), *AIChE J.*, **8**, 93.
Bieber, H. and Gaden, E. L. Jr (1962), Paper no. 102, AIChE Los Angeles Meeting, February.
Nagata, S., Yamaguchi, I., Yabuta, S. and Harada, M. (1960), *Kagaku kogaku*, **24**, 618.
Wilhelm, R. H., Conklin, L. H. and Sauer, T. C. (1941), *Ind. Eng. Chem.*, **33**, 453.
Barker, H. and Treybal, R. E. (1960), *AIChE J.*, **6**, 289.

Table 6.7 — Calculated values of N_{Re} and $N_{Re}/N_{Sc}^{1/3}$ from experimental data obtained under various shear conditions

Experiment no.	$\dot{\gamma}$ (s^{-1})	τ (Pa)	Specific oxygen uptake [10^{-5} kg O_2 kg^{-1} cells s^{-1}]	N_{Re} ($x10^4$)	$N_{Sh}/N_{Sc}^{1/3}$ ($x10^2$)
1	12300	219	5.73	6.52	1.78
2	12300	114	5.52	8.68	2.13
3	11000	98.5	5.64	8.00	2.20
4	12400	66.0	5.43	15.18	2.51
5	6400	47.5	5.54	5.62	2.30
6	6300	40.5	5.57	6.39	2.43
7	4100	24.5	4.89	4.47	2.18
8	3100	16.5	4.22	3.79	1.96
9	5100	14.5	3.29	11.69	1.88
10	3900	13.0	2.93	7.62	1.59
11	3100	11.0	3.49	5.69	1.85
12	2500	10.5	2.87	3.88	1.44
13	1400	7.3	2.79	1.75	1.75
14	1200	4.3	2.91	2.35	2.35
15	1200	3.9	2.55	2.35	2.35
16	1500	173	5.7	0.0848	0.949
17	3500	76	4.95	1.05	1.44
18	1800	55	5.10	0.384	1.32
19	1100	22	5.18	0.358	1.55
20	2200	22	5.02	1.43	1.88
21	3200	20.3	4.98	3.29	2.18
22	2300	14.0	4.12	2.46	1.83
23	3200	15.0	3.50	4.45	1.69
24	2700	12.3	2.76	3.86	1.35
25	1500	9.4	3.65	1.56	1.60

PROBLEM 6.14

Oxygen balance within a spherical pellet of fungus gives the following equation in steady state.

$$\mathscr{D}\left(\frac{d^2 c_L}{dr^2} + \frac{2}{r}\frac{dc_L}{dr}\right) = 2\rho Q_{O_2}\frac{c_L}{K_m + c_L} \tag{6.14.1}$$

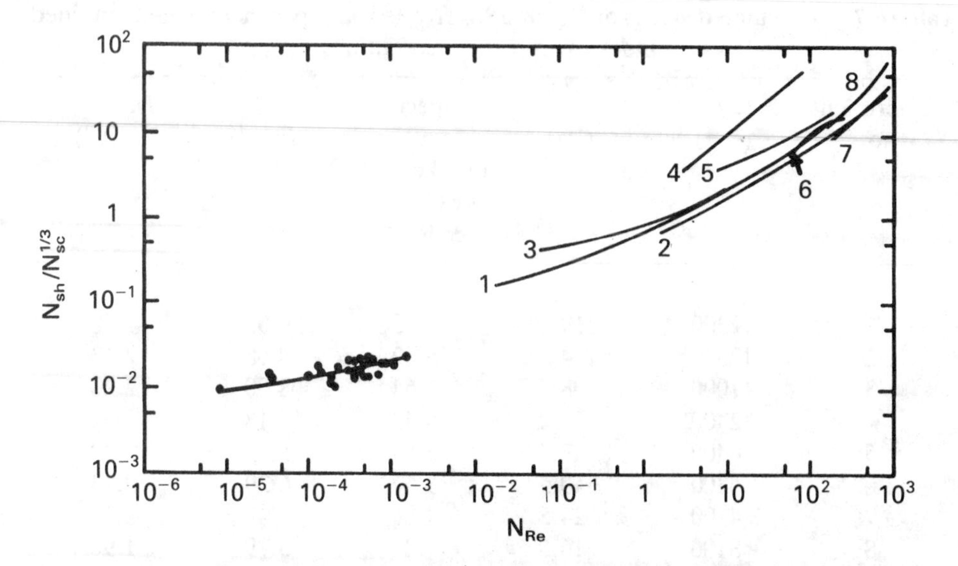

Fig. 6.12 — Correlation between N_{Re} and $N_{Sh}/N_{Sc}^{1/3}$. Closed circles are data of the oxygen transfer around the cells of *Xanthomonas campestris* ATCC 13951 surrounded by a thick layer of xanthan gum. Other curves are from previous studies. **1** and **3**, the data of Harriott (1962) for transport rate of ion exchange resins in glycerin and methocel solution, and in water, respectively; **2**, the data of Harriott (1962) for dissolution in water of boric, benzoic acid particles; **4**, the data of Bieber *et al.* (1962) for adsorption of copper ion on ion exchange resins, **5**, the data of Brian *et al.* (1969) for melting ice spheres and dissolving spheres of pivalic acid suspended in water; **6** and **8**, the data of Nagata *et al.* (1960) and Wilhelm *et al.* (1941), respectively, for transport rate by dissolving sodium chloride in water; and **7**, the data of Barker and Treybal (1960) for dissolution of boric benzoic acid particles in water.

where c_L = dissolved oxygen concentration, $\mathcal{D}$ = molecular diffusivity of oxygen in liquid, r = radial distance from sphere centre, Q_{O_2} = specific respiration rate of fungus, K_m = Michaelis constant, and ρ = density of fungal pellet.

Derive the equation.

Solution

At radius r and $r + \mathrm{d}r$, the flow rate of oxygen will be

$$\mathrm{In} = -4\pi(r+\mathrm{d}r)^2\mathcal{D}\left(\frac{\mathrm{d}c_L}{\mathrm{d}r}\right) \tag{6.14.2}$$

$$\mathrm{Out} = 4\pi(r+\mathrm{d}r)^2\mathcal{D}\left(\frac{\mathrm{d}^2c_L}{\mathrm{d}r^2}\mathrm{d}r + \frac{\mathrm{d}c_L}{\mathrm{d}r}\right) \tag{6.14.3}$$

$$\mathrm{Respiration} = 4\pi r^2\,\mathrm{d}r\rho Q_{O_2} \tag{6.14.4}$$

$$\mathrm{Accumulation} = 4\pi r^2\,\mathrm{d}r\frac{\mathrm{d}c_L}{\mathrm{d}t} \tag{6.14.5}$$

By balance,

$$4\pi r^2\,\mathrm{d}r\frac{\partial c}{\partial t} = -4\pi r^2\mathscr{D}\frac{\mathrm{d}c_{\mathrm{L}}}{\mathrm{d}r} - \left\{-4\pi(r+\mathrm{d}r)^2\mathscr{D}\left(\frac{\mathrm{d}^2c_{\mathrm{L}}}{\mathrm{d}r^2}\mathrm{d}r + \frac{\mathrm{d}c_{\mathrm{L}}}{\mathrm{d}r}\right)\right\} - \\ -\left\{4\pi r^2\mathrm{d}r\rho Q'_{\mathrm{O}_2}\right\} \tag{6.14.6}$$

$$\therefore \frac{\mathrm{d}c_{\mathrm{L}}}{\mathrm{d}t} = \mathscr{D}\left(\frac{\mathrm{d}^2c_{\mathrm{L}}}{\mathrm{d}r^2} + \frac{2}{r}\frac{\mathrm{d}c_{\mathrm{L}}}{\mathrm{d}r}\right) - \rho Q'_{\mathrm{O}_2} \tag{6.14.7}$$

$$Q'_{\mathrm{O}_2} = Q_{\mathrm{O}_2}\frac{c_{\mathrm{L}}}{K_{\mathrm{m}} + c_{\mathrm{L}}} \tag{6.14.8}$$

Q'_{O_2} = specific oxygen consumption rate due to respiration

Q_{O_2} = specific oxygen consumption rate (intrinsic)

$$\therefore \frac{\mathrm{d}c_{\mathrm{L}}}{\mathrm{d}t} = \mathscr{D}\left(\frac{\mathrm{d}^2c_{\mathrm{L}}}{\mathrm{d}r^2} + \frac{2}{r}\frac{\mathrm{d}c_{\mathrm{L}}}{\mathrm{d}r}\right) - 2\rho Q_{\mathrm{O}_2}\frac{c_{\mathrm{L}}}{(K_{\mathrm{m}} + c_{\mathrm{L}})} \tag{6.14.9}$$

At steady state, $\mathrm{d}c_{\mathrm{L}}/\mathrm{d}t = 0$.

$$\therefore \mathscr{D}\left(\frac{\mathrm{d}^2c_{\mathrm{L}}}{\mathrm{d}r^2} + \frac{2}{r}\frac{\mathrm{d}c_{\mathrm{L}}}{\mathrm{d}r}\right) = 2\rho Q_{O_2}\frac{c_{\mathrm{L}}}{K_{\mathrm{m}} + c_{\mathrm{L}}} \tag{6.14.10}$$

PROBLEM 6.15

Many simple bioconversions are carried out using immobilized enzymes. Most immobilized enzymes are charged and so too the substrate they convert to product. In such a system mass transfer with electrostatic interaction is to be considered. For such a system develop external mass transfer equations with an immobilized enzyme reaction which follows the substrate inhibition model and show multiple steady states are possible depending upon the charged interaction between substrate and carrier.

Solution

The flux expression for a charged substrate can be written as

$$J_{\mathrm{s}} = \mathscr{D}\frac{\mathrm{d}s}{\mathrm{d}x} + S\frac{z\mathscr{D}F}{RT}\frac{\mathrm{d}\psi}{\mathrm{d}x} \tag{6.15.1}$$

where $\mathscr{D}$ = diffusivity, $\psi(x)$ = electrostatic potential, T = temperature, z = valency of substrate, F = Faraday's constant, and R = gas constant.

For such a system where electrostatic double layer is much smaller than diffusion layer, the flux expression can be written as

$$J_{\mathrm{s}} = Mk_{\mathrm{c}}(S_{\mathrm{b}} - S_{\mathrm{s}}\mathrm{e}^{\lambda}) \tag{6.15.2}$$

where

$$\lambda = zF\psi(0) \tag{6.15.3}$$

S_b = bulk substrate concentration, S_s = substrate concentration on the surface of IME, $\psi(0)$ = electrostatic charge on the carrier, and k_c = external mass transfer coefficient.

$$M^{-1} = 1 + \frac{1}{\kappa\delta}\sum_{1}^{\infty}\frac{\lambda^n}{n.n!} \tag{6.15.4}$$

and κ = electrostatic double layer.

Equation (6.15.2) gives the external mass transfer rate. It represents a straight line in S_s versus reaction rate coordinate with slope = $-Mk_c e^{\lambda}$ and intercept = $Mk_c S_b$. Fig. 6.13 shows the rate of the substrate inhibited reaction and the rate of

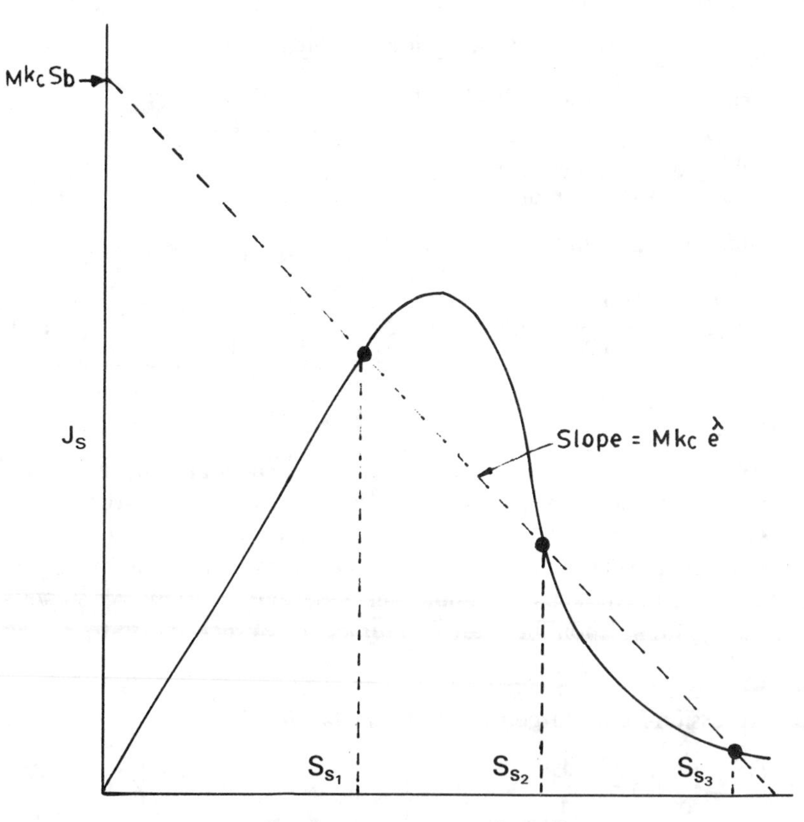

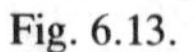
Fig. 6.13.

mass transfer as a function of the substrate concentration at the support surface.

It can be seen that it is possible to have three surface concentrations at which reaction rate is equal to external mass transfer rate. It can also be seen that multiple steady states depend upon the slope of the straight line which depends on λ, the electrostatic interaction parameter.

Reference

Shuler, M. L., Aris, R. and Tsuchiya, H. M. (1972) *J. Theor. Biol.* **35**, 67.

PROBLEM 6.16

Oxygen requirements of activated sludge in an aerator of complete mixing is shown as follows:

$$V\frac{dc_{O_2}}{dt} = aF(c_{s_0} - c_s) + bVx$$

V = aerator volume (m^3), dc_{O_2}/dt = oxygen consumption rate ($kg\,O_2\,m^{-3}\,day^{-1}$), a= empirical constant ($kg\,O_2\,(kg\,BOD)^{-1}$), b = empirical constant ($kg\,O_2(kg\,MLSS)^{-1}\,day^{-1}$), F = flow rate of fresh waste water ($m^3\,day^{-1}$), c_{s0} = BOD in fresh waste water ($kg\,BOD\,m^{-3}$), c_s = BOD in effluent from aerator ($kg\,BOD\,m^{-3}$), and x = sludge concentration ($kg\,MLSS\,m^{-3}$).

In a specific waste treatment facility where the operation is carried out at a BOD loading of 0.4 kg BOD $(kg\,MLSS)^{-1}\,day^{-1}$ with an efficiency of BOD removal of 90% the values of a and b are given below.

$$a = 0.5\,kg\,O_2\,(kg\,BOD)^{-1}$$

$$b = 0.3\,kg\,O_2\,(kg\,MLSS)^{-1}\,day$$

What is the value of oxygen required per kg BOD removed?

Solution

$$V\frac{dc_{O_2}}{dt} = aF(c_{s_0} - c_s) + bVx$$

$$\frac{1}{x}\frac{dc_{O_2}}{dt} = a\frac{F}{Vx}\left(1 - \frac{c_s}{c_{s_0}}\right) + b$$

$$\frac{Fc_{s_0}}{Vx} = \frac{(m^3\,day^{-1})(kg\,BOD\,m^{-3})}{(m^3)(kg\,MLSS\,m^{-3})}$$

$$= kg\,BOD\,(kg\,MLSS)^{-1}\,day^{-1}$$

$$= \text{BOD loading} = 0.4$$

$$\frac{90}{100} = \frac{c_{s_0} - c_s}{c_{s_0}}$$

$$\therefore c_s = 0.1\,c_{s_0}$$

$$\therefore \frac{1}{x}\frac{dc_{O_2}}{dt} = 0.5 \times 0.4(1 - 0.1) + 0.3$$

$$= 0.48\,\text{kg}\,O_2\,(\text{kg MLSS})^{-1}\,\text{day}^{-1}$$

PROBLEM 6.17

Unsteady state current, i_t, of a membrane covered O_2 sensor is given by

$$i_t = i_\infty \left\{ 1 + 2\sum_{n=1}^{\infty} (-1)^n \, e^{-(n^2\pi^2\mathcal{D}_m t)/b^2} \right\}$$

where i_∞ = steady state current (μA), $\mathcal{D}_m$ = O_2 diffusivity in membrane ($m^2 s^{-1}$), b= membrane thickness (m).

What will be the 95% response time of sensor whose characteristics are given below?

$$b = 2.54 \times 10^{-5}\,\text{m}$$

$$\mathcal{D}_m = 1.7 \times 10^{-11}\,\text{m}^2\,\text{s}^{-1}$$

Solution

$$i_t = i_\infty \left\{ 1 + 2\sum_{n=1}^{\infty} (-1)^n \, e^{-(n^2\pi^2\mathcal{D}_m t)/b^2} \right\}$$

$$b = 2.54 \times 10^{-5}\;; \qquad \mathcal{D}_m = 1.7 \times 10^{-11}$$

95% response time meaning $i_t = 0.95 i_\infty$

$$0.95 = 1 + 2\sum_{n=1}^{\infty} (-1)^n \, e^{-(n^2\pi^2\mathcal{D}_m t)/b^2} \Big\}$$

$$= 1 + 2\sum_{n=1}^{\infty} (-1)^n \, e^{-0.26n^2 t}$$

$$\therefore \sum_{n=1}^{\infty} (-1)^n \, e^{-0.26n^2 t} = -0.025$$

$$-e^{-0.26t} + (e^{-0.26t})^4 - (e^{-0.26t})^9 + \ldots + (-1)^n (e^{-0.26t})^{n^2} = -0.025$$

$$0.025 = e^{-0.26t} - (e^{-0.26t})^4 + \ldots + (-1)^n (e^{-0.26t})^{n^2}$$

Neglecting higher order terms,

$$e^{-0.26t} = 0.025$$

$$\therefore t = \underline{14.2\,s}$$

PROBLEM 6.18

A technique for oxygen absorption studies in a microbial slime system resulted the data given in Table 6.8 and Table 6.9. Estimate from these data, the

Table 6.8 — Dissolved oxygen concentration in a microbial slime bathed in $0.55\,\text{kg}\,\text{m}^{-3}$ nutrient broth at 26°C

Distance (m)	Oxygen tension (c_L) ($\text{kg}\,\text{m}^{-3}$)	
	Above film	In film
0	4.7×10^{-3}	4.7×10^{-3}
2.5×10^{-5}	6.0×10^{-3}	3.2×10^{-3}
5×10^{-5}	6.9×10^{-3}	1.9×10^{-3}
10^{-4}	7.5×10^{-3}	0.4×10^{-3}
1.25×10^{-4}	—	0.3×10^{-3}
1.5×10^{-4}	—	0.1×10^{-3}

Table 6.9 — Change in dissolved oxygen concentration after nutrient flow stopped with probe positioned in 500 ppm nutrient broth at 25 μm from slime surface

Time (min)	Oxygen tension (c_L) (kg m^{-3})
0.9	7.8×10^{-3}
1.0	5.6×10^{-3}
1.1	1.0×10^{-3}
1.2	2.3×10^{-3}
1.4	1.5×10^{-3}
2.0	1.0×10^{-3}
3.0	0.7×10^{-3}
4.0	0.6×10^{-3}
4.5	0.6×10^{-3}
5.0	0.6×10^{-3}

diffusivity of oxygen in:

(a) the nutrient medium at a thickness of 2.5×10^{-5} m above the slime film;

(b) the slime film at a depth of 2.5×10^{-5} m (assume oxygen uptake rate by the slime at this depth as $2.5 \times 10^{-5}\,\text{kg}\,\text{m}^{-3}\,\text{s}^{-1}$.

Solution

(a) We have the relation

$$\frac{dc_L}{dt} = -\mathcal{D}\frac{d^2c_L}{dx^2}$$

To find the value of initial rate of oxygen absorption, the DO values are plotted against time (from Table 6.9) as shown in Fig. 6.14. From this the following is ascertained:

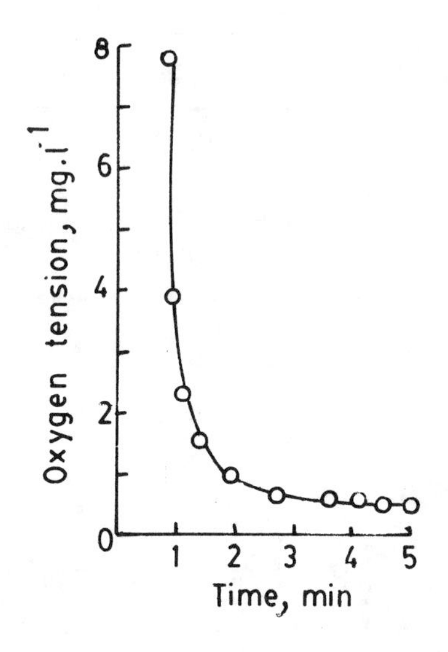

Fig. 6.14.

$$\text{Initial rate of } O_2 \text{ absorption}, \quad \frac{dc_L}{dt} = 6.24 \times 10^{-4}\,\text{kg}\,\text{m}^{-3}\,\text{s}^{-1}$$

Now the determination of d^2c_L/dx^2 is made as follows: c_L is plotted against the corresponding values x (from Table 6.8). The plot is shown in Fig. 6.15. Slopes of this curve at various points give the values of dc_L/dx at various dx values, next, dc_L/dx values are plotted against the corresponding x values resulting in Fig. 6.16. The slopes of this curve predict d^2c_L/dx^2 at any depth.

From Fig. 6.16 we see that the slope at a depth of 25 μm above the slime film $= 7.55 \times 10^6\,\text{kg}\,\text{m}^{-3}\,\text{m}^{-2}$.

So,

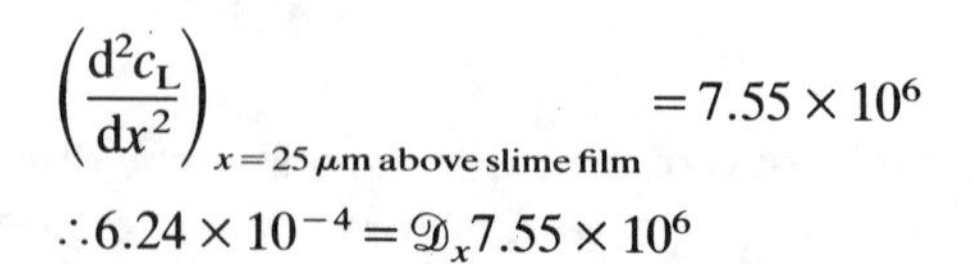

$$\left(\frac{d^2c_L}{dx^2}\right)_{x=25\,\mu\text{m above slime film}} = 7.55 \times 10^6$$

$$\therefore 6.24 \times 10^{-4} = \mathcal{D}_x 7.55 \times 10^6$$

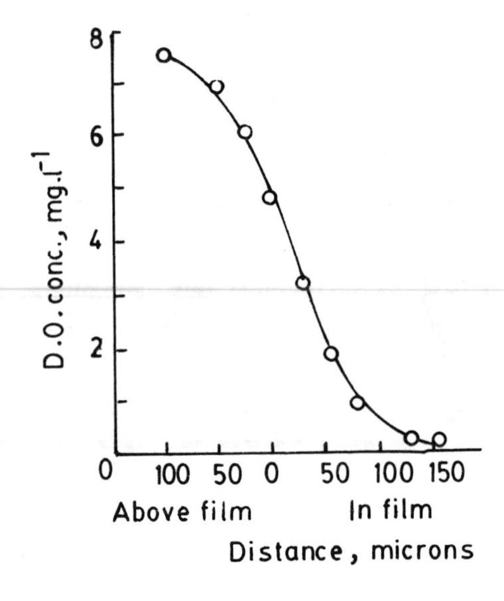

Fig. 6.15.

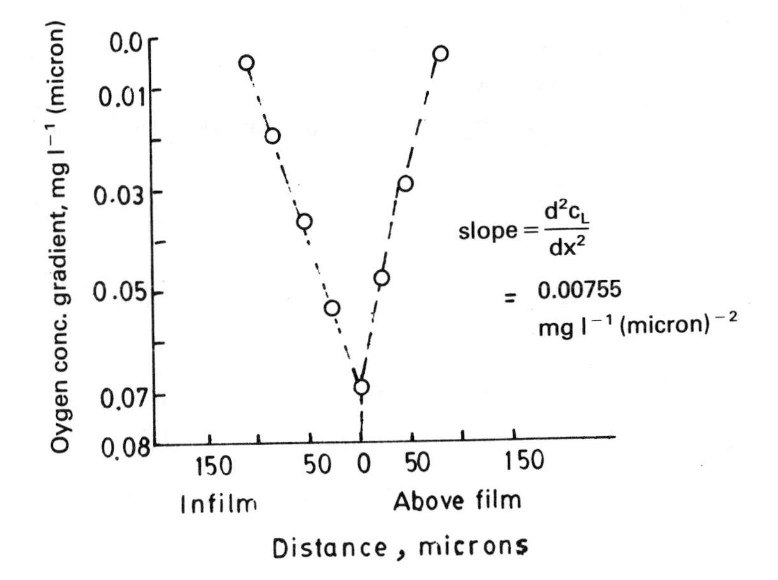

Fig. 6.16.

$$\mathscr{D} = \frac{6.24 \times 10^{-4}}{7.55 \times 10^{6}} = \underline{8.265 \times 10^{-11}\,\mathrm{m^2\,s^{-1}}}$$

(b) Inside the film under steady state:

$$\frac{dc_L}{dx} = \frac{-q_{O_2}}{\mathscr{D}}x + K$$

where q_{O_2} represents the oxygen uptake rate in the film.

From Fig. 6.16, the slope of the plot of dc_L/dx against x (depth in the film) gives the value of $-q_{O_2}/\mathscr{D}$ and its value is

$$\text{Slope} = \frac{-q_{O_2}}{\mathscr{D}} = \underline{-6.3 \times 10^{5}\,\mathrm{kg\,m^{-3}\,m^{-2}}}$$

PROBLEM 6.19

In an oxygen absorption study the following data were obtained on the oxygen transfer capacity of an air diffusion unit:

Air flow rate = $9.439\,\mathrm{m^3\,s^{-1}}$

Temperature = 12°C

Liquid depth = 4.27 m

Average bubble diameter = $2.5 \times 10^{-3}\,\mathrm{m}$

Average bubble velocity = $0.3\,\mathrm{m\,s^{-1}}$

The dissolved oxygen concentration measurement was tabulated as in Table 6.10.

Table 6.10

Time (min)	c_L ($\mathrm{kg\,m^{-3}}$)
3	0.6×10^{-3}
6	1.6×10^{-3}
9	3.1×10^{-3}
12	4.3×10^{-3}
15	5.4×10^{-3}
18	6.0×10^{-3}
21	7.0×10^{-3}

From this information, compute

(a) $k_L a$ and k_L;

(b) mass of oxygen per hour transferred per 28.317 $\mathrm{m^3}$ (1000 $\mathrm{ft^3}$) at 20°C and zero dissolved oxygen concentration and the oxygen transfer efficiency;

(c) how much oxygen will be transferred to a waste with an α of 0.80, a temperature of 32°C and an operating dissolved oxygen of $15 \times 10^{-3}\,\mathrm{kg\,m^{-3}}$? Assume the saturation concentration of oxygen in the liquid at 12°C to be $1.08 \times 10^{-2}\,\mathrm{kg\,m^{-3}}$.

Solution

(a) The mean oxygen saturation concentration c_{s_m} in the aeration tank assuming 10% oxygen absorption is given by the following empirical relation:

$$c_{s_m} = c_s\left(\frac{p_b}{29.4} + \frac{O_t}{42}\right)$$

where, p_b = absolute pressure at the depth of air release (in psi), O_t = concentration of O_2 in the air leaving the aeration tank in percent, and c_s = saturation of O_2 at the experimental temperature.

Thus,

$$c_s = 1.08 \times 10^{-2}\,\mathrm{kg\,m^{-3}};\ O_t = \frac{18.9}{18.9 + 79} \times 100 = 19.3\%$$

$$p_b = 20.2 \text{ psi} = 1.42 = \text{atm}$$

$$c_{s_m} = 10.8 \times 10^{-3}\left(\frac{20.2}{29.4} + \frac{19.3}{42}\right) = 0.0123\,\mathrm{kg\,m^{-3}}$$

From Table 6.10 we produce Table 6.11.

Table 6.11

Time (min)	$c_s - c_L$ ($\mathrm{kg\,m^{-3}}$)
3	11.7×10^{-3}
6	10.7×10^{-3}
9	9.2×10^{-3}
12	8.0×10^{-3}
15	6.9×10^{-3}
18	6.3×10^{-3}
21	5.7×10^{-3}

From the plot of log $(c_L - c_L)$ against time we get the curve as shown in Fig. 6.17. The slope of the curve is

$$k_L a = [\{2.3 \log(12/8)\}/10] \times 60$$
$$= 2.60\,\mathrm{h^{-1}}$$

The interfacial area per unit volume is given by:

$$\frac{A}{V} = \frac{6FH_L}{d_B U_B V}$$

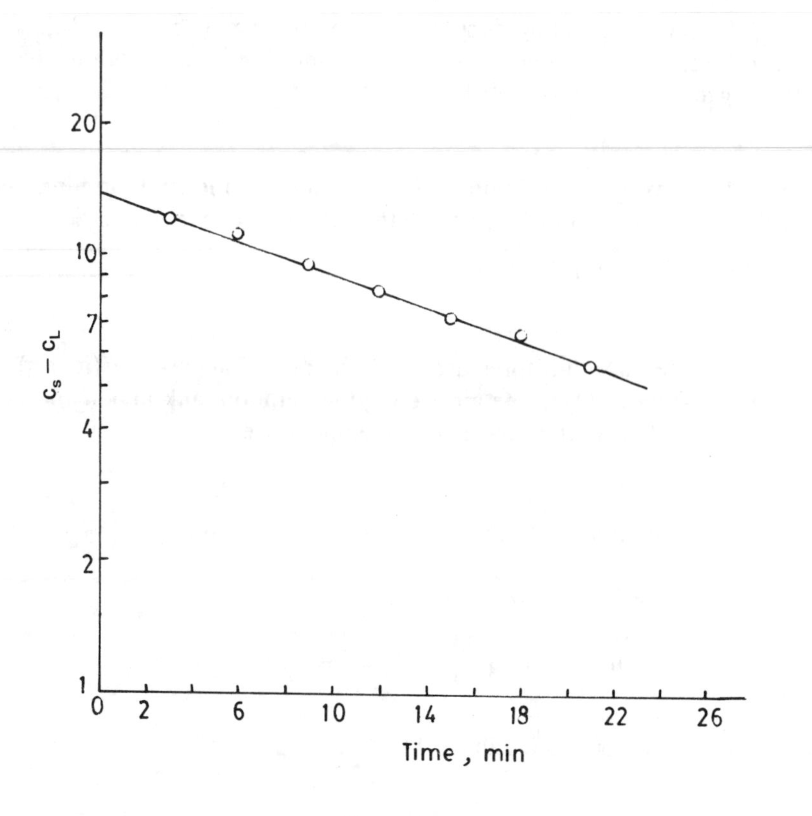

Fig. 6.17.

where F = air flow rate, H_L = liquid depth, d_B = bubble diameter, U_B = bubble velocity, and V = volume of liquid.

$$\therefore \frac{A}{V} = \frac{6 \times 9.439 \times 4.27}{(2.5 \times 10^{-3}) \times 0.3 \times V}$$

$$= 9.7416\,\mathrm{m}^{-1}$$

$$\therefore k_L = \frac{k_L a}{AV} = \frac{2.6}{9.7416} = \underline{0.265\,\mathrm{m\,h}^{-1}}$$

(b) $k_L a(T) = k_L a(20°\mathrm{C}) \times 1.02.$

$$\therefore k_L a(20°\mathrm{C}) = \frac{k_L a(t)}{1.02^{(t-20)}}$$

$$= \frac{2.60}{1.02^{(12-20)}} = \underline{3.04\,\mathrm{h}^{-1}}$$

$$c_s(20°\mathrm{C}) = 10.7 \times 10^{-3}\,\mathrm{kg\,m}^{-3}$$

So, the amount of oxygen transferred per hour

$$= k_L a_{20°C}(c_{s(20°C)} - c_L)$$
$$= k_L a c_s \qquad \text{Since } c_L = 0$$
$$= 3.04 \times 10.7 \times 10^{-3} \times 28.317 = 0.921\,\text{kg}\,\text{h}^{-1}$$

(c) The amount of oxygen transferred at 32°C is:

$$N = N_0 \left[\frac{c_{s(32°C)} - c_{L(32°C)}}{c_{s(20°C)}} \right] \alpha (1.02)^{(t-20)}$$

$$= 2.0 \left[\frac{(7.35 \times 1.18) - 1.5}{10.7} \right] 0.80 \times (1.02)^{12}$$

$$= \underline{0.617\,\text{kg}\,\text{h}^{-1}}$$

PROBLEM 6.20

In course of estimation of oxygen uptake rates and respiratory quotient RQ of a yeast suspension by Warburg constant volume respirometer the data given in Table 6.12 were obtained. Manometer readings were taken at 5-min intervals. The volume

Table 6.12 — Warburg manometer readings

Flask 1		Flask 2	
Left arm	Right arm	Left arm	Right arm
15	15	15	15
15	15	15	15
15	15	14.6	15.4
14.5	15.5	14.8	15.2
14.8	15.2	15	15
13.2	15.8	15	15

of the manometer filled with mercury was in both flasks $1.8 \times 10^{-6}\,\text{m}^3$.

Experimental temperature = 35°C .

Contents of the flasks were

Flask 1: 3 ml of yeast suspension with alkali in the central cup.
Flask 2: 3 ml of yeast suspension (washed cells).
Flask 3: Distilled water + 0.3 ml KOH in the cup.

From these data, calculate
(a) the oxygen uptake rate of the yeast suspension and
(b) the respiratory quotient, RQ

Solution
Flask constant for flask 1

$$k_{L_{O2}} = \frac{V_g \frac{273}{T} + V_f \alpha}{P_O}$$

where

$h = V_f = 13.2 \times 10^{-6}\,m^3$ = observed manometer reading.

V_g = total volume of manometer − volume of the liquid in the arm

$= 18 - 13.2 - = 4.8$ ml $= 4.8 \times 10^{-6} m^3$

$T = 35 + 273 = 308\,K$

$\alpha = 1.3$ mm gas ml^{-1} of liquid in flask

$= 0.0013$ mm $(\mu l)^{-1}$

$P_O = \frac{760 \times 13.6}{1.033} = 10{,}000$ mm of liquid having density of 1.033 g ml^{-1}

$$k_{L_{O2}} = \frac{4800 \frac{273}{308} + 13200 \times 0.0013}{10{,}000}$$

$= 0.427$

$k_{L_{CO2}}$ determination: data given:

$V_f = 15 \times 10^{-6} m^3$

$V_g = 3 \times 10^{-6} m^3$

$\alpha = 0.001$ mm gas ml^{-1} of liquid in flask

$$k_{L_{CO2}} = \frac{300 \frac{273}{308} + 15000 \times 0.001}{10000}$$

$= 0.2664$

For flask 2:

$V_f = 15 \times 10^{-6} m^3$

$V_g = 3 \times 10^{-6} m^3$

$T = 308°K$

$\alpha = 1.3$ mm ml^{-1} of liquid in flask

$$P_O = 10{,}000\,\text{mm}$$

so,

$$k_{2O_2} = \frac{300\dfrac{273}{308} + 15000 \times 0.0013}{10000}$$

$$= 0.205$$

Hence in flask 1, the amount of oxygen absorbed $= X_{O_2} = h_{O_2} \times k_{LO_2}$

$$h_{O_2} = \text{observed manometer reading}$$
$$= 15 - 13.2 = 4.8$$

so,

$$X_{O_2} = 4.8 \times 0.427 = 2.05$$

Amount of CO_2 released:

$$X_{CO2} = k_{2CO_2}\left[h - \frac{X_{O_2}}{k_{2O_2}}\right]$$

$$k_{2O_2} = 0.2664$$

$$h_2 = 4.8\,\text{ml}$$

$$X_{O_2} = 2.05$$

$$k_{2O_2} = 0.205$$

$$X_{CO_2} = 0.2664\left[4.8 - \frac{2.05}{0.205}\right]$$

$$X_{CO_2} = 0.2664\,(-10 + 4.8)$$
$$= -1.38$$

So,

$$RQ = \frac{X_{CO_2}}{X_{O_2}} = \frac{1.38}{2.05} = \underline{0.67}$$

PROBLEM 6.21

A mixed liquor was sampled from an aeration basin in a sewage treatment plant. The respiration rate, Q_{res} was then measured polarographically. The value Q_{res} was $15.6 \times 10^{-3}\,\text{kg}\,\text{m}^{-3}\,\text{h}^{-1}$. Concentration of dissolved oxygen, $\bar{c}$ and saturation, c^* were determined polarographically as $\bar{c} = 5.1 \times 10^{-3}\,\text{kg}\,O_2\,\text{m}^{-3}$ and $c^* = 7.8 \times 10^{-3}\,\text{kg}\,O_2\,\text{m}^{-3}$ (at 24.5°C). Calculate k_La value, assuming that supply and demand for oxygen is balanced. MLSS (mixed liquor suspended solids) was $2\text{kg}\,\text{m}^{-3}$. What is the Q_{res} value of this sludge? Based on the value of k_La assessed so far, calculate the efficiency of aeration (efficiency is defined as the ratio of actual oxygen transferred to that theoretically possible). The aeration rate was $113\,\mathrm{m^3\,min^{-1}}$ at 24.5°C and $1.01325 \times 10^5\,\mathrm{N\,m^{-2}}$. The working volume of the aeration basin was $5200\,\mathrm{m^3}$.

Solution

$$\frac{\mathrm{d}c}{\mathrm{d}t} = k_\mathrm{L}a(c^* - \bar{c}) - Q_\mathrm{res}X$$

since supply = demand, $\mathrm{d}c/\mathrm{d}t = 0$

$$k_\mathrm{L}a(c^* - \bar{c}) = Q_\mathrm{res}X = 15.6 \times 10^{-3}\,\mathrm{kg\,O_2\,m^{-3}\,h^{-1}}X$$

$$k_\mathrm{L}a = \frac{15.6 \times 10^{-3}}{7.8 \times 10^{-3} - 5.1 \times 10^{-3}} = 5.78\,\mathrm{h^{-1}}$$

$$Q'_\mathrm{res} = Q_\mathrm{res}X$$

$$Q_\mathrm{res} = \frac{Q'_\mathrm{res}}{X} = \frac{15.6 \times 10^{-3}}{2} = 7.8 \times 10^{-3}\,\mathrm{h^{-1}}$$

Amount of oxygen entering = $113\,\mathrm{m^3\,min^{-1}} \times \mathrm{O_2^{-1}}$

$$PV = nRT$$

$$m = \frac{PVM}{RT} = \frac{1 \times 1 \times 32}{0.032\,(273 + 24.5)}$$

$$= 1.312\,\mathrm{kg\ for\ 1\,m^3\ of\ O_2}$$

$$= 113 \times 0.2 \times 1.312$$

$$= 1.78 \times 10^3\,\mathrm{kg\,h^{-1}}$$

Total volume of tank = $5200\,\mathrm{m^3}$.

For each $1\,\mathrm{m^3}$ of tank

$$\frac{1.78 \times 10^3}{5200} = 0.342\,\mathrm{kg\,O_2\,h^{-1}\,m^{-3}}\ \text{(theoretical)}$$

$$\text{Efficiency } \eta = \frac{15.6}{342} = \underline{4.56\%}$$

7

Bioreactor systems

Contributors: Tarun Ghose, John Villadsen, K. B. Ramachandran

PROBLEM 7.1

A 20 m^3 working volume bioreactor with the following dimensions is available in a plant for the production of penicillin. What initial concentration of the limiting nutrient (sugar) you will choose if the oxygen transfer rate for the antibiotic production is not to be limiting?

Data given:
Tank diameter = 2.4 m
Impeller diameter = 0.8 m and 3 Nos
Impeller speed = 2.5 rps
Number of blades = 8
Type of impeller = turbine
Viscosity of broth = 1 mpa s
Density of broth = 1.2×10^3 kg m^{-3}
Ratio of gassed to ungassed power = 0.4
Aeration rate = 1 vvm
Driving force for oxygen transfer = 6×10^{-3} kg m^{-3}
Specific oxygen uptake rate of cells = 0.65 millimoles O_2 (kg cell)$^{-1}$
Maximum specific growth rate of cells = 0.5 h^{-1}
Specific sugar consumption rate of cells = 1.0 kg(kg cell)$^{-1}$ h^{-1}
Volumetric mass transfer coefficient

$$k_L a = 2 \times 10^{-3} \left(\frac{P_g}{V}\right)^{0.6} (V_s)^{0.667} \tag{7.1.1}$$

where power input per unit volume $\left(\frac{P_g}{V}\right)$ (=) hp m^{-3} and superficial gas velocity V_s(=) cm min^{-1}. The unit of $k_L a$ is s^{-1}.

Reference

Bailey, J. E. & Ollis, D. O. (1985) *Biochemical Engineering Fundamentals*, 2nd edn., McGraw Hill, New York.

Solution

The power number in the turbulent flow is constant and is equal to 6.0. For this system, Reynolds' number is

$$N_{Re} = \frac{n\, D_i^2\, \rho}{\mu} \tag{7.1.2}$$

$$= \frac{2.5 \times (0.8)^2 \times (1.2 \times 10^3)}{0.1}$$

$$= 1.92 \times 10^4 \text{ (this is turbulent flow)}$$

$$\therefore N_p = 6 = \frac{P\, g_c}{\rho\, n^3\, D^5} \tag{7.1.3}$$

$$\therefore \text{Ungassed power } P = \frac{6\, \rho\, n^3\, D^5}{g_c} \tag{7.1.4}$$

$$= \frac{6 \times 1.2 \times 10^3 \times (2.5)^3 \times (0.8)^5}{9.81}$$

$$= 3.75 \times 10^3 \text{ kg m s}^{-1} = 49.3 \text{ hp}$$

Correction factor for non-geometrical similarity is

$$f_c = \sqrt{\frac{3 \times 5.53}{3 \times 3}} = 1.36 \tag{7.1.5}$$

$$\therefore P = 3 \times 1.36 \times 49.3 = 200 \text{ hp} \tag{7.1.6}$$

$$\therefore P_g = P \times 0.4 = 80 \text{ hp} \tag{7.1.7}$$

Aeration rate = 20 m^3 min^{-1}

$$\therefore V_s = \frac{20}{\left(\frac{\pi}{4}\right) \times (2.4)^2} \tag{7.1.8}$$

$$= 4.42 \text{ m min}^{-1}$$

$$= 442 \text{ cm min}^{-1}.$$

$$\therefore k_L a = 2 \times 10^{-3} \left(\frac{80}{20}\right)^{0.6} (442)^{0.667} \tag{7.1.9}$$

$$= 0.267 \text{ s}^{-1}.$$

$$\therefore \text{oxygen transfer rate} = k_L a(c^* - c)$$
$$= 0.267 \times 6 \times 10^{-3}$$
$$= 1.603 \times 10^{-3} \text{ kg m}^{-3} \text{ s}^{-1} \tag{7.1.10}$$

The maximum cell concentration we can have without oxygen deficiency is

$$xq_{O_2} = 1.603 \times 10^{-3} \text{ kg m}^{-3} \text{ s}^{-1}$$

$$\therefore x = \frac{1.603 \times 10^{-3}}{q_{O_2}}.$$

$$q_{O_2} = 0.65 \times 10^{-3} \times 32 \times 10^{-3} = 20.8 \times 10^{-6} \text{ kg } O_2 \text{ (kg cell)}^{-1} \text{ s}^{-1}.$$

$$\therefore x = \frac{1.603 \times 10^{-3}}{20.8 \times 10^{-6}} = 77 \text{ kg m}^{-3}$$

Stationary phase population is given by Bailey & Ollis

$$x_s = x_0 + \frac{\mu}{q_s} c_s \tag{7.1.11}$$

where $x_s = 77$ kg m^{-3} (stationary cell population)
x_0 = initial cell population
μ = specific growth rate of cells
c_s = limiting substrate (sugar) concentration
q_s = specific substrate uptake of the cells.
Assuming $x_s \gg x_0$

and
$$s_0 = \frac{x_s\, q_s}{\mu} \tag{7.1.12}$$
$$= 77 \times \quad 1.0 \times \frac{1}{0.5} = \underline{154 \text{ kg m}^{-3}}$$

Therefore, initial substrate concentration should be 154 kg m^{-3}.

PROBLEM 7.2

The nominal holding time t_p of liquid in a continuous reactor wherein cell concentration is enhanced from $x = x_1$ to $x = x_2$ is

$$t_p = \int_{x_1}^{x_2} \frac{dx}{\left(\dfrac{dx}{dt}\right)_{\text{growth}}} \tag{7.2.1}$$

in which the flow pattern is of plug flow. What will be the holding time, t_c, if the flow is assumed to be of complete mixing (back-mixing)? If the batch growth rate data of a microorganism could be used in continuous culture, how do you estimate the value of t_p and t_c?

Solution

Refer to Fig. 7.1.

(i) For plug flow system:

$$Fx + dV\left(\frac{dx}{dt}\right)_{growth} = F(x + dx) \tag{7.2.2}$$

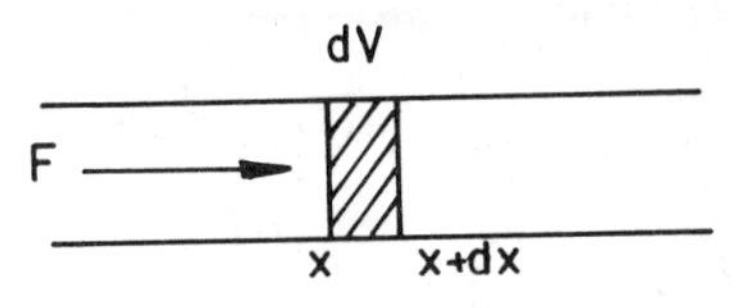

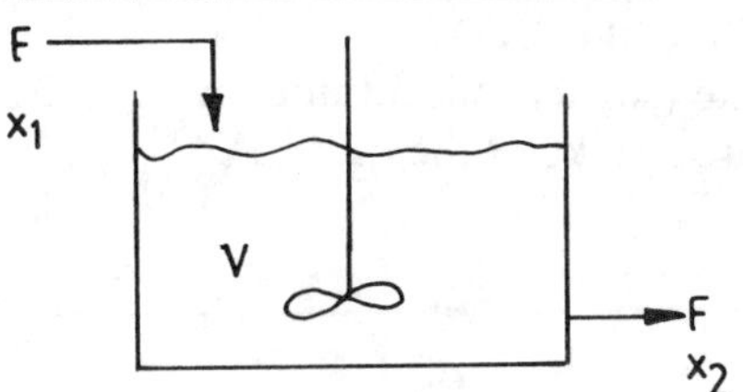

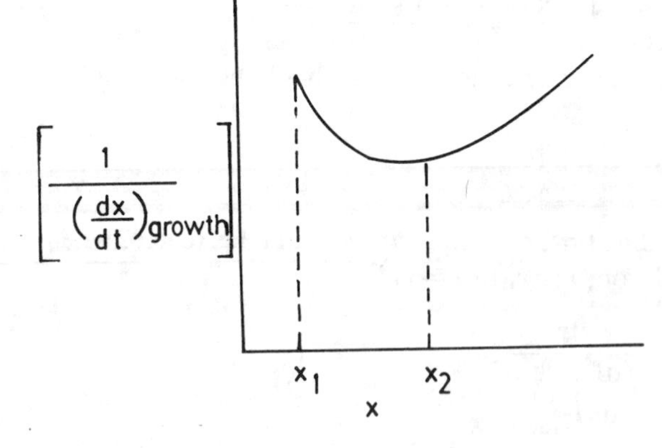

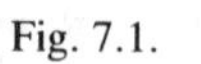

Fig. 7.1.

We may write the differential mass balance equation for the system as:

$$dt_p = \frac{dV}{F} = \frac{dx}{(dx/dt)_{growth}} \quad (7.2.2)$$

$$\therefore t_p = \int_{x_1}^{x_2} \frac{dx}{(dx/dt)_{growth}} \quad (7.2.3)$$

(ii) For the completely mixed system:

$$Fx_1 + V\left(\frac{dx}{dt}\right)_{growth} = Fx_2 \quad (7.2.4)$$

and the mass balance equation (differential) for the system

$$t_c = \frac{V}{F} \quad (7.2.5)$$

$$= \frac{x_2 - x_1}{\left(\frac{dx}{dt}\right)_{growth}}$$

PROBLEM 7.3

Show that the residence time curve, $R(\phi)$ of a laminar flow of a fermentation broth through a straight pipe can be expressed by

$$R(\phi) = \left(\frac{1}{2\phi}\right)^2 \quad (7.3.1)$$

where

$\phi = \frac{\theta}{\theta_T} = \frac{L/u}{L/\bar{u}}$
L = pipe length
u = liquid velocity
$\bar{u}$ = mean liquid velocity.

Solution
For laminar flow (refer to Fig. 7.2)

$$u = u_0\left\{1 - \left(\frac{r}{r_0}\right)^2\right\} \quad (7.3.2)$$

$$R(\phi) = 1 - \frac{\int_0^r u\, 2\pi r\, dr}{\pi\, r_0^2\, \bar{u}} \quad (7.3.3)$$

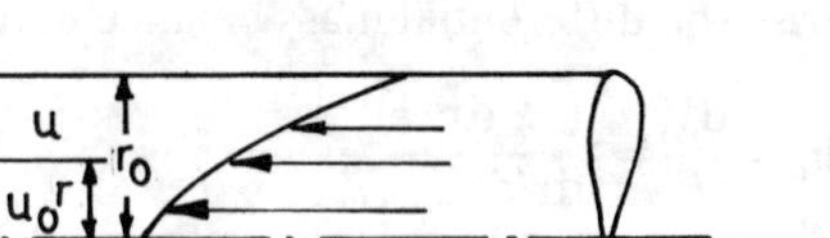

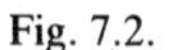
Fig. 7.2.

$$= 1 - \frac{1}{\pi\, r_0^2\, \bar{u}} \int_0^r u_0 \left\{1 - \left(\frac{r}{r_0}\right)^2\right\} 2\pi r \, \mathrm{d}r$$

$$= 1 - \frac{2u_0}{r_0^2 \bar{u}} \int_0^r \left\{r - \frac{r^3}{r_0^2}\right\} \mathrm{d}r$$

$$= 1 - \frac{2u_0}{r_0^2\, \bar{u}} \left[\frac{r^2}{2} - \frac{r^4}{4r_0^2}\right]$$

$$= 1 - \frac{u_0}{\bar{u}} \left(\frac{r}{r_0}\right)^2 + \frac{u_0}{2\bar{u}} \left(\frac{r}{r_0}\right)^4$$

$$= 1 - 2\left(\frac{r}{r_0}\right)^2 + \left(\frac{r}{r_0}\right)^4 \left[\text{because } \frac{u_0}{\bar{u}} = 2\right]$$

$$= \left[1 - \left(\frac{r}{r_0}\right)^2\right]^2$$

$$= \left\{\frac{u_0}{2\bar{u}} \left[1 - \left(\frac{r}{r_0}\right)^2\right]\right\}^2$$

$$= \left(\frac{u}{2\bar{u}}\right)^2$$

$$= \left[\frac{1}{2} \frac{(L/\bar{u})}{(L/\bar{u})}\right]^2$$

$$= \left(\frac{1}{2} \frac{\theta_T}{\theta}\right)^2$$

$$= \left(\frac{1}{2\phi}\right)^2$$

PROBLEM 7.4

A bacterial fermentation was carried out in a reactor containing a medium having a density of 1.2×10^3 kg m^{-3} and a viscosity of 0.02 N s m^{-2}. The broth was agitated with the help of an impeller having a speed of 90 rpm and air was introduced through the sparger at the rate of 0.4 vvm. The fermenter, equipped with two sets of flat blade turbines and four baffle plates, was having the following dimensions:

diameter of the fermenter	= 4 m
diameter of the impeller	= 2 m
baffle plate width	= 0.4 m
liquid depth in fermenter	= 6.5 m

For designing the overall fermentation process, it is required to determine:

(a) power requirement, P, for ungassed system,
(b) power consumption, P_g, for gassed system,
(c) volumetric oxygen transfer coefficient, k_La, and
(d) hold up of bubbles, H_0.

Solution

(a) We have $\frac{D_t}{D_i} = \frac{4}{2} = 2.0$

$$\frac{H_L}{D_i} = \frac{6.5}{2} = 3.25$$

$$n = \frac{90}{60} = 1.5 \text{ rps}$$

$$N_{Re} = \frac{n\, d_i^2\, \rho}{\mu} = \frac{1.5 \times (2)^2 \times 1.2 \times 10^3}{2 \times 10^{-2}}$$

$$= 3.6 \times 10^5$$

From Fig. 6.5 in Aiba *et al.* (1973, p. 174)

$$N_{Re} = 3.6 \times 10^5$$

$$N_P = \frac{P\, g_c}{N^3 D_i^5 \rho} = 6$$

So, for the ungassed system, the power requirement will be

$$P = \frac{\rho n^3\, D_i^5\, N_P}{g_c} = 8.35 \times 10^3 \text{ kg m}^{-1}\text{ s}^{-1} = 81885.67 \text{ watts}$$

Since the geometrical ratios, $(D_t/D_i)^*$ and $(H_L/D_i)^*$ of this problem deviate from (D_t/D_i) and (H_L/D_i) in Fig. 6.4 in Aiba *et al.* (1973, p. 174), a correction factor f_c, which can be calculated from the relation

$$f_c = \sqrt{\frac{(D_t/D_i)^*(H_L/D_i)^*}{(D_t/D_i)(H_L/D_i)}} = \sqrt{\frac{2 \times 3.25}{3 \times 3}} = 0.85$$

must be multiplied with the power calculated to get the actual power requirement by the one set of impellers. Thus,

$$P^* = Pf_c = 81{,}885.67 \times 0.85$$
$$= 69{,}602.82 \text{ W}$$

The power requirement by the two sets of impellers will be then

$$P^{**} = P^* \sqrt{2} = 69{,}602.82 \times \sqrt{2}$$
$$= 98{,}139.98 \text{ W}$$

(b) Aeration number, $N_a = \dfrac{F}{N\,D_i^3}$

$$= 3.5 \times 10^{-2}$$

Asuming that curve A in Fig. 6.6 Aiba *et al.* (1973, p. 176) can be used in this case, we get

$$N_a = 3.5 \times 10^{-2},\ (P_g/P^{**}) = 0.76$$

so,

$$P_g = 0.76 \times 98{,}139.98$$
$$= 74{,}202.73 \text{ W}$$

(c)
$$F = 0.4 \times \frac{\pi}{4} \times 4^2 \times 6.5$$
$$= 0.1 \times 3.14 \times 16 \times 6.5 = 32.8 \text{ m}^3 \text{ min}^{-1}$$
$$V_s = \frac{F}{\frac{\pi}{4} D_t^2} = \frac{32.80}{3.14 \times 4} = 2.6 \text{ m min}^{-1}$$
$$= 156 \text{ m h}^{-1}$$

Now, we have the relationship

$$k_L a = 0.0635\,(P_g/V)^{0.95}(V_s)^{0.67}$$

where P_g is in horse power (hp)
or, $P_g = 74{,}202.73 \text{ W} = 99.56 \text{ hp}$

$$\therefore k_L a = 0.0635 \times \left[\frac{99.56}{\frac{\pi}{4} \times 4^2 \times 6.5}\right] \times (156)^{0.67}$$

$$= 2.51 \text{ kg mol m}^{-3} \text{ h}^{-1} \text{ atm}^{-1}$$

In this case, $H_L/D_t = \dfrac{6.5}{4} = 1.62$. Now, from Fig. 6.9 Aiba *et al.* (1973, p. 181), we see that the correction factor for this H_L/D_t ratio which is to be multiplied with the calculated $k_L a$ value to get the actual $k_L a$ value is:

$$f_c = 1.25$$

Therefore, the actual volumetric oxygen transfer coefficient $(k_L a)^*$ becomes:

$$(k_L a)^* = k_L a f_c = 2.51 \times 1.25$$
$$= \underline{3.16 \text{ kg mol m}^{-3} \text{ h}^{-1} \text{ atm}^{-1}}$$

(d) For the calculation of hold up, H_0, we have to make use of Fig. 6.7 of Aiba *et al.* (1973, p. 178). Thus,

$$(P/V)^{0.4} V_s^{0.5} = \frac{131}{\left(\frac{\pi}{4} \times 4^2 \times 6.5\right)}^{0.4} \times (156)^{0.5}$$
$$= 1.2 \times 12.49$$
$$= 14.98$$

Now, from Fig. 6.7 Aiba *et al.* (1973, p. 178), the corresponding value of bubble hold up, H_0, becomes

$$H_0 = \underline{13.8\%}$$

Reference
Aiba, S., Humphrey, A. E. and Millis, Nancy F. (1973) *Biochemical Engineering*, 2nd edn, University of Tokyo Press.

PROBLEM 7.5

The data given in Table 7.1 were obtained by agitating a CMC solution with a specific impeller in a fermenter.

Table 7.1

Flow behaviour of CMC used		Rotational speed of	Power number impeller	Reynolds number
n'	$K(\text{kg m}^{-1}\text{ s}^{2-n'})$	$n(\text{s}^{-1})$	$= \dfrac{P g_c}{n^3 D_i^5 \rho}$	$= \dfrac{n D_i^2 \rho}{\mu_a}$
0.48	223	3.0	9.05	19.0
0.48	223	4.0	7.26	29.0
0.50	160	1.5	16.0	8.6
0.50	160	2.0	11.1	13.8
0.54	91	1.0	17.4	7.8
0.54	91	2.5	7.27	29.5

given: n' = flow behaviour index, K = consistency index, μ_a = apparent viscosity.

The Reynolds number shown in Table 7.1 was estimated from the power number versus Reynolds number obtained separately by using another viscous and Newtonian liquid in the same vessel. It is desired to assess the proportionality constant, in the following equation

$$\left(\frac{\mathrm{d}v}{\mathrm{d}r}\right)_{\mathrm{av}} = \gamma n$$

where $\left(\frac{\mathrm{d}v}{\mathrm{d}r}\right)_{\mathrm{av}}$ = average value of shear rate in the vessel. Take $D_{\mathrm{i}} = 0.155$ m (impeller diameter) and $\rho = 10^3$ kg m^{-3} (density of CMC solution).

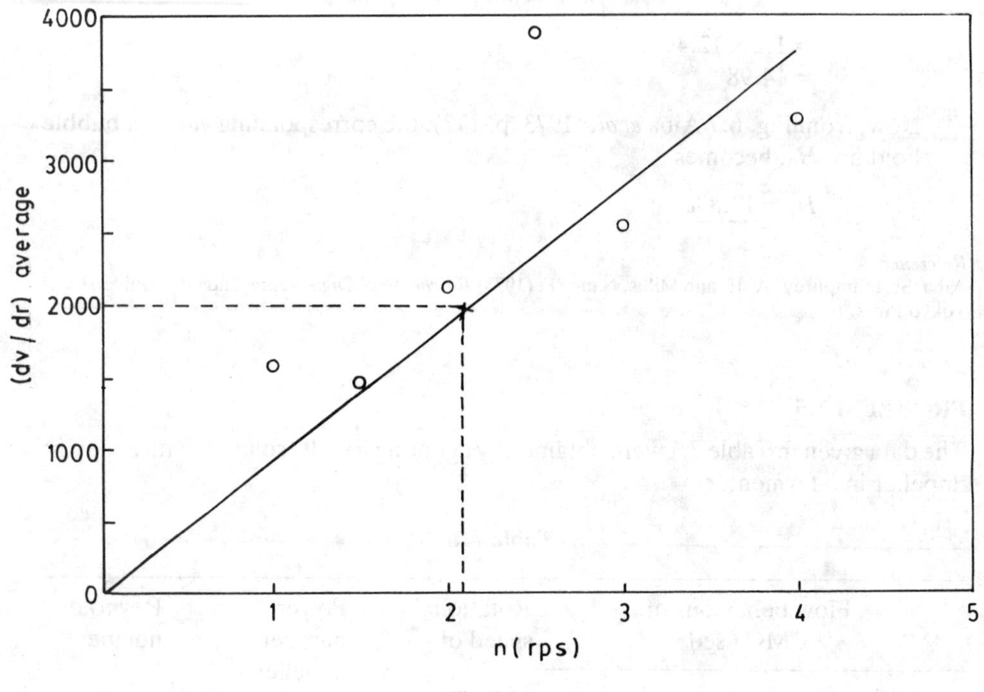

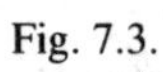
Fig. 7.3.

Solution

$\left(\frac{\mathrm{d}v}{\mathrm{d}r}\right)$ can be expressed as

$$\mu_{\mathrm{a}} = \frac{\tau g_{\mathrm{c}}}{\left(\frac{\mathrm{d}v}{\mathrm{d}r}\right)}$$

for CMC solution, $\tau g_{\mathrm{c}} = K\left(\frac{\mathrm{d}v}{\mathrm{d}r}\right)^n$.

$$\therefore \mu_{\mathrm{a}}\left(\frac{\mathrm{d}v}{\mathrm{d}r}\right) = K\left(\frac{\mathrm{d}v}{\mathrm{d}r}\right)^n \text{ or, } \left(\frac{\mathrm{d}v}{\mathrm{d}r}\right) = \left(\frac{K}{\mu_{\mathrm{a}}}\right)^{1/(1-n)}$$

$$\therefore \left(\frac{\mathrm{d}v}{\mathrm{d}r}\right)_{\mathrm{av.}} = \left(K\frac{n\,D_{\mathrm{i}}^2\rho}{\mu_{\mathrm{a}}}\frac{1}{n\,D_{\mathrm{i}}^2\rho}\right)^{1/(1-n)}$$

From Fig. 7.3, the following data are obtained:

n (rps)	$\left(\frac{dv}{dr}\right)_{av.}$ Average velocity gradient (shear rate)
3.0	2526
4.0	3276
1.5	1457.9
2.0	2111.6
1.0	1572.8
2.5	3868.3

The value of γ is found as 2,000

PROBLEM 7.6

Batch rate data of alcohol fermentation are given in Table 7.2. Assuming that the data can be used for designing continuous culture in a single vessel (working volume $= 0.1\ m^3$) determine the alcohol concentration in the broth when operated at $F = 4 \times 10^{-3}\ m^3\ h^{-1}$.

Table 7.2

Culture time (h)	Alcohol concentration in the broth ($kg\ m^{-3}$)
12	2
15	4
18	6.2
21	10.0
27	16.1
33	30.5
42	56.5
48	75.7
51	85.0
54	96.0
57	103.5
60	107.3
66	110.5
72	113.6

Solution

The rate data of Table 7.2 are plotted in Fig. 7.4:

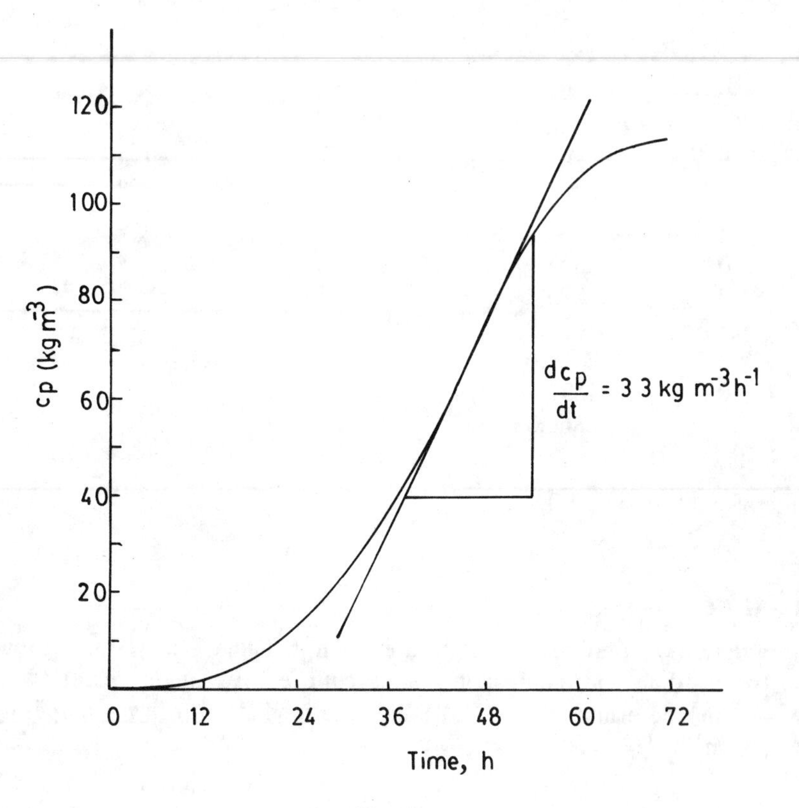

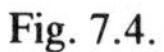
Fig. 7.4.

It gives an exponential value as:

$$\frac{dc_p}{dt} = 3.3 \text{ kg m}^{-3}\text{ h}^{-1} \times 0.1 \text{ m}^3$$

$$= 0.33 \text{ kg h}^{-1}$$

$$\frac{dc_p}{dt} = \frac{F}{V}(c_{p0} - c_p) + r_p \text{ (product mass balance)}$$

At steady state,

$$\frac{dc_p}{dt} = 0; \; c_{p_0} = 0$$

or, $$c_p = \frac{V}{F}\left(\frac{dc_p}{dt}\right)$$

and $$\frac{c_p}{V} = \frac{1}{F}\left(\frac{dc_p}{dt}\right)$$

$$= \frac{0.33}{4 \times 10^{-3}} = \underline{82.5 \text{ kg m}^{-3}}$$

PROBLEM 7.7

Kirpekar and Kirwan (1985) report on experiments with a mutant P of *N. lactamdurans*. When P grows on sodium-L-glutamate monohydrate (MSG) with glucose and NH_4^+ as extra substrates it produces the pharmaceutical Cephamycin C. Unfortunately P will eventually revert to a non-producing wild strain B. The *total* biomass production (concentration x g m^{-3}, rate r_x) does not change, but the concentration of B, relative to P increases. After about 4–5 residence times in a chemostat $f = b/(b+p) = b/x \rightarrow 1$. Let the ratio between specific growth rates of B and P be $c = \mu_b/\mu_p = 1.25$. P is inherently unstable and reverts to B with rate $r_{bp} = \alpha p$. α may be a constant K or a simple function of f and μ_p. Thus, there are two reasons for the eventual washout of P: It grows more slowly than B ($c > 1$), and it reverts to B. Fig. 7.4 shows how f increases from $f(t=0) = 0.05$ towards 1 as a function of $tD = \theta$, the number of residence times, in chemostat experiments (constant total cell density x) with $D = 0.036$ h^{-1} (circles) and $D = 0.045$ h^{-1} (squares). $\mu_b = 0.064$ h^{-1}, $\mu_p = 0.051$ h^{-1} ($c = 1.25$).

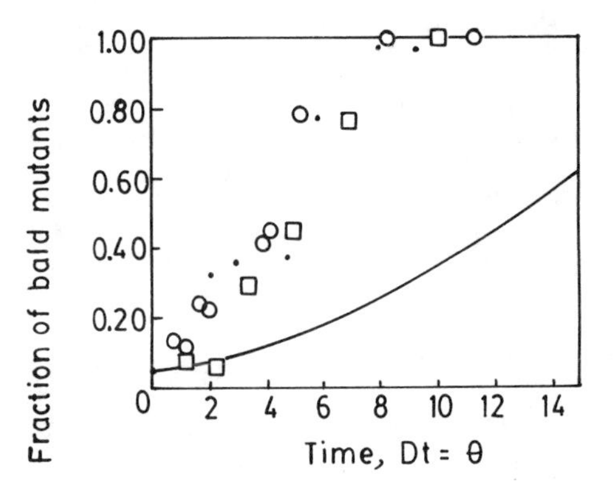

Fig. 7.5 — Fraction of bald mutants as a function of the number of residence times $(\tau - \tau_0)$. τ_0 corresponds to $f_0 = 0.05$. Solid line represents prediction of equation (7.7.1) with $\alpha = 0$.

(1) Write mass balances for P, B and X. Use $dx/dt = 0$ to find μ_b as a function of f, c and D.

Show that

$$\frac{df}{dt} = \frac{f}{b}\frac{db}{dt}$$

and next

$$\frac{df}{d\theta}=\frac{(c-1)f(1-f)}{cf+1-f}+\frac{(1-f)\alpha}{D} \quad (7.7.1)$$

where $f=f_0$ at $t=0$.

(2) The full line on Fig. 7.5 shows the solution of (7.1.1) for $\alpha=0$, i.e. washout due to the difference in growth rates alone. Solve (7.1.1) and show that the curve correctly predicts $f=0.20$ and $f=0.60$.

(3) Obviously washout of P is much faster than predicted in (2). Thus α must be $\neq 0$. Let α = constant K, integrate (7.1.1) and show that the experiments for $D=0.036\ \mathrm{h}^{-1}$ are reasonably well simulated when $K=0.0035\ \mathrm{h}^{-1}$ is used in the solution.

(4) The authors also use other models for α ($r_{pb}=\alpha p$):
(a) $\alpha=K\mu_p$, (b) $\alpha=kf$ and (c) $\alpha\ Kf\mu_p$.
What is (qualitatively) the biological explanation for each of these reversion rate expressions? Is (1.7.1) more difficult to solve analytically with one of these expressions for α.

Reference
Kirpekar, A. C. & Kirwan, D. J. (1985)*Biotechnology Progress*, **1**, 231–236.

Solution

(1) Mass balances for B, P and X in a chemostat:

$$\mathrm{B:}\frac{\mathrm{d}}{b\mathrm{d}t}=-b\mu_b+p\alpha-bD$$

$$\mathrm{P:}\ \frac{\mathrm{d}p}{\mathrm{d}t}=p\mu_p-p\alpha-pD=-\frac{\mathrm{d}b}{\mathrm{d}t}$$

$$\mathrm{X:}\ \frac{\mathrm{d}x}{\mathrm{d}t}=b\mu_b+p\mu_p-xD=0$$

From the last equation

$$f\mu_b+(1-f)\ \mu_p-D=0 \text{ or with } c=\frac{\mu_b}{\mu_p}$$

$$fc+(1-f)-\frac{cD}{\mu_b}=0$$

i.e.

$$\mu_b=\frac{cD}{fc+1-f}$$

Also,

$$\frac{df}{dt}=\frac{\mathrm{d}\left(\frac{b}{b+p}\right)}{\mathrm{d}t}=\frac{1}{(b+p)}\frac{\mathrm{d}b}{\mathrm{d}t}-\frac{b}{(b+p)^2}\frac{\mathrm{d}x}{\mathrm{d}t}=\frac{f}{b}\frac{\mathrm{d}b}{\mathrm{d}t}$$

Inserting $\frac{\mathrm{d}b}{\mathrm{d}t}$ and μ_b yields

$$\frac{1}{D}\frac{df}{dt}=\frac{df}{d\theta}=\frac{fc}{fc+1-f}+\frac{(1-f)}{D}\alpha-f=\frac{f(c-1)(1-f)}{fc+1-f}+\frac{(1-f)}{D}\alpha$$

(2) For $\alpha=0$ and $c=1.25$, one obtains

$$\frac{df}{d\theta}=\frac{f(1-f)}{f+4}\rightarrow d\theta=\left\{\frac{4}{f}+\frac{5}{(1-f)}\right\}df\rightarrow 4\ln\frac{f}{f_0}-5\ln\left(\frac{1-f}{1-f_0}\right)=\theta$$

For $f=0.20$; $u=6.40$ and for $f=0.60$; $\theta=14.26$. Both fit well with the full curve on Fig. 7.5. Note that $f(\theta)$ is independent of D. This is of course not true for $f(t)$.

(3) For $\alpha=K$:

$$d\theta=\frac{f+4}{(1-f)[f(1+K/D)+4K/D]}df$$

$$=\left\{\frac{5}{1+5\,K/D)}\frac{1}{(1-f)}+\frac{4}{(1+5\,K/D)}\frac{1}{(1+\,K/D)\,f+4\,K/D}\right\}df$$

$$\theta=-\frac{5}{(1+5\,K/D)}\ln\left(\frac{1-+f}{1-f_0}\right)+$$

$$+\frac{4}{(1+5\,K/D)\,(1+K/D)}\ln\frac{(1+K/D)\,f+4\,K/D}{(1+\,K/D)\,f_0+4\,K/D}$$

For $K=0.035\ h^{-1}$ and $D=0.036\ h^{-1}$ one obtains $f=0.60\rightarrow\theta=5.016$, $f=0.80\rightarrow\theta=7.815$ (i.e. much faster growth of f as a function of time).

(4) See the authors' explanation in the paper. None of the other expressions are more difficult to handle analytically.

PROBLEM 7.8

For a fermentation in a chemostat,

$$g_x=\frac{4}{3}\frac{v_{sx}}{c_s+4}\frac{\text{g cells}}{\text{m}^3\text{h}}$$

We wish to compare the performance of two reactor types, $c_{s_f}=60\ \text{g m}^{-3}$, $x_f=0$, $Y_{x/s}=0.1$.

(a) Let the plug flow reactor have avolume $V=1\ \text{m}^3$. Determine the recirculation ratio R for which maximum cell production Q_x is obtained. Compare with the corresponding result for the chemostat.

(b) Determine the optimal value of R and the corresponding value of Q_x, if the exit substrate concentration is chosen as $c_s=3\ \text{g m}^{-3}$. Compare with the corresponding result for a chemostat.

(c) Consider a reactor system with a chemostat and a plug flow reactor in series. $v_0=2.5\ \text{m}^3\ \text{h}^{-1}$, $c_{s_f}=60\ \text{g m}^{-3}$, $x_f=0$, and $c_s=3\ \text{g m}^{-3}$ at exit. Determine the volume of each reactor so that the total volume is as small as possible. Again compare with the reactor configuration, which could give the maximum cell productivity per m³ reactor. (See Fig. 7.6.)

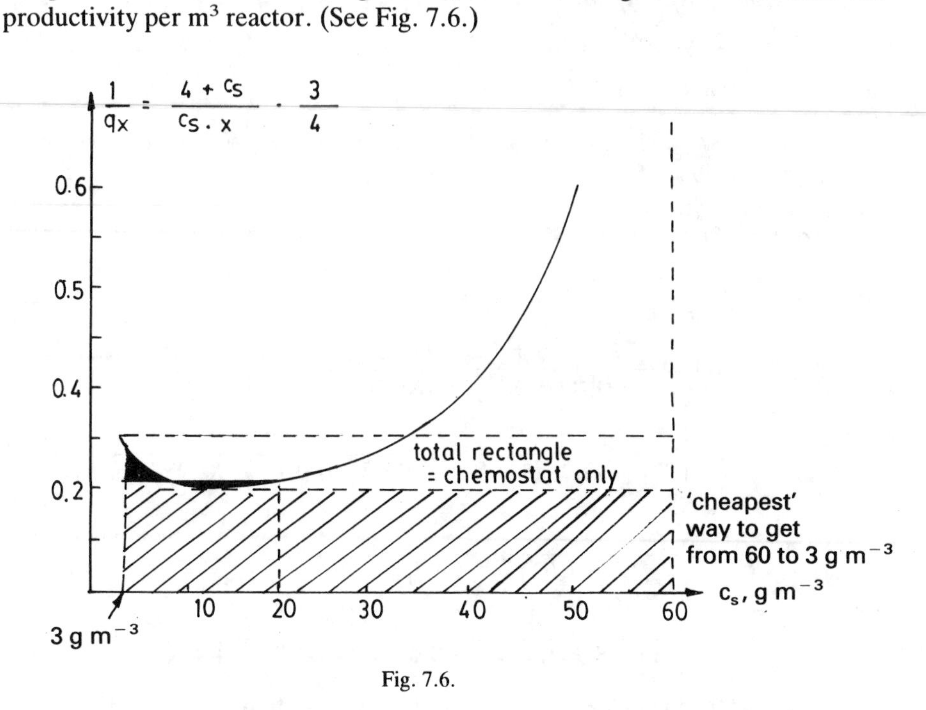

Fig. 7.6.

Solution

(a) Maximum cell production is obtained in a chemostat using

$$S = -a + \sqrt{a + a^2} = \frac{1}{15} + \sqrt{\frac{1}{15} + \frac{1}{225}} = \frac{1}{5}$$

(as seen in Problem 3.23). Hence, R should be infinity. No plug flow reactor can compete with the best chemostat if the objective is to obtain the highest possible cell productivity.

(b) If one insists on having $c_s = 3\ \mathrm{g\,m^{-3}}$ the very best Q_x cannot be obtained but now a recirculation reactor is much better than a chemostat.

In question (a),

$$\tau_{opt} = \tau(s = \frac{1}{5}) = \frac{\frac{4}{2} \times 12}{4 + 12} = 1\ \mathrm{h}$$

For the chemostat with $c_s = 3\ \mathrm{g\,m^{-3}}$,

$$\tau = \frac{S \times 7}{10\,\frac{4}{3}\,\frac{S \times 7 \times 3}{4 + 3}}$$

$$= \frac{7}{4}\ \mathrm{h}$$

The recirculation rate is determined from

$$\frac{K_m}{s_f} \ln \frac{c_{s_f} + Rc_s}{\mathbf{R}c_s} + \ln \frac{R+1}{R} = \frac{R+1}{R} \frac{K_m}{f+R} + \frac{1}{R}$$

or

$$\underset{\text{(LHS)}}{\frac{1}{15} \ln \frac{20+R}{R}} + \ln \frac{R+1}{\text{R}} = \underset{\text{(RHS)}}{\frac{R+1}{\text{R}} \frac{4/3}{(20+R)}} + \frac{1}{R}$$

R	2	2.5	3.0
LHS	0.5653	0.4830	0.4235
RHS	0.5909	0.4830	0.4106

Accept $R = R_{opt} = 2.5$.

Now use the design formula for a recirculation reactor:

$$\tau k = (R+1)\left[\ln \frac{R+1}{R} + \frac{K_m}{f} \ln \frac{c_{s_f} + Rc_s}{R}\right]$$

$$\tau = \frac{3}{4} \times 3.5 \left[\ln \frac{7}{5} + \frac{1}{15} \ln \left(\frac{20+2.5}{2.5}\right)\right]$$

$$= \underline{1.26 \text{ h}}$$

This is obviously better than the chemostat result $\tau = \frac{7}{4} = 1.75$ h but not as good as the optimal result, $\tau = 1$ h.

(c) For $v_0 = 2.5 \text{ m}^3 \text{ h}^{-1}$ we need a reactor (chemostat) with volume $V_0 = 2.5 \text{ m}^3$ to bring c_s from 60 g m^{-3} to 12 g m^{-3}. Next we use a plug flow reactor to bring c_s to 3 g m^{-3}.

Design formula

$$k\tau = (1+F) \ln \frac{X}{X_F} - F \ln \frac{c_s}{c_{s_f}}$$

where

$$F = \frac{K_m}{c_{s_f} + Y_{x/s} X_f} = \frac{4}{12+48}$$

$$\tau = \frac{3}{4}\left[\left(1 + \frac{1}{15}\right) \ln \frac{5.7}{4.8} - \frac{1}{15} \ln \frac{3}{12}\right] = 0.2068 \text{ h}$$

which corresponds to $V = 2.5 \times 0.2068 = \underline{0.5170\ \text{m}^3}$.

Total space time used to obtain $c_s = 3\ \text{g m}^{-3}$ is

$$1 + 0.2068 = \underline{1.2068\ \text{h}}$$

Conclusions

(1) Maximum productivity is obtained when a chemostat working at $c_s = \underline{12\ \text{g m}^{-3}}$ is used.

(2) If c_s has to be $3\ \text{g m}^{-3}$ it is best to combine a chemostat and a plug flow reactor. The chemostat brings the substrate concentration to $12\ \text{g m}^{-3}$, the plug flow reactor takes c_s to $3\ \text{g m}^{-3}$.

$$\text{Productivity} = \frac{0.1(60-3)}{1.2068} = \underline{4.723\ \text{g m}^{-3}\ \text{h}^{-1}}$$

(3) A recirculation reactor should work at $R = 2.5$ when $c_s = 3\ \text{g m}^{-3}$.

$$\text{Productivity} = \frac{0.1(60-3)}{1.26} = \underline{4.52\ \text{g m}^{-3}\ \text{h}^{-1}}$$

(4) A single chemostat which takes substrate concentration from 60 to $3\ \text{g m}^{-3}$ has a productivity

$$Q_x = \frac{5.7}{1.75} = \underline{3.26\ \text{g m}^{-3}\ \text{h}^{-1}}$$

PROBLEM 7.9

The rate of cell production q_x and the rate of sustrate consumption q_s are given by (in units of $\text{kg m}^{-3}\ \text{h}^{-1}$).

$$q_x = \frac{kc_s x}{K_m + c_s};\ q_s = Y_{x/s}(q_x + k_e x)$$

The feed is $v\ \text{m}^3\ \text{h}^{-1}$ to a reactor of volume $V\ \text{m}^3$. The feed is sterile and substrate feed concentration is $c_{s_f}\ \text{kg m}^{-3}$.

k_e is a maintenance constant (h^{-1}); $k_e/k = \text{b}$. $b = 0$. $a = K_m/c_{s_f}$.

We wish to work with the two variables $S = c_s/c_{s_f}$ and $X = xY_{x/s}/c_{s_f}$

In the transient problem, we use $S = Y_1 = c_s/c_{s_f}$ where x_0 is the value of x at $t = 0$ (or at $\theta = tD = 0$, where $D = v/V$).

(1) For $k_e = 0$ find an expression for $X(S)$ in a chemostat. Determine the value of S for which the maximum productivity of cells Q ($\text{kg h}^{-1}\ \text{m}^{-3}$ reactor) is obtained. Show on the same graph X and $G = Q/Y_{s/x}kc_{s_f}$ as functions of S for $a = 0.2$.

(2) In a steady state chemostat experiment with $D = 1\ \text{h}^{-1}$ we obtain $S = 0.8$. $a = 0.2$, $k_e = 0$. Determine the maximum growth rate k and the largest permissible value of D.

(3) Consider a transient experiment in the chemostat.

For $t<0$ $s\sim 0$. What is the corresponding value of $x=x_0$ and of X_0? For $t>0$ the space time V/v is changed to 1 h.

What is the expected final value of X?

After 24 min $x/x_0=0.8$. Does this check with the value of k determined in question (2)? How long will it take for x/x_0 to get to within 10% of its final value?

(4) Assume that $V/v = 1$ h for the chemostat. A plug flow reactor is used after the chemostat. Determine the necessary τ if $S=0.01$ after the plug flow reactor. ($a=0.2$, $k=1.25\ \text{h}^{-1}$ and $k_e=0$).

What is the minimum τ (total) for a chemostat followed by a plug flow reactor in which S is reduced to 0.01?

(5) Now consider $k_e>0$.

Determine $X(S)$ and find the S value for which X is maximum. Consider $a=0.2$ and $b=0.1$. Find S_{opt} and X_{max}. Show graphically $X(S)$.

On the same graph show $G=P/k\,Y_{x/s}c_{sf}$ and prove that the maximum value of G is obtained for $(1+b)S^3+2a(1+2b)S^2-(1+2b-3ab)aS-2a^2b=0$.

(6) For the non-steady-state behaviour of the chemostat with $k_e>0$ show that

$$\frac{dY_1}{d\theta}=1-Y_1-\frac{Y_{x/s}x_0}{c_{s_f}}(a+1)n\left(\frac{Y_1Y_2}{Y_1+a}+bY_2\right);\ \theta=tD$$

$$\frac{dY_2}{d\theta}=\ -Y_2\,n(a+1)n\,\frac{Y_1Y_2}{Y_1+a}\quad n=\tau/\tau_{crit}=k\tau/(a+1)$$

Let Y_1 and Y_{10} for $t=0$. Determine $X_0=Y_{x/s}x_0/c_{s_f}$.

Assume that $\tau(t<0)$ is so large that $X_0\sim 0$. Solve the set of coupled o.d.e. analytically.

Compare with the solution of the true set of o.d.e. when $Y_{10}=10^{-6}$, $a=0.2$, $b=0.1$ and $n=1.4$. What is the expected value of Y_1 and of Y_2 for $t\to\infty$?

Solution

(1)

$$\tau=\frac{c_s+K_m}{kc_s}=\frac{S+a}{kS}=\frac{x}{q_x}=\frac{c_s-c_{s_f}}{q_s}=\frac{c_{s_f}(1-S)}{Y_{x/s}q_x}$$

i.e.

$$X=\frac{xY_{x/s}}{c_{sf}}=\underline{1-S}$$

$$Q_2=\frac{xv}{V}=\frac{c}{\tau}=q_x$$

$$G=\frac{Q}{Y_{s/x}kc_{s_f}}=\frac{q_x}{Y_{s/x}kc_{s_f}}=\frac{qx}{c_s+K_m}\frac{1}{Y_{s/x}c_{s_f}}=X\frac{S}{(S+a)}=\frac{S(1-S)}{S+a}$$

The rate of cell production is a maximum when

$$\frac{\mathrm{d}G}{\mathrm{d}S}=0$$

$$\frac{\mathrm{d}G}{\mathrm{d}S}=\frac{(-2S\mathrm{H})(S+a)-S(1-S)}{(S+a)^2}\rightarrow S^2+2as-a=0$$

$$S=-a+\sqrt{a^2+a}$$

$$G|_{S_{\text{opt}}}=1+2a-2\sqrt{a^2+a}=\{\sqrt{1+a}-\sqrt{a}\}^2$$

For $a=0.2$ we obtain $(S,G_{\text{opt}})=(0.290,0.420)$ $X(S)$ and $G(S)$ are shown on Fig. 7.7 for $a=0.2$.

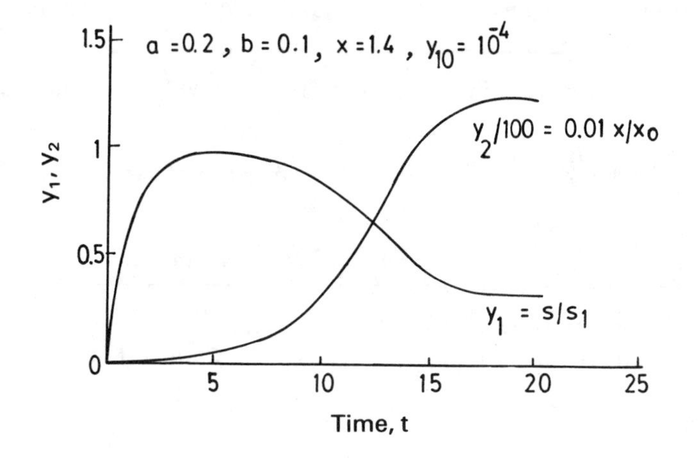

Fig. 7.7 — Chemostat transient with maintenance.

(2)

$$\tau=1\ \mathrm{h}=\frac{0.8+0.2}{0.8\ k}\rightarrow k=1.25\ \mathrm{h}^{-1}$$

$$\tau_{\text{min}}(=1/D_{\text{max}})=\frac{1+0.2}{1.25}=\frac{24}{25}=0.96\ \mathrm{h}$$

$$D_{\text{max}}=\underline{1.042\ \mathrm{h}^{-1}}$$

(3) Since $c_s\approx 0$, $X\simeq 1=X_0$ before start of the transient. As τ is changed to 1 h x must decrease to 0.2 for $t\rightarrow\infty$ corresponding to the S value of 0.8 used in (2) above.

$$\text{Consequently}\ \frac{x(t\to\infty)}{x_0}=0.2;\ n=\frac{\tau}{\tau_{\text{orit}}}=\frac{1}{0.96}$$

$$kt=(a+1)\frac{n}{(n-1)}\left(\ln\frac{x}{x_0}-\frac{na}{n-1+na}\ln\frac{b-x/x_0}{b-1}\right)$$

$$b=\frac{c_{sf}\,Y_{s/x}}{x_0}\frac{(a+1)(n-1)}{(n-1+na)}=1\,\frac{\frac{6}{5}\,\frac{1}{24}}{\frac{1}{24}+\frac{5}{24}}=\frac{1}{5}=X_\infty$$

as shown above

$$kt=30\left(\ln\frac{x}{x_0}-\frac{5}{6}\ln\frac{0.2-x/x_0}{-0.8}\right)=0.4977$$

$$k=0.4977\,\frac{60}{24}=1.244\ \text{h}^{-1}$$

which checks quite well.

For $x/x_0=0.22$, we get

$$1.25t=30\left(\ln 0.22-\frac{5}{6}\ln\frac{0.02}{0.8}\right)$$

$$\underline{t=37.44\ \text{h.}}$$

The uncomfortably large transient indicates how difficult it is to obtain the true steady state measurements of the rate in the chemostat operating under 'normal' fermentation conditions.

(4) Plug the flow reactor formula:

$$k\tau=(1+F)\ln\frac{x}{x_f}-F\ln\frac{c_s}{c_{s_f}}$$

$$F=\frac{K_m}{c_{s_f}+Y_{x/s}\,x_f}.$$

where c_{s_f}, x_f are inlet conditions to the plug flow reactor. If the feed to the plug flow reactor comes from another unit which has processed an original feed from $(c^o_{s_f}, x^o_f=0)$; then with $S_1=\frac{c_{c_f}}{c^o_{s_f}}$; $F=\frac{a}{S_1+X_1}=a$ since $S_1+X_1=1$ when there is no maintenance;

$$a=\frac{K_m}{f^o}=0.2;\ S_1=0.8\ \text{if}\ \tau_{\text{chemostat}}=1\ \text{h}$$

Consequently

$$1.25\,\tau=(1+\text{a})\ln\frac{0.99}{0.2}-a\ln\frac{0.01}{0.8}$$

and $\tau = 2.24$ h.

We can do no better than can the chemostat at its highest rate q_x and then sweep up the remaining conversion in a plug flow reactor.

Chemostat:

$$S_{opt} = -a + \sqrt{a^2 + a} = 0.2899$$

$$\tau_{chemostat} = \frac{0.2899 + 0.2}{1.25 \times 0.2899} = 1.352 \text{ h}$$

Plug flow reactor:

$$\tau = \frac{1}{1.25}\left(1.2 \ln \frac{0.99}{0.71} - 0.2 \ln \frac{0.01}{0.29}\right)$$

$$= 0.858 \text{ h}$$

Total $\tau = 1.352 + 0.858 = \underline{2.210 \text{ h} = \tau_{minimum}}$

(5) For $k_e = bk\, q_s = \frac{xY_{x/s}k}{c_{c_f}}\left(\frac{S}{a+S} + b\right) c_{s_f}$

$$\tau = \frac{x}{qx} = \frac{s+a}{kS} = \frac{c_{s_f} - c_s}{q_s} = \frac{1-S}{Xk\left(\frac{S}{a+S} + b\right)}$$

$$X = \frac{S(1-S)}{S + b(S+a)} = \frac{S(1-S)}{(b+1)S + ab}$$

X is maximum for $(1-2S)\{S(b+1) + ab\} - (b+1)S(1-S) = 0$

$$S_{opt} = \frac{ab}{b+1}\left(-1 + \sqrt{1 + \frac{b+1}{ab}}\right)$$

$$X_{max} = \frac{1 - S_{opt}}{b + 1 + 2ab + (b+1)S_{opt}}$$

for $b = 0.1$, $a = 0.2$; $S_{opt} = 0.1179$ and $X_{max} = 0.6948$.

On Fig. 7.8, X is shown as a function of S for $a = 0.2$, $b = 0.1$. Note that $X = 0$ at both ends of the S interval (0,1).

$$G = \frac{q_x}{Y_{x/s}\, kc_{s_f}} = \frac{S}{S+a} X = \frac{S^2(1-S)}{(S+a)[(b+1)S + ab]}$$

$G(S)$ is also shown on the figure. G_{max} is at $S \approx 0.30$. A numerical scheme for calculating S_{opt} follows by differentiation of G:

$$\frac{dG}{dS} = 0$$

or,

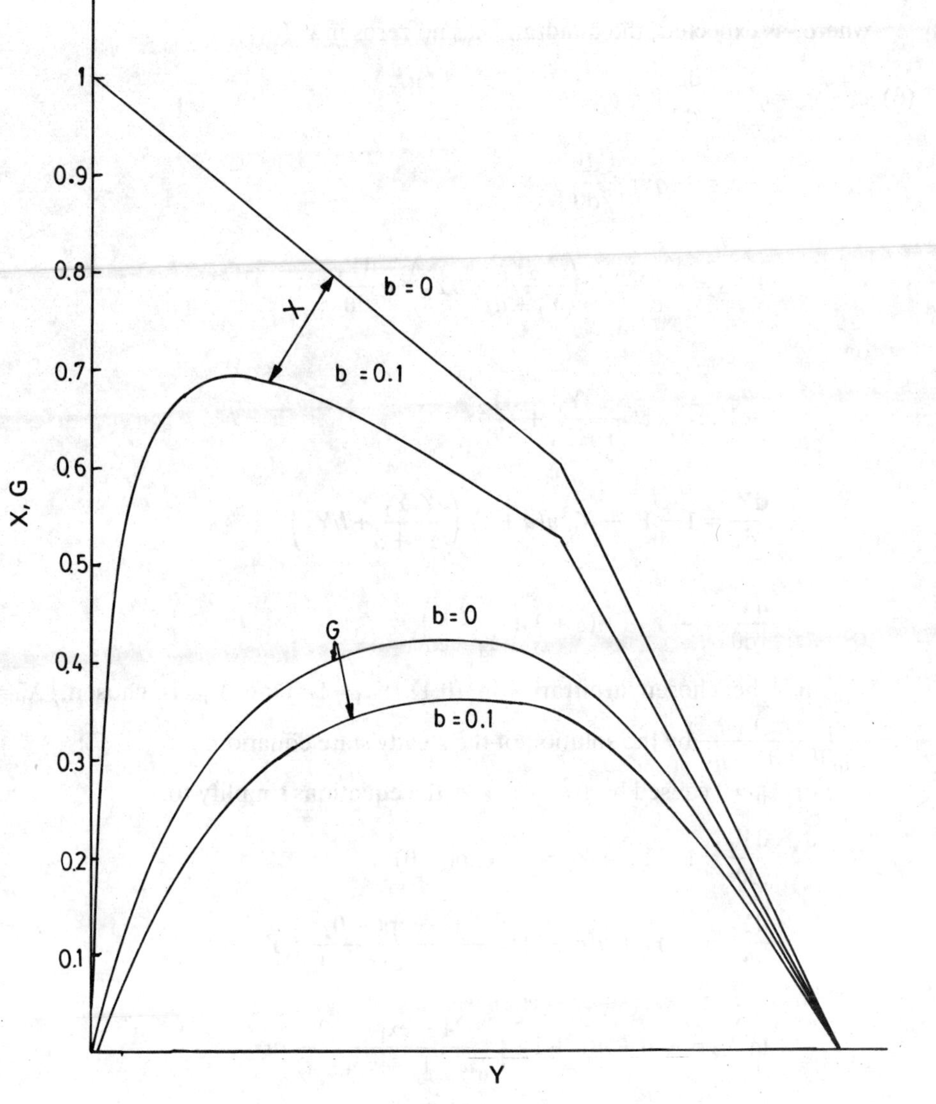

Fig. 7.8.

$$(2S - 3S^2)(a + s)[(1 + b)s + ab] - S^2(1 - s)[2(1 + b)s + a(1 + 2b)] = 0$$

or

$$(1 + b)s^2 - 2a(1 + 2b)s^2 - (1 + 2b - 3ab)as - 2a^2b = 0$$

for $(a,b) = (0.2,0.1)$, one obtains

$$1.1s^3 + 0.48s^2 - 0.228s - 0.008 =$$
$$= (s - 0.3094348)(s^2 + 0.8204s + 0.0258)$$

where, as expected, the quadratic has no zeros in $s \in (0,1)$.

(6) $c_{s_f} - c_s = q_s \tau + \frac{dc_s}{d\theta} \quad \theta = t/\tau$

$$-x = -qx\tau + \frac{dx}{d\theta}$$

$$1 - s = \frac{Y_{x/s}\, x_0}{c_{s_f}} k\tau\left(\frac{Y_2 Y_1}{Y_1 + a} + bY_2\right) + \frac{dY_1}{d\theta} = 1 - Y_1$$

or,

$$-Y_2 = -\tau k \frac{Y_2 Y_1}{a + Y_1} + \frac{dY_2}{d\theta}$$

$$\frac{dY_1}{d\theta} = 1 - Y_1 - X_0\, n(a+1)\left(\frac{Y_2 Y_1}{Y_1 + a} + bY_2\right)$$

$$\frac{dY_2}{d\theta} = -Y_2 + n(a+1)\left(\frac{Y_2 Y_1}{Y_1 + a}\right)$$

Y_{10} may be chosen arbitrarily in $(0,1)$ $Y_{20} \equiv 1$. One Y_{10} is chosen, $X_0 = \frac{Y_{10}(1 - Y_{10})}{Y_{10}0b + 1) + ab}$ by the solution of the steady state equation.

For $X_0 \approx 0$ caused by $\tau(t<0) \to \infty$ the equations simplify to

$$\frac{dY_1}{d\theta} = 1 - Y_1 \to Y_1 = 1 - \exp(-\theta)$$

$$\frac{dY_2}{d\upsilon} = -Y_2 + n(a+1)Y_2 \frac{1 - \exp(-\theta)}{a + 1 - \exp(-\theta)}; \; Y_{20} \equiv 1$$

$$\ln Y_2 = -\theta + n(a+1)\int_0^{\theta} \frac{1 - \exp(-t)}{a + 1 - \exp(-t)}\, dt$$

$$= -\theta + n(a+1)\int_1^{\exp(-\theta)} \frac{(1-t)}{(a+1-t)}\left(\frac{-1}{t}\right) dt; \; t > \exp(-\theta)$$

$$-\theta + n(a+1)\int_1^{\exp(-\theta)} \frac{1}{(a+1)}\left(\frac{a}{a+1-t} - \frac{1}{t}\right) dt$$

$$\ln Y_2 = (n-1)\theta - na \ln \frac{a + 1 - \exp(-\theta)}{a}$$

For small θ:

$$1-\exp(-\theta)\sim\theta \text{ and } \ln\frac{a+1-\exp(-\theta)}{a}\sim\ln\left(1+\frac{\theta}{a}\right)\approx\frac{\theta}{a}$$

Hence for small θ: $\ln Y_2 \approx (n-1)\theta - n\theta = -\theta$. This means that Y_2 decreases for small θ. When θ increases, the first term increases $\alpha\theta$ while the second becomes constant. Hence Y_2 has a minimum for some small θ value. The ever-increasing Y_2 by the approximate formula must be wrong. We know that $(Y_1,Y_2)_\infty$ can be found by solution of

$$1-Y_1 = X_0 n(a+1)\left(\frac{Y_1}{a+Y_1}+b\right)Y_2$$

and

$$Y_2 = n(a+1)\frac{Y_2Y_1}{a+Y_1}$$

$$Y_1=\frac{a}{n(a+1)-1};\ Y_2=\frac{1}{X_0}\frac{(n-1)\ (a+1)}{[n(a+1)-1][1+bn(a+1)]}$$

$$(Y_2X_0)_\infty=\left(\frac{xY_{s/x}}{s_f}\right)_\infty=X_\infty=\frac{(n-1)(a+1)}{[n(a+1)-1][1+bn(a+1)]}$$

For $\begin{cases} a=0.2 \\ b=0.1, \end{cases} n=1.4$

$$Y_1=\frac{0.2}{1.4\times 1.2-1}=0.2941;\ X_\infty=0.6044$$

For very small Y_{10}: $x_0=\dfrac{Y_{10}}{ab}$ and $Y_2(\infty)\sim\dfrac{\text{ab}(0.6044)}{Y_{10}}$

Thus for $Y_{10}=10^{-6}$ the expected value of $Y_2(\infty)$ is 1.2087×10^4, while Y_1 is 0.2941 independently of b and Y_{10}.

X and Y_1 are of course much easier to understand than Y_2 and Y_1.

The computer output and Fig. 7.7 show $Y_1(\theta)$ and $Y_2(\theta)$. The accuracy of the approximate method is indicated.

θ	Y_1	Y_2	Approximate method	
			Y_1	Y_2
0.02	0.0199	0.9817		
0.08	0.0768	0.9426		
0.20	0.1844	0.9034		
0.37	0.3068	0.8922		
0.61	0.4569	0.9150		
0.99	0.6231	0.9954		
1.48	0.7668	1.1577	0.7724	1.161
2.06	0.8647	1.4172		
2.74	0.9253	1.8282	0.9354	1.840
3.56	0.9576	2.5096		

4.41	0.9684	3.5084	0.9878	3.544
5.24	0.9677	4.8569		
6.06	0.9604	6.6847	0.9977	6.840
6.87	0.9479	9.1618		
7.68	0.9298	12.5153		
8.49	0.9047	17.0465		
9.30	0.8707	23.1481		
10.16	0.8251	31.3141		
11.02	0.7645	42.1098		
11.92	0.6862	56.0115		
12.86	0.5905	72.8620		
13.84	0.4878	90.7095		
14.83	0.3990	105.7557		
15.91	0.3363	115.8809		
17.61	0.3002	121.0771		
18.97	0.2946	121.6468		
20.50	0.2939	121.6294		
22.89	0.2940	121.5603		
28.76	0.2941	121.5473		
39.38	0.2941	121.5472		
50.00	0.2941	121.5472		

$a = 0.2$, $b = 0.1$, $n = 1.4$, $Y_{10} = 10^{-4}$

PROBLEM 7.10

A typical rate expression for growth of *Saccharomyces cerevisiae* on glucose is

$$q_x = \frac{kS}{a+S}\frac{b}{b+S}x$$

$$q_s = Y_{s/x}\, kx\left(\frac{S}{a+S}\frac{b}{b+S} + c\right)$$

Here c is the maintenance term while b describes the substrate inhibition. $S = c_s/c_{s_f}$ and $a = K_m/c_{s_f}$ as in Problem 7.9.

(1) Determine (as in Problem 7.9) X and

$$UX = \frac{DX}{k} \text{ as functions of } S,\ a,\ b \text{ and } c$$

For $c = 0$ (no maintenance) $a = 0.2$ and $b = 2$, show that maximum cell production rate is obtained for $S = 1/4$. Find the corresponding values of X and U.

(2) Next consider start-up of the chemostat ($c > 0$) from an original V_0 with $c_s = c_{s0}$ and $x = x_0$. Fore $t > 0$ feed ($c_{s_f}, x = 0$) is supplied at a rate $v(t)$. There is no product removal, i.e. the total mass of cells in the reactor at time $t > 0$ is $V(t)x(t)$. Show that the start-up is described by:

$$\frac{d(RX)}{d\theta} = \frac{S}{a+S}\frac{b}{b+S}RX$$

$$\frac{d(RS)}{d\theta} = U - \left(\frac{S}{a+S}\frac{b}{b+S} + c\right) RX$$

$$\frac{dR}{d\theta} = U$$

$$R = \frac{V}{V_0}, \cdot U = \frac{v}{V_0 k}; \ \theta = kt$$

(3) Show that independent of the value of c the maximum cell mass in the reactor is obtained when $S(t)$ is constant at $\sqrt{ab}$ (or $S = 1$ if $ab > 1$) for $t > 0$.
Show that

$$RX = X_0 \exp\left[\left(1 - \sqrt{\frac{a}{b}}\right)^{-2} \theta\right] \text{ when } S(t) = \sqrt{ab}$$

Finally determine $U(t)$ and $R(t)$.

(4) Discuss a suitable start-up procedure if a given steady state (e.g. the one discussed in question (1) is to be obtained in the best possible way.

Solution

(1) $D = \frac{q_x}{x} = \frac{q_s}{c_{sf} - c_s}$

$$\therefore x = \frac{q_x(c_{s_f} - c_s)}{q_s} = \frac{c_{sf}\, Sb(1-S)}{Y_{s/x}[Sb + c(a+S)(b+S)]}$$

$$X = \frac{xY_{s/x}}{c_{s_f}} = \frac{bS(1-S)}{Sb + c(a+S)(b+S)}$$

$$UX = \frac{1}{k}\frac{q_x}{x}X = \frac{b^2S^2(1-S)}{[Sb + c(a+S)(b+S)](a+S)(b+S)}$$

For $c = 0$

$$UX = \frac{bS(1-S)}{(a+S)(b+S)}$$

$$\frac{d(UX)}{dS} = 0 \Rightarrow -(a+b+1)S^2 - 2abS + ab = 0$$

or,

$$S = \frac{ab}{a+b+1}\left[-1 + \sqrt{1 + \frac{a+b+1}{ab}}\right] = S_{opt}$$

For $a=0.2$ and $b=2$

$$S_{\text{opt}}=\frac{0.4}{3.2}[-1+\sqrt{1+8}]=\frac{1}{4}.$$

$$X=1-S=\frac{3}{4} \text{ and } U=\frac{bS}{(a+S)(b+S)}=\frac{40}{81}$$

(2)

$$\frac{d(Vx)}{dt}=q_xV; \qquad \frac{d(RX)}{d\theta}=\frac{Sb}{(a+S)(b+S)}RX$$

$$\frac{d(Vc_s)}{dt}=vc_{s_f}-Y_{s/x}(q_x+k_ex)V$$

$$\frac{d(RS)}{d\theta}=U-\left(\frac{S}{a+S}\frac{b}{b+S}+c\right)RX.$$

$$\frac{dV}{dt}=\upsilon \Rightarrow \frac{dR}{d\theta}=U$$

(3) From the cell mass balance:

$$\frac{d[\ln(RX)]}{d\theta}=\frac{bS}{(a+S)(b+S)}$$

$\ln(RX)$ is a monotonously increasing function of Vx. Hence, if at any time RX is as large as possible this is also the case for Vx.

The rate of increase of RX is as large as possible if

$$\frac{d}{dS}\left[\frac{S}{(a+S)(b+S)}\right]=0 \Rightarrow -bS^2+b^2a=0 \Rightarrow S=\sqrt{ab}$$

hence, whatever the value of X_0 we obtain the fastest increase of Vx, and consequently the largest value of (cells m^{-3}) when S is kept at $\sqrt{ab}$ for all $t>0$. If $ab>1 \Rightarrow S1$

Since S is constant, $\frac{dS}{d\theta}=0$

$$\frac{dS}{d\theta}=\frac{1}{R}\left[\frac{d(RS)}{d\theta}-S\frac{dR}{d\theta}\right]=\frac{1}{R}\left[\frac{d(RS)}{d\theta}-SU\right]=0$$

$$U(1-S)=\left[\frac{S}{(a+S)}\frac{b}{(b+S)}+c\right]RX \text{ with } S=\sqrt{ab}$$

i.e.

$$U=\frac{(1+\sqrt{a/b})^{-2}+c}{1-\sqrt{ab}}RX$$

$$\frac{d(\ln RX)}{d\theta} = \left\{\frac{bS}{(a+S)(b+S)}\right\}_{S=\sqrt{ab}} = \left(1+\sqrt{\frac{a}{b}}\right)^{-2}$$

$$RX = 1 \times X_0 \exp\left[\left(1+\sqrt{\frac{a}{b}}\right)^{-2}\theta\right]$$

and new

$$U = X_0 \exp\left[\left(1+\sqrt{\frac{a}{b}}\right)^{-2}\theta\right] \frac{\left(1+\sqrt{\frac{a}{b}}\right)^{-2}+c}{1-\sqrt{ab}}$$

Finally

$$\frac{dR}{d\theta} = U \Rightarrow R = 1 - X_0\left\{1-\exp\left[\left(1+\sqrt{\frac{a}{b}}\right)^{-2}\theta\right]\right\}\frac{1+c\left(1+\sqrt{\frac{a}{b}}\right)^{2}}{1-\sqrt{ab}}$$

(4) Note that $X(\theta \to \infty) = \left(\frac{RX}{R}\right)_{\theta\to\infty} = \frac{1-\sqrt{ab}}{1+c(1+\sqrt{a/b})^2}$. Thus, the largest attainable X is $(1-\sqrt{ab})$ if we follow the procedure described under (3) for $c = 0$.

For $a = 0.2$, $b = 2$, X_{max} becomes $1 - 0.6325 = 0.3675$ and we cannot reach the steady state of (1) ($X = 0.75$) following the procedure of (3) alone. Assume that we stop the fed-batch procedure when $X = 0.25$. θ is found from the expression for $\left(\frac{RX}{R}\right)$

$$0.25 = \frac{X_0\beta}{1+X_0(\beta-1)2.72}; \quad \beta = \exp(0.5772\theta)$$

or for $X_0 = 10^{-5}$, $\theta = 19.518$ and $R = 1 + 2.125 = 3.125$, i.e. $V = 3.125\ V_0$

Next with *no feed* S is allowed to convert to X.

$$\frac{dX}{dt} = \frac{q_x Y_{s/x}}{c_{sf}} = -\frac{dS}{dt} \rightarrow X + S = 0.25 + 0.6325 = 0.8825$$

$$\frac{dS}{d\theta} = -\frac{S}{(a+S)}\frac{b}{(b+S)}(0.8825 - S)$$

with $S = 0.6325$ for $\theta = 19.518$.

$$-\frac{1}{b}\left[\frac{A+S}{(0.8825-S)} + \frac{B}{S}\right] dS = d\theta \text{ with } A = a + b + \left(\frac{0.8825}{ab}\right)^{-1}$$

and,

$$B = \left(\frac{0.8825}{ab}\right)^{-1}$$

$$= \left(\frac{1}{2} - \frac{1}{2}\frac{3.5357}{(0.8825 - S)} - \frac{0.4532}{2S}\right) \mathrm{d}S$$

$$\theta = 19.518 + \frac{1}{2}\left\{(S - S_0) + 3.5357 \ln\left(\frac{0.8825 - S}{0.8825 - S_0}\right) - \ln\frac{S}{S_0}\, 0.4532\right\}$$

$$= \left\{19.201 + \frac{1}{2}S + 1.7678 \ln\left(\frac{0.8825 - S}{0.25}\right) - \right.$$

$$\left. - 0.2266 \ln\frac{S}{0.6325}\right\}$$

Assume that we wish to end up with V_{final}. We follow the batch procedure until S has reached a value where dilution with a volume V_f of feed with $S = 1$ brings X to 0.75 and S to 0.25. Then we start feed again, following the solution to question 1 discussed earlier.

$$VS + V_+ \times 1 = V_{final}(0.25).\ (V = 3.125\ V_0)$$

$$V(0.8825 - S) + V_+ \times 0 = V_{final}\ (0.75).$$

$$V + V_+ = V_{final}$$

The solution of this set of equations yields V (and hence V_0), V_f, the pulse added for the final balancing of S and X, and S. Then from the equation for $\theta(S)$ above one may determine θ where the continuous reactor operation may start.

This mixed procedure (fed batch $\rightarrow$ batch $\rightarrow$ pulse addition of fresh feed) can, of course only yield an ad hoc solution to an optimal start-up procedure. The switch between fed-batch and batch operation is seen to be arbitrarily set (at $X = 0.25$). Other switch points might be explored or one might compare with a batch operation from $\theta = 0$ until addition of a feed pulse brings (with some luck) X and S to their desired value at $V = V_{final}$. Lots of ideas may be explored!

8

Bioprocess simulation and control

Contributor: T. Shioya

PROBLEM 8.1

Deduce the general system equations including cell mass, substrate and product concentrations for batch, fed-batch and continuous stirred tank reactors. Use the normalized dimensionless variables which represent the system characteristics.

Solution

From the mass balance, the following equations can be derived.

$$\frac{d(Vx)}{dt} = \mu Vx - F_o x \tag{8.1.1}$$

$$\frac{d(VC_s)}{dt} = -\mu Vx/Y_{x/s} - \rho Vx/Y_{p/s} + F_i C_{s_f} - F_o C_s \tag{8.1.2}$$

$$\frac{d(VC_p)}{dt} = \rho Vx - F_o C_p \tag{8.1.3}$$

$$\frac{dV}{dt} = F_i - F_o \tag{8.1.4}$$

where x = cell concentration in the reactor ($kg\,m^{-3}$)
c_s = substrate concentration in the reactor ($kg\,m^{-3}$)
c_p = product concentration in the reactor ($kg\,m^{-3}$)
V = liquid volume (m^3)
F_i = inlet flow rate ($m^3\,h^{-1}$)
F_o = outlet flow rate ($m^3\,h^{-1}$)
c_{sf} = substrate concentration in the feed ($kg\,m^{-3}$)
$Y_{x/s}$ = yield coefficient for cell growth (kg cell kg^{-1} substrate)
$Y_{p/s}$ = yield coefficient for product (kg product kg^{-1} substrate)
μ = specific growth rate (h^{-1})
ρ = specific production rate of the product (kg product kg^{-1} cell h^{-1})

By arranging $F_i = F_o = 0$ for a batch culture, $F_i \neq 0$, $F_o = 0$ for a fed-batch culture and $F_i = F_o \neq 0$ for a continuous stirred tank reactor (CSTR), equations (8.1.1)–(8.1.4) can be used for every operation of the fermenter. For equations (8.1.1)–(8.1.4), we can rewrite the system equations to equations (8.1.5)–(8.1.8) introducing the normalized variables such as,

$$x' = x/x(0) \ , \quad s = Y_{x/s}C_s/x(0) \ , \quad s_f = Y_{x/s}C_{s_f}/x(0) \ ,$$

$$s(0) = Y_{x/s}c_s(0)/x(0) \ , \quad p = c_p/x(0) \ , \quad \gamma = (Y_{x/s})/(Y_{p/s}) \ ,$$

$$\tau = \mu_m t$$

$$\mu = \mu/\mu_m \ , \quad \rho = \rho/\mu_m \ , \quad \upsilon = V/V(0)$$

$$D_i = F_i/V(0)\mu_m \ , \quad D_o = F_o/V(0)\mu_m \ ,$$

where, $x(0)$, $V(0)$, μ_m are nominal cell concentration, liquid volume and specific growth rate, respectively. Usually $x(0)$, $V(0)$ are taken as initial value and μ_m is taken as maximum specific growth rate.

$$\frac{d(\upsilon x')}{d\tau} = \mu\upsilon x' - D_o x' \tag{8.1.5}$$

$$\frac{d(\upsilon s)}{d\tau} = -\mu\upsilon x' - \gamma\rho\upsilon x' + D_i s_f - D_o s \tag{8.1.6}$$

$$\frac{d(\upsilon p)}{d\tau} = \rho\upsilon x' - D_o p \tag{8.1.7}$$

$$\frac{d\upsilon}{d\tau} = D_i - D_o \tag{8.1.8}$$

Finally, equations (8.1.5)–(8.1.8) are a general experession of the batch and continuous cultures to be considered hereafter.

PROBLEM 8.2

When the specific growth rate is represented by a Monod-type equation for the batch cultivation of *Saccharomyces cerevisiae* which assimilates ehtanol as a carbon source (Mor & Fiechter, 1968).

(a) Calculate the final cell concentration for given operation time τ_f and initial condition $x(0)$ $(=1)$, and $c_s(0)$.
(b) Check the linearity, that is, when initial concentration $x(0)$ and $c_s(0)$ becomes twice, does $x(t_f)$ becomes two times larger than that in case (a)?
(c) Establish the above things by numerical calculations where $K_s = 3$(mmol/l), $\mu_m = 0.18(\mathrm{h}^{-1})$, $Y_{x/s} = 0.1$ (g cell/mmol).

Reference
Mor, J. R. & Fiechter, A. (1968) *Biotechnol. & Bioeng.*, **10**, 159.

Solution

(a) According to the normalized general equations (8.2.1) and (8.2.2),

$$\frac{d(\upsilon x')}{d\tau} = \mu\upsilon x' - D_0 x' \tag{8.2.1}$$

$$\frac{d(\upsilon s)}{d\tau} = -\mu\upsilon x' - \gamma\rho\upsilon x' + D_i s_f - D_0 s \tag{8.2.2}$$

system equations can be derived by setting $D_o = D_i = 0$ and $\gamma = 0$. For the Monod equation, μ can be written as

$$\mu = \frac{s}{k+s} \tag{8.2.3}$$

where $k = K_s Y_{x/s}/x(0)$ and K_s is Monod constant. Equations (8.2.1) and (8.2.2) becomes

$$\frac{dx'}{d\tau} = \frac{s}{k+s} x' \tag{8.2.4}$$

$$\frac{ds}{d\tau} = -\frac{s}{k+s} x' \tag{8.2.5}$$

From equations (8.2.4) and (8.2.5)

$$\frac{dx'}{d\tau} + \frac{ds}{d\tau} = 0 \tag{8.2.6}$$

then,

$$s(\tau) = x'(0) + s(0) - x'(\tau) = z_o - x'(\tau) \tag{8.2.7}$$

where

$$z_o \equiv x'(0) + s(0) = 1 + s(0) \ . \tag{8.2.8}$$

Substituting equation (8.2.8) into (8.2.4),

$$\frac{dx'}{d\tau} = \frac{z_o - x'}{k + z_o - x'} \tag{8.2.9}$$

From equation (8.2.9)

$$\left\{ \frac{k/z_o}{z_o - x'} + \frac{(1 + k/z_o)}{x'} \right\} dx' = d\tau \tag{8.2.10}$$

Integrating equation (8.2.10)

$$\ln \left\{ \left(\frac{x'_f}{(z_o - x'_f)/s(0)} \right)^{k/z_o} x'_f \right\} = \tau_f \tag{8.2.11}$$

Equation (8.2.11) is the solution of equations (8.2.4) and (8.2.5).

(b) If setting initial concentrations to $\tilde{x}(0)=2x(0)$ and $\tilde{C}_s(0)=2C_s(0)$, the normalized variables become as:

$$\tilde{s}(0)=s(0)\ , \qquad \tilde{z}_o=z_o\ , \qquad x'(0)=1\ , \qquad \tilde{k}=k/2\ .$$

Then, at τ_f, $\tilde{x}(\tau_f)=2x(\tau_f)$ does not hold because

$$\tau_f \neq \ln\left\{\left(\frac{\tilde{x}_f}{(\tilde{z}_o-\tilde{x}_f)/\tilde{s}(0)}\right)^{k/2\tilde{z}_o}\tilde{x}_f\right\}$$

This means the system equation (8.2.4) and (8.2.5) are not linear but non-linear.

(c) Now for initial condition: $x(0)=0.1\ \mathrm{kg\ m^{-3}}$, $c_s(0)=10\ \mathrm{kg\ m^{-3}}=0.217(\mathrm{M})$ are taken. Then normalized initial conditions and parameters are given as:

$$Y_{x/s}=0.1(\text{g cell/mmol})=100(\text{g cell/mol}).\ x'(0)=1,$$
$$s(0)=c_s(0)Y_{x/s}/x(0)=0.217\times 100/0.1=217.$$
$$k=K_sY_{x/s}/x(0)=3\times 0.1/0.1=3\ , \qquad z_0=x'(0)+s(0)=218\ .$$

(i) First, when x'_f is given as $x_f=200$ $\{x(\tau_f)=20(\mathrm{kg\ m^{-3}})\}$, that is, $s(\tau_f)=18$, batch operation time τ_f can be calculated from equation (8.2.11)

$$\tau_f=\ln\left\{\left(\frac{200}{(218-200)/217}\right)^{3/218}200\right\}$$

$$=5.40 \qquad (t_f=30.0\ (\mathrm{h}))$$

(ii) Next, when initial conditions are increased two times larger than the first cases ($x(0)=0.2\ \mathrm{kg\ m^{-3}}$, $C_s(0)=20\ \mathrm{kg\ m^{-3}}$, normalized initial condition do not change such as: $\tilde{s}(0)=0.434\times 100/0.2=217$, $\tilde{z}_o=218$. But the parameter $\tilde{k}$ should be changed as

$$\tilde{k}=K_sY_{x/s}/x(0)=3\times 0.1/0.2=1.5\ .$$

Then the operation time τ_f required for achieving $x'_f=200$ $\{x(\tau_f)=40\ \mathrm{kg\ m^{-3}}\}$ is calculated as

$$\tau_f=\ln\left\{\left(\frac{200}{(218-200)/217}\right)^{1.5/218}200\right\}$$

$$=5.35 \qquad (t_f=29.7\ \mathrm{h})$$

Then, x'_f becomes larger than that in case (i) when $\tau_f=5.40$. That is, when initial condition $x(0)$ and $c_s(0)$ become twice, $x(\tau_f)$ is more than two times. For another final concentration x'_f, the required time τ_f is given depending on the initial concentration as follows:

$$x'_f=217.5\ ; \qquad x(0)=0.1\ , \quad c_s(0)=10\ , \quad \tau_f=5.54\ .$$
$$x(0)=0.2\ , \quad c_s(0)=20\ , \quad \tau_f=5.46\ .$$

Then non-linearity will appear more strongly in larger x'_f. In any case the non-linearity of the system equations (8.2.4) and (8.2.5) was also shown numerically.

PROBLEM 8.3

Find an optimum condition for

(a) maximum productivity of the cell and
(b) maximum productivity of the products.
(c) Compare the cell productivity obtained here with the one in batch culture where the specific growth is given as a Monod-type equation. Use the same parameters as for problem 8.2.

Solution

(a) As already shown in general equation in a CSTR, the system equation becomes as,

$$\frac{dx'}{d\tau} = (\mu - D)x' \tag{8.3.1}$$

$$\frac{ds}{d\tau} = -\mu x' + D(s_f - s) \tag{8.3.2}$$

where

$$\mu = \frac{s}{k+s} \tag{8.3.3}$$

And cell productivity Q_x will be calculated as,

$$Q_x = Dx' \tag{8.3.4}$$

From equations (8.3.1) and (8.3.2)

$$s = \frac{kD}{1-D}$$

$$x' = s_f - s = \frac{s_f - D(s_f + k)}{1-D} \tag{8.3.5}$$

Then Q_x becomes as

$$Q_x = \frac{Ds_f - D^2(s_f + k)}{1-D} \tag{8.3.6}$$

In order to get maximum value of Q_x with respect to D, differentiating Q_x by D,

$$\frac{\partial Q_x}{\partial D} = \frac{(-2D(k+s_f) + s_f)(1-D) - D^2(k+s_f) + s_f D}{(1-D)^2}$$

$$= \frac{D^2(k+s_f) - 2D(k+s_f) + s_f}{(1-D)^2} \tag{8.3.7}$$

As the necessary condition of maximizing Q_x is $\partial Q_x/\partial D=0$, the numerator of equation (8.3.7) should be zero. Then D is given by

$$D=\frac{(k+s_f)-\sqrt{(k+s_f)^2-(k+s_f)s_f}}{(s_f+k)}$$

$$=1-\sqrt{k/(s_f+k)} \tag{8.3.8}$$

because D should be less than 1. And s is written as

$$s=-k+\sqrt{k(s_f+k)} \tag{8.3.9}$$

By confirming $\partial^2 Q_x/\partial D^2<0$, D given by equation (8.3.8) becomes the optimal condition which maximizes the cell productivity.

(b) For the production, the system equation becomes as

$$\frac{dx'}{d\tau}=(\mu-D)x'=0 \tag{8.3.10}$$

$$\frac{ds}{d\tau}=-\mu x'-\gamma\rho x'+D(s_f-s)=0 \tag{8.3.11}$$

$$\frac{dp}{d\tau}=\rho x'-Dp=0 \tag{8.3.12}$$

$$Q_p=Dp \tag{8.3.13}$$

Using equation (8.3.10), (8.3.11) and (8.3.12),

$$s_f=x'+\gamma p+s \tag{8.3.14}$$

From equation (8.3.12)

$$p=\rho x/D \tag{8.3.15}$$

Then,

$$x'=(s_f-s)/(1+(\rho/D)\gamma) \tag{8.3.16}$$

That is,

$$Q_p=Dp$$

$$=\rho\frac{(s_f-s)}{(1+(\rho/\mu)\gamma)}$$

$$=\frac{\mu\rho(s_f-s)}{\mu+\rho\gamma} \tag{8.3.17}$$

By solving the following equation

$$\frac{\partial Q_p}{\partial s}=0 \tag{8.3.18}$$

we will get a necessary condition for the maximum productivity. If ρ can be represented by $\rho = K\mu$, the optimal condition for maximum productivity coincides with one given by equation (8.3.8).

(c) By comparing the cell productivity of the fed-batch culture, at first, cell productivity in the CSTR is calculated as,

$$\begin{aligned} Q_x|_{CSTR} &= (1-\sqrt{k/(s_f+k)})(s_f+k-\sqrt{k(s_f+k)}) \\ &= s_f+k+k-\sqrt{(s_f+k)k}-\sqrt{k(s_f+k)} \\ &= (\sqrt{s_f+k}-\sqrt{k})^2 \end{aligned} \tag{8.3.19}$$

if $C_{sf} \gg k$,

$$Q_x|_{CSTR} \doteqdot s_f \tag{8.3.20}$$

On the other hand, cell productivity in the batch culture is,

$$Q_x|_{batch} = \frac{x(\tau_f)-1}{\tau_f+\tau_b} \tag{8.3.21}$$

where τ_b is processing time for batch operation, i.e. which is required for washing, sterilization, feeding and so on. If batch operation time τ_f is large, $x'(\tau_f)-1 = s(0)$ and if initial substrate concentration of batch culture, $s(0)$ is equal to s_f,

$$Q_x|_{batch} \doteqdot s_f/(\tau_f+\tau_b) \ .$$

Then

$$Q_x|_{CSTR}/Q_x|_{batch} \doteqdot \tau_f+\tau_b \tag{8.3.22}$$

The processing time, τ_b, is an important factor in the rational operation of the batch fermenter when τ_b becomes large. CSTR then becomes superior to batch operation. For the numerical calculations, the same parameters are used. That is, $s_f = s(0) = 217$, $k = 3$, $\tau_b = 0.18$, $\tau_f = 5.54$ (optimal batch operation time). Then,

$$Q_x|_{CSTR} = (\sqrt{220}-\sqrt{3})^2 = 171.6$$

$$Q_x|_{batch} = (217.5-1)/(5.54+0.18) = 37.85$$

$$Q_x|_{CSTR}/Q_x|_{batch} = \underline{4.53}$$

That is, cell productivity in CSTR is about 4.5 times higher than that in the batch operation.

PROBLEM 8.4

Check the stability condition for an equilibrium point of CSTR in the following cases:

(a) yield coefficient is constant
(b) yield coefficient is a function of substrate S.

Solution

The stability of the equilibrium point can be evaluated generally using a linear dynamic system theory. Let us consider the following differential equation,

$$\dot{\mathbf{X}} = \mathbf{f}(\mathbf{X}) \tag{8.4.1}$$

where $\mathbf{X}$ is an n-dimensional state vector. Around the equilibrium point $\mathbf{X}^*$ which satisfies equation (8.4.2)

$$f(\mathbf{X}^*) = \mathbf{0} \tag{8.4.2}$$

the deviation of $\mathbf{X}$ which is rewritten as $\mathbf{Y}$ can be represented by the following linear differential equation

$$\dot{\mathbf{Y}} = \mathbf{AY} \tag{8.4.3}$$

where

$$\mathbf{A} \triangleq \left(\frac{\partial \mathbf{f}}{\partial \mathbf{X}}\right)_{\mathbf{X}=\mathbf{X}^*}$$

When the system starts from an arbitrary initial state $\mathbf{Y}(0) = \mathbf{0}$, if $\mathbf{Y}$ approaches to the origin $\mathbf{Y} = \mathbf{0}$, that is, to the equilibrium point, the equilibrium point $\mathbf{X}^*$ is called as stable one (mathematically, strictly speaking, it is asymptotically stable). The condition for stable equilibrium point is that all eigenvalues of matrix $\mathbf{A}$ have negative real parts or that matrix $\mathbf{A}$ is negative definite.

(a) For the system equations of CSTR, $\mathbf{A}$ becomes as

$$\mathbf{A} = \begin{bmatrix} \mu - D, & \mu_s x' \\ -\mu, & -\mu_s x' - D \end{bmatrix}$$

$$= \begin{bmatrix} 0, & \mu_s x' \\ -\mu, & -\mu_s x' - D \end{bmatrix} \tag{8.4.4}$$

because $\mu = D$, where μ_s means derivative of μ with respect to s. Eigenvalue λ of $\mathbf{A}$ should satisfy,

$$\lambda^2 + \lambda(\mu_s x' + D) + \mu_s \mu x' = 0 \tag{8.4.5}$$

Then,

$$\lambda_i = \frac{-(\mu_s x' + \mu) \pm \sqrt{(\mu_s x' + \mu)^2 - 4\mu_s \mu x'}}{2}$$

$$= \frac{-(\mu_s x' + \mu) \pm (\mu_s x' - \mu)}{2}$$

$$= -\mu \quad \text{or} \quad -\mu_s x' \tag{8.4.6}$$

From equation (8.4.6), if $\mu_s > 0$, eigenvalues $\lambda_i (i = 1,2)$ are negative. That is, if μ_s is positive, as is true in Monod-type equation, the equilibrium state ($\mu = D$) is stable. For another equilibrium point $x' = 0$, $s = s_f$, which is called as wash-out point, $\mathbf{A}$ becomes,

$$\mathbf{A} = \begin{bmatrix} \mu - D, & 0 \\ -\mu, & -D \end{bmatrix} \tag{8.4.7}$$

Then, the characteristic equation of **A** becomes

$$(\mu - D - \lambda)(-D - \lambda) = 0 \tag{8.4.8}$$

that is,

$$\lambda = \mu - D \quad \text{or} \quad -D$$

If $D > \theta(c_{s_f})$, the equilibrium point is stable, that is, the wash-out steady state is stable.

On the other hand, if μ is represented by

$$\mu = \frac{s}{(k_s + s + s^2/k_I)} \tag{8.4.9}$$

which is called as substrate inhibition kinetics, we have three equilibrium points. One is trivial wash-out steady state which is stable if

$$D > \frac{\sqrt{k_s k_I}}{2k_s + \sqrt{k_s k_I}}$$

In terms of the other two equilibrium points, larger one is unstable because $\mu_s < 0$ which means the eigenvalue is positive. However, the smaller one is stable as seen in equation (8.4.6) because $\mu_s > 0$.

(b) If the yield coefficient is written as a function of s, as

$$Y_{x/s} = as + b \tag{8.4.10}$$

where μ is given by equation (8.4.9), the system equation has limit cycle at a certain D and also have bifurcation points. For example, for $a = 0.1$, $b = 0.05$, $k_I = 15$, $D = 0.18$, $k = 0.5$, $s_f = 5$, it can be shown that the system has a limit cycle.

PROBLEM 8.5

Find the optimal start-up policy which minimizes the start-up time required for moving the state from the appropriate initial state to the final state (x'_t, s_t). It is assumed that the specific growth rate μ is represented by a substrate inhibition type equation and is written as follows in dimensionless form,

$$\mu = bs - as^2 \tag{8.5.1}$$

It is also assumed that the initial state always satisfies the following relationship:

$$x'(0) + s(0) = s_f \tag{8.5.2}$$

Also, the total amount of $\upsilon(0)x'(0)$ available at the initial time is fixed. Then one of $\upsilon(0)$, $x'(0)$ and $s(0)$ can be arbitrarily chosen and the other two variables will be determined to satisfy equation (8.5.2) and $\upsilon(0)x'(0)$ = given constant. And the flow rate D has upper and lower limit

$$0 \leqslant D \leqslant D_u \tag{8.5.3}$$

Parameters employed here are:

$$a = 2.0 \ , \quad b = 1.6 \ , \quad s_f = 0.8 \ ,$$

$$s(\tau_f) = 0.2 \ , \quad \upsilon(\tau_f) = 1.0 \ , \quad D_u = 0.4 \ , \quad \upsilon(0)x'(0) = 0.04 \ .$$

Solution

The system equation of the fed-batch culture can be simplified using the condition of equation (8.5.2) as,

$$\frac{ds}{d\tau} = (-\mu + D/\upsilon)(s_f - s) \tag{8.5.4}$$

$$\frac{d\upsilon}{d\tau} = D \tag{8.5.5}$$

because for any time, x' can be written by,

$$x' = s_f - s \tag{8.5.6}$$

according to the condition of equation (8.5.2).

Objective function J is defined as

$$J = -\int_0^{\tau_f} dt \tag{8.5.7}$$

The problem can be deduced to: to find the initial condition of $s(0)$ and $\upsilon(0)$ and the control policy of $D(\tau)$ $(0 < \tau < \tau_f)$ so as to arrive at the final state (s_f, υ_f) taking the minimum time, which means to maximize J in equation (8.5.7). Time optimal control can be solved by the well-known Maximum principle (Pontryagin, 1962). The theory can be simply summarized as follows.

Let us consider the following system

$$\dot{\mathbf{X}} = \mathbf{f}(\mathbf{X}, \mathbf{u}) \tag{8.5.8}$$

$$\mathbf{u}_{min} \leqslant \mathbf{u} \leqslant \mathbf{u}_{max}$$

$$\mathbf{X}(\mathbf{O}) = \text{given}$$

$$\mathbf{X}(\tau_f) = \text{given}$$

and the objective function to be maximized is written as,

$$J = \int_0^{\tau_f} g(\mathbf{X}, \mathbf{u})\, dt \tag{8.5.9}$$

For the time optimal control case, $g(\mathbf{X}, \mathbf{u}) = -1$. Introducing the Hamiltonian, H, defined as

$$H = g(\mathbf{X}, \mathbf{u}) + \boldsymbol{\lambda}^{\mathrm{T}}\mathbf{f} \tag{8.5.10}$$

where $\boldsymbol{\lambda}$ is the adjoint vector which satisfies the following equation.

$$\frac{\mathrm{d}\boldsymbol{\lambda}^{\mathrm{T}}}{\mathrm{d}\tau} = -\frac{\partial H}{\partial \mathbf{X}} \tag{8.5.11}$$

where superscript T means transposition of the vector, the optimal control **u** should maximize the Hamiltonian, H. The maximum principle is named after this condition. Usually the condition can be deduced that the optimal **u** satisfies

$$\frac{\partial H}{\partial \mathbf{u}} = 0 \tag{8.5.12}$$

However, when H is a linear function of **u**, then **u** should take $\mathbf{u}_{\max}$ or $\mathbf{u}_{\min}$ depending on the coefficient of **u** in the Hamiltonian, H. This control is called bang-bang control. On the other hand, if the coefficient of **u** becomes zero within a certain time, the maximization of H cannot give any information. This case is called singular control and the other discussion will be needed for decision of control **u** (Kelly *et al.*, 1967).

Based on the Maximum principle, the Hamiltonian in our problem becomes,

$$H = -1 + \lambda_1(-\mu + D/\upsilon)(s_{\mathrm{f}} - s) + \lambda_2 D \tag{8.5.13}$$

and the adjoint system λ_{i} should satisfy the following equation,

$$\frac{\mathrm{d}\lambda_1}{\mathrm{d}\tau} = -\frac{\partial H}{\partial s} = (\mu_{\mathrm{s}}(s_{\mathrm{f}} - s) + (-\mu + D/\upsilon))\lambda_1 \tag{8.5.14}$$

$$\frac{\mathrm{d}\lambda_2}{\mathrm{d}\tau} = -\frac{\partial H}{\partial \upsilon} = \lambda_1 D(s_{\mathrm{f}} - s)/\upsilon^2 \tag{8.5.15}$$

where the suffix ‘s’ means the derivative with respect to s and initial and terminal conditions of λ_{i} are all unknown. Then the time optimal control, D, becomes

$$D = \begin{cases} 0 & \text{if } \phi < 0 \\ D_{\mathrm{u}} & \text{if } \phi > 0 \\ D_{\mathrm{sing}} & \text{if } \phi \equiv 0 \end{cases} \tag{8.5.16}$$

for an initial condition $\upsilon(0)$ and $s(0)$, where ϕ is a coefficient of D in equation (8.5.13) given as,

$$\phi \triangleq \lambda_1(s_{\mathrm{f}} - s)/\upsilon + \lambda_2 \tag{8.5.17}$$

Singular control D_{sing} will be derived by setting $\phi \equiv 0$. It means that $\phi = \dot{\phi} = \ddot{\phi} = \dddot{\phi} = \ldots = 0$. From $\dot{\phi} = 0$, the following equation is derived.

$$\lambda_1 \mu_{\mathrm{s}}(s_{\mathrm{f}} - s)^2/\upsilon = 0 \tag{8.5.18}$$

Then,

$$\mu_s = 0 \text{ that is } s^* = b/2a \tag{8.5.19}$$

is a singular state and it corresponds to the value which maximizes μ. After all, D_{sing} should be exponentially increasing feed rate (exponential fed-batch culture) in order to keep the substrate concentration s at s^*. Then,

$$D = \upsilon(0)\mu^* \exp(\mu^*\tau) \tag{8.5.20}$$

where

$$\mu^* = b^2/4a$$

According to the optimal control policy given by equation (8.5.16) we should solve basically two point boundary value problem because $s(0)$, $\upsilon(0)$, $s(\tau_f)$ and $\upsilon(\tau_f)$ are given but $\lambda_i(0)$, $\lambda_i(\tau_f)$ are unknown, and compare it with the singular control. In our case, $s(0)$ can be chosen arbitrarily. Then the initial state can be taken at singular state. After due consideration of the singular control and bang-bang control, time optimal control can be derived as follows: (1) Start from the singular state ($s = s^* = 0.4$, $\upsilon(0) = 0.1$) and follow the singular control which is the exponential fed-batch culture to keep $s = s^* = 0.4$. The duration time of singular control is 7.2.(2) At the point $\upsilon(\tau) = \upsilon(\tau_f) = 1$, fed-batch culture should be changed to batch culture ($D = 0$). (3) The minimum start-up time τ_f required to get the final state ($s(\tau_f) = 0.2$, $\upsilon(\tau_f) = 1.0$) is 8.6. For the sake of contrast, let us consider the case that the batch culture is taken as the start-up policy. In this case $\upsilon(0) = 1.0$, then $x(0) = 0.04$ and $s(0) = 0.76$. By the batch culture, the required time for assimilating s from initial to 0.2 is about 18. Then, the start-up time required by the batch operation is about twice as long as by the optimal one. Using the technique employed here, several dynamical optimization problems have been solved; for example, see Takamutsu *et al.* (1975). However, it is still difficult to achieve optimal control in high dimensional systems.

References

Pontryagin (1962) *The Mathematical Theory of Optimal Process*, (translated by K. N. Trirogoff), Interscience, New York.

Kelly, H. J., Kopp, R. E. and Moger, H. G. (1967) Singular Extremals, in *Topics in Optimization* (ed. G. Leitmann, Vol. 63, Academic Press, New York.

Takamatsu, T., Hashimoto, I., Shioya, S. and Ohno, H. (1975) *Automatica* **11**, 141.

PROBLEM 8.6

In a two-stage CSTR, find an optimal volume of each tank υ_i which gives the minimum total volume of two tanks in order to get the desired final cell concentration when the specific growth rate is written by first order equation such as

$$\mu = as \tag{8.6.1}$$

The parameters employed hereafter are: input substrate concentration $s_f = 200$, desired cell concentration in second tank $x_2' = 190$ and $a = 0.001$. The Lagrangian method for deducing the optimal condition is recommended here for the demonstration.

Solution

In a two-stage CSTR, the system equation becomes as

$$\frac{dx_i'}{d\tau} = \mu_i x_i' + D_i(x_{i-1}' - x_i') = 0 \qquad i = 1, 2 \tag{8.6.2}$$

$$\frac{ds_i}{d\tau} = -\mu_i x_i' + D_i(s_{i-1} - s_i) = 0 \tag{8.6.3}$$

$$\mu_i = a s_1$$

where $x_0' = 0$, $s_o = s_f$ (substrate concentration in the feed). As the objective function, J, is the total tank volume, J is written as,

$$J = V_1 + V_2 = 1/D_1 + 1/D_2 \tag{8.6.4}$$

As the final cell concentration x_2' is given, we have four equations given in equations (8.6.2) and (8.6.3) and five unknown variables, x_1', s_1, s_2, D_1, D_2. Then the degree of freedom is one and it can be determined as as to minimize J in equation (8.6.4). From equations (8.6.2) and (8.6.3) of the first stage tank,

$$\mu_1 = D_1 = a s_1 \tag{8.6.5}$$

$$x_1' + s_1 = s_f \tag{8.6.6}$$

are derived. Also for the second stage tank,

$$D_2(s_1 - s_2) = \mu_2 x_2' = a s_2 x_2' \tag{8.6.7}$$

$$x_2' + s_2 = s_f \tag{8.6.8}$$

are satisfied.

From equation (8.6.5), $s_1 = D_1/a$ is derived. Substituting this relation into equation (8.6.7) and using equation (8.6.8),

$$D_2(D_1/a - s_2) = a s_2(s_f - s_2) \tag{8.6.9}$$

is obtained. Here s_2 is fixed. Then, equation (8.6.9) is a constraint existing between D_1 and D_2. The problem can be deduced to: find optimal D_1 and D_2 which minimizes J in equation (8.6.4) where the constraint described by equation (8.6.9) should be satisfied. For this type of optimization problem, the Langrangian method will be utilized. That is, the Lagrangian, L, is defined as

$$L = \frac{1}{D_1} + \frac{1}{D_2} + \lambda\left[a s_2(s_f - s_2) - D_2\left(\frac{D_1}{a} - s_2\right)\right] \tag{8.6.10}$$

where λ is the Lagrangian multiplier and the problem can be converted to a constraint-free minimization problem of L with respect to D_1, D_2 and λ. The necessary conditions for this optimization problem are,

$$\frac{\partial L}{\partial D_1} = \frac{\partial L}{\partial D_2} = \frac{\partial L}{\partial \lambda} = 0 \tag{8.6.11}$$

From these conditions

$$\frac{\partial L}{\partial D_1} = -\frac{1}{D_1^2} - \frac{\lambda D_2}{a} = 0 \tag{8.6.12}$$

$$\frac{\partial L}{\partial D_2} = -\frac{1}{D_2^2} - \lambda\left(\frac{D_1}{a} - s_2\right) = 0 \tag{8.6.13}$$

$$\frac{\partial L}{\partial \lambda} = as_2(s_f - s_2) - D_2\left(\frac{D_1}{a} - s_2\right) = 0 \tag{8.6.14}$$

From equations (8.6.12) and (8.6.13), by eliminating λ, the following relation

$$\frac{D_1^2}{a} = D_2\left(\frac{D_1}{a} - s_2\right) \tag{8.6.15}$$

is derived. Substituting equation (8.6.15) into equation (8.6.14)

$$D_1^2 = a^2 s_2(s_f - s_2) \tag{8.6.16}$$

is reduced. Then

$$D_1 = a\sqrt{s_2(s_f - s_2)} \tag{8.6..17}$$

That is the optimal solution becomes

$$D_1 = 0.001\sqrt{10 \times 190} = 0.044 \ , \qquad D_2 = 0.056$$

$$x_1' = 156 \ , \qquad x_2' = 190 \ , \qquad s_1 = 44 \ , \qquad s_2 = 10$$

The Lagrangian method is frequently used for getting the necessary condition for optimal design. And several numerical optimization techniques such as the Simplex method, the Gradient method, the Davison & Fletcher Powell method and so on are mostly utilized for practical static optimization problems.

9

Downstream processing and bioseparation

Contributors: Tarun Ghose, K. Suga, V. S. Bisaria, Subhash Chand, Purnendu Ghosh

PROBLEM 9.1

A small laboratory column of inner diameter 0.02 m contained a packed bed of 1 m high Sephadex G-100 (total volume of gel used was 0.065 kg with water regain value of $0.003\ m^3\,kg^{-1}$ of dry Sephadex). It was used to separate a mixture of bovine serum albumin (BSA) from glucose. The sample volume of $2 \times 10^{-6}\,m^3$ contained $5\,kg\,m^{-3}$ BSA and $10\,kg\,m^{-3}$ glucose. With water as eluent, the elution volume of BSA and glucose were found to be $1.05 \times 10^{-4}\,m^3$ and $2.95 \times 10^{-4}\,m^3$ respectively. The void volume of the column using blue dextran 2000 was found to be $1.05 \times 10^{-4}\,m^3$.

It is proposed to scale up the operation to the size of a packed bed of Sephadex with column size of 3 m height and 0.06 m diameter such that the ratios of (a) void volume to total bed volume and (b) inner gel volume to total bed volume of the bed remain the same as that of the laboratory column. Determine the elution volume of BSA and glucose from the bigger column assuming that the flow pattern is the same in both the columns.

Solution

Elution volume is given by the expression

$$V_e = V_0 + K_P V_i$$

where V_0 = void volume, K_p = partition coefficient and V_i = inner gel volume.

V_i can be calculated from

$$\begin{aligned} V_i &= aW_r \\ &= 0.065 \times 0.003 \\ &= 1.95 \times 10^{-4}\,m^3 \end{aligned}$$

where a = amount of gel and W_r = water regain value.

The elution volume of BSA is $1.05 \times 10^{-4}\,m^3$; hence K_p for BSA is

$$1.05 \times 10^{-4} = (1.05 \times 10^{-4}) + (1.95 \times 10^{-4})K_p$$

or, $K_p = 0$.

The K_p for glucose is given by

$$2.95 \times 10^{-4} = (1.05 \times 10^{-4}) + (1.95 \times 10^{-4})K_p$$

or

$$K_p = 0.974$$

The total bed volume of small laboratory column

$$V = \pi r^2 l$$
$$= 3.14 \times (0.01)^2 \times 1.0 = 3.14 \times 10^{-4}\,\text{m}^3$$

and of bigger column is

$$= 3.14 \times (0.03)^2 \times 3.0$$
$$= 8.48 \times 10^{-3}\,\text{m}^3$$

The ratios of void to total bed volume are same in the two columns. For small column

$$\frac{V_0}{V_t} = \frac{1.05 \times 10^{-4}}{3.14 \times 10^{-4}} = \frac{1}{3}$$

Hence, the void volume of the bigger column

$$V_0' = \frac{8.48 \times 10^{-3}}{3} = \underline{2.826 \times 10^{-3}\,\text{m}^3}$$

Similarly, the ratios of inner gel volume to total bed volume in the two columns are the same. For small column:

$$\frac{V_i}{V_t} = \frac{1.95 \times 10^{-4}}{3.14 \times 10^{-4}} = 0.621$$

Hence, inner gel volume of bigger column is

$$V_i' = 0.621 \times 8.48 \times 10^{-3} = \underline{5.266 \times 10^{-3}\,\text{m}^3}$$

(i) Elution volume of BSA is

$$V_e' = V_0' + K_p' V_i'$$
$$= \underline{2.826 \times 10^{-3}\,\text{m}^3}$$

(as K_p does not depend on the geometry of the column and is zero for BSA in the smaller column).

(ii) Elution volume of glucose is

$$V_e' = 2.826 \times 10^{-3} + 0.974 \times 5.266 \times 10^{-3}$$
$$= \underline{7.955 \times 10^{-3}\,\text{m}^3}$$

Thus, in the bigger column, the elution volumes are

for BSA $= 2.826 \times 10^{-3}\,\mathrm{m}^3$
for glucose $= 7.955 \times 10^{-3}\,\mathrm{m}^3$

Reference

Fischer, L. (1980) Gel filtration chromatography, in *Laboratory Techniques in Biochemistry and Molecular Biology*, Vol. 1, Part 2, Work, T. S. and Burdson, R. H. (eds), Elsevier, Holland.

PROBLEM 9.2

Bacterial cells are to be separated from a culture both at a rate of $3.34 \times 10^{-3}\,\mathrm{m^3\,s^{-1}}$. Assuming that the cells are spherical and their average diameter is $1\,\mu\mathrm{m}$, select a Sharples centrifuge that can perform the separation. Density of the bacterial cell and viscosity of both are taken as 110% and 300% of those for water at 25°C respectively.

Solution

Physical properties of water at 25°C:

$$\rho_f = 997\,\mathrm{kg\,m^{-3}}$$

$$\mu = 0.894 \times 10^{-3}\,\mathrm{N\,s\,m^{-2}}$$

$$\rho_m = \rho_f\ ; \qquad \rho_y = 1.1\rho_f\ ; \qquad \mu_m = 3\mu$$

$$U_{co} = \frac{g z_c \mathrm{d}p^2(\rho_y - \rho_m)}{18\mu} = z_c V_g$$

or,

$$z_c = \frac{r\omega^2}{g}$$

$$x = U_{co}t = z_c V_g \frac{V}{\bar{F}}$$

$$\frac{V z_c}{x} = \frac{\bar{F}}{V_g} = S$$

$$S = \frac{\bar{F}}{V_g} = \frac{\bar{F}\,18\mu}{\mathrm{d}p^2(\rho_y - \rho_m)g}$$

$$= \frac{3.34 \times 10^{-3} \times \frac{1}{60} \times 18 \times 0.894 \times 10^{-3} \times 3}{(1.0 \times 10^{-6})^2 \times [1.1 \times 997 - 997] \times 9.8}$$

$$= \underline{1.65 \times 10^5\,\mathrm{m}^2}$$

$$2\Sigma = S$$

$$\therefore \Sigma = \underline{8.25 \times 10^4\,\mathrm{m}^2}$$

Reference

Aiba, S., Humphrey, A. E. & Millis, Nancy F. *Biochemical Engineering* (2nd edn. 1973), University of Tokyo Press, p. 349.

PROBLEM 9.3

A mixed liquor from an aeration basin proceeds to the secondary sedimentation to obtain a clear effluent to be discharge into a river. When the mixed liquor is charged into the clarifier at a rate of $10{,}000\,m^3\,h^{-1}$, what is the surface area of the basin required? Physical properties of the liquid are those of water at 20°C. The floc density is taken as $1090\,kg\,m^{-3}$. An equivalent diameter of the floc is 700 micrometres. It is assumed that the discharge rate of sludge from the bottom of clarifier can be disregarded in the calculation. See Fig. 9.1

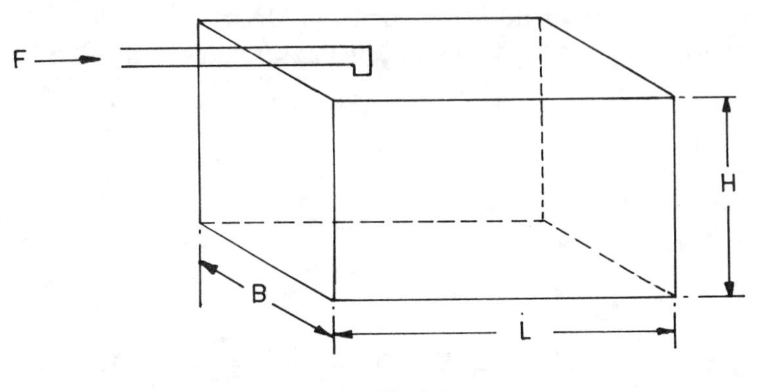

Fig. 9.1.

Solution

Physical properties of water at 20°C:

$$\rho_m = 998.2\,kg\,m^{-3}$$

$$\mu_m = 1.005 \times 10^{-3}\,N\,s\,m^{-2}$$

By Stoke's law,

$$3\pi\mu\, dp u_0 = (\pi/6)\, dp^3(\rho_y - \rho_m)g$$

therefore,

$$U_0 = \frac{dp^2(\rho_y - \rho_m)g}{18\mu}$$

$$= \frac{(700 \times 10^{-6})^2(1090 - 998.2) \times 9.8}{18 \times 1.005 \times 10^{-3}}$$

$$= 0.0244\,m\,s^{-1}$$

$$\frac{H}{u_0} \leqslant \frac{1}{(F/BH)}$$

$$\therefore \frac{1}{u_0} \leqslant \frac{BL}{F}$$

$$\therefore BL\ (=\text{ surface area}) \geqslant \frac{F}{u_0}$$

$$= \frac{10{,}000}{2.44 \times 10^2 \times 60^2}$$

$$= \underline{113.8\,\text{m}^2}$$

PROBLEM 9.4

A mixed liquor is separated in a sedimentation basin. Estimate the recycle ratio in order to secure the steady-state operation of the activated sludge process, where the MLVSS value in the aeration basin is kept at 4000 ppm. See Fig. 9.2.

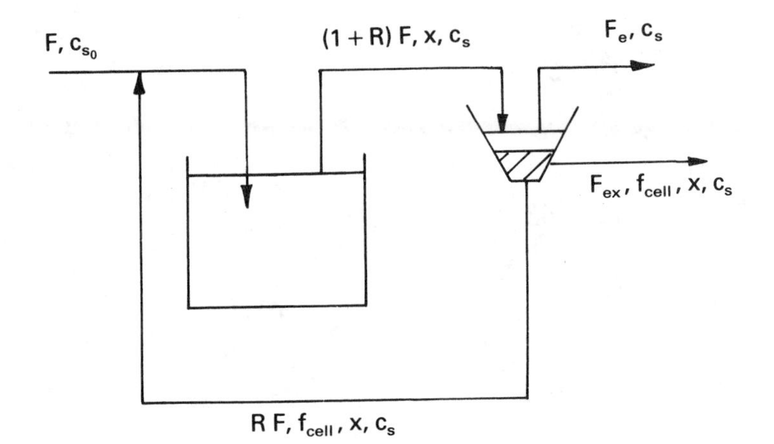

Fig. 9.2.

Solution

$$V\frac{dx}{dt} = RFf_{cell}x - (1+R)Fx + V\mu x \tag{9.4.1}$$

$$V\frac{dc_s}{dt} = Fc_{s_0} + RFc_s - (1+R)Fc_s - V\frac{\mu x}{Y_{x/s}} \tag{9.4.2}$$

$$D = \frac{F}{V}$$

At steady state

$$\frac{dx}{dt} = 0$$

$$\therefore \mu = \{(1+R) - Rf_{cell}\}D \tag{9.4.3}$$

$$\mu x = D(c_{s_0} - c_s)Y_{x/s} \tag{9.4.4}$$

$$\therefore x = \frac{(c_{s_0} - c_s)}{1 + R - Rf_{cell}} Y_{x/s} \tag{9.4.5}$$

$$(1+R)Fx = RFx + F_e x x \tag{9.4.6}$$

$$cx = \frac{1}{SVI}$$

$$x = \frac{(RF + F_e x)f_{cell} x}{(1+R)F} = \frac{RF + F_e x}{(1+R)F(SVI)}$$

$$RF \gg F_e x$$

$$x = \frac{R}{(1+R)(SVI)}$$

$$R = \frac{(SVI)x}{1 - (SVI)x}$$

$$= \frac{(65)(4000 \times 10^{-6})}{1 - (65)(4000 \times 10^{-6})} = 0.35$$

$$\underline{R = 0.35}$$

PROBLEM 9.5

A bacterial suspension (bacterial size dp = 1 micrometre; cell density ρ_y = 1030 kg m^{-3}; volume fraction of the cells, $c = 0.3$) in a buffer solution (density ρ_m = 1000 kg m^{-3}; viscosity $\mu = 1.01 \times 10^{-3}$ N s m^{-2}) is treated with a suspended-type centrifuge. It is desired to estimate the time, t, required for the interface of the suspension to travel (hindered settling) from $r_1 = 0.09$ m to $r_2 = 0.1$ m from the rotation axis when the centrifuge is operated at a speed of $N = 5000$ rpm.

Solution

$$u_0 = \frac{dp^2(\rho_y - \rho_m)g}{18\mu}$$

$$= \frac{(1.0 \times 10^{-6})^2(1030 - 1000)9.81}{18 \times 1.01 \times 10^{-3}}$$

$$= 1.62 \times 10^{-8}\ \mathrm{m\,s^{-1}}$$

$$u = \frac{u_0}{1 + \alpha' c^{1/3}}$$

(hindered setting $\alpha' = 1 + 229\,c^{3.43}$)

$$= \frac{1.62 \times 10^{-8}}{1 + \{1 + 229 \times (0.3)^{3.43}\}(0.3)^{1/3}}$$

$$= 3.9 \times 10^{-9}\,\mathrm{m\,s^{-1}}$$

Interface substance rate

$$1 \cdot u = (1 - f_{\mathrm{cell}})u$$

$$= (1 - 0.3) \times (3.9 \times 10^{-9})$$

$$= 2.73 \times 10^{-9}\,\mathrm{m\,s^{-9}}$$

$$u_c = \frac{\mathrm{d}r}{\mathrm{d}t} = u\left(\frac{r\omega^2}{g}\right)$$

$$= \frac{ur(2\pi n)^2}{g}$$

$$\mathrm{d}t = \frac{g}{u(2\pi n)^2}\frac{\mathrm{d}r}{r}$$

$$t_{(0.09 \to 0.1)} = \frac{9.8}{(2.73 \times 10^{-9})\left\{2 \times 3.14 \times \dfrac{5000}{60}\right\}^2} \ln\left(\frac{0.1}{0.09}\right)$$

$$= \underline{1379.5 \text{ seconds}} = \underline{23 \text{ minutes}}$$

Reference

Aiba, S., Humphrey, A. E. & Millis, Nancy F. *Biochemical Engineering* (2nd edn. 1973), University of Tokyo Press, p. 349.

PROBLEM 9.6

(i) Calculate the rate of settling of a particle (diameter = 5 μm) in water at 20°C. (ii) What is the rate of settling at r = 0.15 m when the suspension is centrifuged employing a speed of 3000 rpm. Viscosity of water at 20°C is $1.01 \times 10^{-3}\,\mathrm{N\,s\,m^{-2}}$ and the particle density is taken as $1100\,\mathrm{kg\,m^{-3}}$.

Solution

(i)

$$u_g = \frac{dp^2(\rho y - \rho m)g}{18\mu}$$

$$= \frac{(5 \times 10^{-6})^2(1100 - 1000)9.81}{18 \times 1.01 \times 10^{-3}}$$

$$= 1.35 \times 10^{-6}\,\mathrm{m\,s^{-1}}$$

(ii)

$$z_c = \frac{r\omega^2}{g} = \frac{r(2\pi n)^2}{g}$$

$$= \frac{0.15 \times \left(2x3.14 \times \frac{3000}{60}\right)^2}{9.81}$$

$$= 1.51 \times 10^3$$

$$V = z_c u_g$$

$$= (1.51 \times 10^3)(1.35 \times 10^{-4}) = \underline{2.03 \times 10^{-3}\,\mathrm{m\,s^{-1}}}$$

PROBLEM 9.7

An actinomycetes broth supplemented with a filter aid of a particular kind to an extent of 5% is filtered by a porotype filter, where the filter area, A, and pressure difference, ΔP, are:

$$A = 35 \times 10^{-4}\,\mathrm{m}^2 \;; \qquad \Delta P = 99{,}990\,\mathrm{N\,m^{-2}}$$

The data obtained are

t (s)	V (m^3)
20	9.5×10^{-6}
40	16.5×10^{-6}
60	22.0×10^{-6}
120	35.0×10^{-6}
180	45.0×10^{-6}
300	61.0×10^{-6}
420	74.5×10^{-6}

Assess the filtration constants k and V_0 of this broth. See Fig. 9.3.

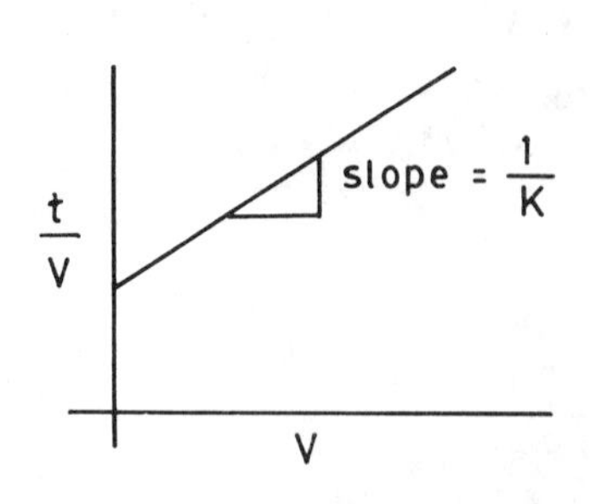

Fig. 9.3.

Solution

$$V^2 = 2V \quad V_0 = kt$$

$$V + 2V_0 = k\frac{t}{V}$$

$$k = \frac{k}{A^2} = \frac{1.79 \times 10^{-11}}{(3.5)^2 \times 10^{-6}} = \underline{1.46 \times 10^{-6}\,\mathrm{m^2\,s^{-1}}}$$

$$V_0 = \frac{V_0}{A} = \frac{1.34 \times 10^{-5}}{3.5 \times 10^{-3}} = \underline{3.83 \times 10^{-3}\,\mathrm{m^3}}$$

PROBLEM 9.8

Amongst various methods for disintegration of cells, dynomill is one employed to break the cell structure for release of proteins. Batch disruption data for baker's yeast slurry (45% wet weight volume^{-1} and containing 0.1 kg protein kg^{-1} yeast cell) in a $6 \times 10^{-4}\,\mathrm{m^3}$ mill at 5°C and using a stainless steel impeller are noted in Table 9.1. If the same mill is to be used in a continuous mode for the same cell suspension,

Table 9.1

Time (s)	Weight of protein released per unit weight of packed yeast (kg protein kg^{-1} yeast)
15	0.055
30	0.080
45	0.091
60	0.096

compute the volumetric feed rate and power consumption to achieve 90% release of the proteins. Also given are, impeller speed = 10 rpm, impeller diameter = 0.04 m, apparent viscosity of the yeast suspension = 0.84 kg m^{-1} s^{-1}, and density = 1100 kg m^{-3}.

Reference
Limon-Lason, J., Hoare, M., Orsborn, C. B., Doyle, D. J. & Dunnill, P. (1979) *Biotechnol. Bioeng.*, **21**, 745.

Solution

Cell disruption in an agitator mill (e.g. dyno mill) has been found to be a first order process (Hethering *et al.*, 1971). Therefore,

$$\frac{dR_P}{dt} = k_1(R_P^M - R_P) \tag{9.8.1}$$

R_p = weight of protein released/unit weight of packed yeast ($kg\,kg^{-1}$), R_p^M = total protein/unit weight of packed yeast = $0.1\,kg\,kg^{-1}$, k_1 = rate constant for the disintegration process.

Integrating equation (9.8.1)

$$\int_0^{R_p} \frac{dR_p}{R_p^M - R_p} = \int_0^t k_1\,dt \tag{9.8.2}$$

or

$$\ln\left[\frac{R_p^M}{R_p^M - R_p}\right] = k_1 t \tag{9.8.3}$$

Plotting $\ln\left[\frac{R_p^M}{R_p^M - R_p}\right]$ versus, t, the first order rate constant 'k_1' can be determined by the slope of the straight line. Such a plot for the batch data given (Table 9.2) is

Table 9.2 — Batch data for disintegration of baker's yeast cells

t (s)	$R_P(kg\,kg^{-1})$	$\left[\frac{R_P^M}{R_P^M - R_P}\right]$	$\ln\left[\frac{R_P^M}{R_P^M - R_P}\right]$
0	0	1	0
15	0.055	2.22	0.798
30	0.080	5.00	1.609
45	0.091	11.11	2.408
60	0.096	25.00	3.218

shown in Fig. 9.4. This figure confirms the applicability of first order kinetics with 'k_1' = $0.0535\,s^{-1}$. Assuming the continuous operation of dynomill to the analogous to a CSTR:

$$\tau = \frac{X R_p^M}{(-dR_p/dt)}$$

where, τ = residence time = VF^{-1}, V = volume of the vessel and F = volumetric flow rate and X = fraction of cells disintegrated.

$$= \frac{X R_p^M}{k_1(R_p^M - R_p)}$$

Here,

$$X = 0.9\ , \qquad R_p^M = 0.1\,kg\,kg^{-1}$$

$$k_1 = 0.0535\,s^{-1}$$

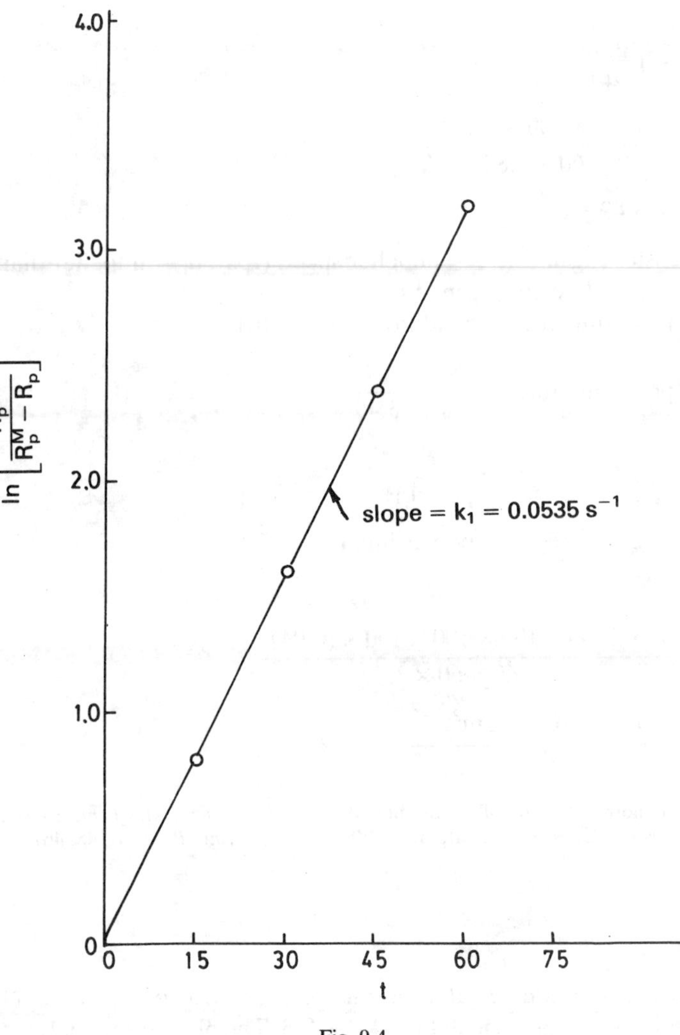

Fig. 9.4.

$$\tau = \frac{0.9 \times 0.1}{0.0535 \times 0.01}$$

$$= 168.2\,\text{s}$$

since,

$$V = 6 \times 10^{-4}\,\text{m}^3$$

$$F = \frac{6 \times 10^{-4}}{168.2} = 1.66 \times 10^{-7}\,\text{m}^3\,\text{s}^{-1}$$

Now, the power input is dependent on the Reynolds' number (N_{Re}) (Rushton *et al.* 1950),

$$N_{Re} = \rho \frac{nd_i^2}{\mu_a}$$

$$= \frac{1100 \times 100 \times (0.04)^2}{60 \times 0.84}$$

$$= 3.49$$

where ρ = density ($kg\,m^{-3}$), n = impeller speed (rps), d_i = impeller diameter (m) and μ_a = apparent viscosity ($kg\,m^{-1}\,s^{-1}$).

Hence, under laminar flow conditions ($N_{Re} < 10$),

$$N_P\,(\text{power number}) = \frac{64}{N_{Re}} = \frac{64}{3.49}$$

$$= \frac{P}{\rho n^3 (d_i)^5}$$

$$= \text{power input}$$

therefore

$$P = \frac{64 \times 1100 \times 100 \times 100 \times 100 \times (0.04)^5}{60 \times 60 \times 60 \times 3.49}$$

$$= \underline{5.96 \times 10^{-3}\,kg\,m^2\,s^{-3}}$$

References

Hethering, P. J., Follows, M., Dunnill, P. & Lilly, M. D. (1971) *Trans. Inst. Chem. Engrs.*, **49**, 142.
Rushton, J. H., Costich, E. W. & Everett, H. J. (1950) *Chemical Eng. Progress*, **46**, 467.

PROBLEM 9.9

Laboratory filtration conducted at a constant pressure drop of $19.6 \times 10^4\,N\,m^{-2}$ on a slurry of mycelial broth gave the data of Table 9.3. The filter area was $1.1 \times 10^{-2}\,m^2$,

Table 9.3

Time (s)	Filtrak volume (m^3)
9.0	1.0×10^{-4}
46.0	2.0×10^{-4}
126.0	3.0×10^{-4}
240.0	4.0×10^{-4}
405.0	5.0×10^{-4}
610.0	6.0×10^{-4}
860.0	7.0×10^{-4}

the mass of mycelial solids per unit volume of filtrate was 40 kg m^{-3} and the viscosity of the slurry was 2×10^{-3} N s m^{-2}. The filtration unit must be shut down periodically in order to remove the accumulated filter cake. The constant pressure performance during one filtration run can be represented by (McCabe and Smith, 1985):

$$V_l^2 = \frac{2\Delta p A^2 t}{w_l \alpha \mu}$$

where V_l = volume of filtrate collected up to time t, Δp = pressure drop across the cake, A = total filtering area, w_l = mass of dry cake solids per unit volume of filtrate, α = specific cake resistance and μ = viscosity of filtrate.

Estimate the value of specific cake resistance. How many hours should each filtration run last between shutdowns in order to obtain the maximum amount of filtrate per day? The filter is operated 24 hours per day and the total time taken for each cleaning shutdown may be taken as 0.5 hours.

Reference

McCabe, W. L. & Smith J. C. (1985) *Unit Operations of Chemical Engineering*, McGraw-Hill, New York.

Solution

The performance of the filtration is represented as

$$V_l^2 = \frac{2\Delta p A^2}{w_l \alpha \mu} t$$

$$= Kt$$

where

$$K = \frac{2\Delta p A^2}{w_l \alpha \mu}$$

A plot of t/V_l against V_l obtained from data points is given in Fig. 9.5. All the data points do not lie on a straight line indicating that the mycelial cake is compressible. However, assuming that the relation between t/V_l and V_l is nearly linear, the slope gives the volume of K and α.

Thus, from Fig. 9.5,

$$\underline{K = 36 \times 10^{-7}\ \mathrm{m^6\,h^{-1}}}$$

and

$$36 \times 10^{-7} = \frac{2 \times 2 \times (110/100 \times 100)^2}{40 \times \alpha \times 2 \times 10^{-3}}$$

$$\underline{\alpha = 1680.5\ \mathrm{m\,kg^{-1}}}$$

If t is the filtering time for one operating cycle and V_l is the amount of filtrate collected in one cycle:

Total number of cycles per 24 hours (considering 0.5 hours of cleaning shotdown period after each cycle)

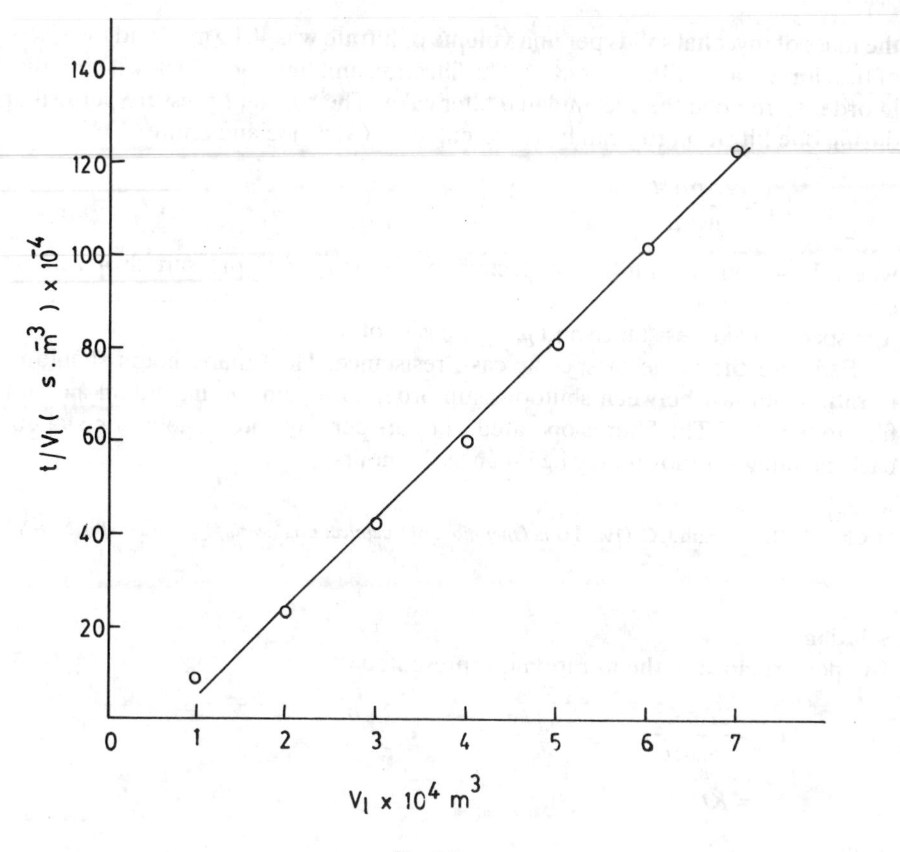

Fig. 9.5.

$$= \frac{24}{(t+0.5)}$$

And, total volume of filtrate collected in 24 hours

$$= (Kt)^{1/2}(24/(t+0.5))$$

The value of t at $|(Kt)^{1/2}[24/(t+0.5)]|_{max}$ represents the optimum length of one operating cycle.

The value of t at the maximum of $(Kt)^{1/2}\dfrac{24}{(t+0.5)}$ can be obtained from

$\dfrac{d}{dt}\left(\dfrac{t^{1/2}}{t+0.5}\right) = 0$ or, $t = 0.5$ h. Thus 24 cycles per day are possible.

PROBLEM 9.10

Ultrafiltration membranes are used for separation and concentration of enzyme from fermentation broth. Table 9.4 shows the rejection of proteins and salts

Table 9.4 — Characteristics of ultra-filtration membranes (Michael, 1968)

Membrane type	Water permeability at 0.686 MPa ($m\,s^{-1}$)	Rejection characteristics	
		Compound (MW)	Rejection (%)
PM-10	5.0×10^{-4}	Salts	0
		Protease (35,000)	99
PM-30	1.33×10^{-3}	Salts	0
		Protease (35,000)	80

using two different membranes. The 10^{-2} m^3 of solution containing 0.1 $kg\,m^{-3}$ of protease and 4.0×10^2 $mol\,m^{-3}$ Na_2CO_3 is concentrated in a batchwise operation, after continuously desalting.

(a) Determine the total volume of distilled water required for 99% desalting.
(b) Calculate the total loss of the protein when it is concentrated to ten times of its original level after 99% desalting using PM-10 and PM-30, respectively.

Reference

Michael, A. S. (1968) *Chem. Eng. Progress*, **64**, p.31.

Solution

(a) The material balance of component i in the separator gives

$$V\frac{dc_i}{dt} = -F_F c_i(1 - R_i) \tag{9.10.1}$$

where

c_i = concentration of component i in the separator ($kg\,m^{-3}$ or $mol\,m^{-3}$),
V = volume of solution (= constant) (m^3),
F_F = filtration rate ($m^3\,s^{-1}$)
R_i = rejection for component i(—).

Integrating equation (9.10.1) and putting $V_F = \int_0^t F_F\,dt$, we obtain

$$\frac{c_i}{c_{i_0}} = \exp\left\{-\frac{V_F}{V}(1 - R_i)\right\} \tag{9.10.2}$$

where c_{i_0} = initial concentration of component i ($kg\,m^{-3}$ or $mol\,m^{-3}$).

Initial concentration of salt (Na_2CO_3) c_{s_0} is

$$c_{s_0} = 4.0 \times 10^2 \ (mol\,m^{-3})$$

and rejection for salt, $R_s = 0$

The total volume, V_F of distilled water required for 99% desalting is given from equation (9.10.2)

$$V_F = \frac{V}{(1-R_i)} \ln \frac{c_{i_0}}{c_i}$$

$$V_F = \frac{10^{-2}}{(1-0)} \ln \left(\frac{0.4}{0.4(1-0.99)} \right) = 4.6 \times 10^{-2} \, m^3$$

(b) The loss of protein, c_{PL} at desalting step using PM-10 is given

$$c_{PL} = c_{i_0} V \left[1 - \exp \left\{ -\frac{V_d}{V}(1-R_i) \right\} \right]$$

$$= (0.1)(10^{-2}) \left[1 - \exp \left\{ -\frac{4.6 \times 10^{-2}}{10^{-2}}(1-0.99) \right\} \right]$$

$$= 4.5 \times 10^{-5} \, kg$$

The relationship between the filtrate volume and concentration of component i in the filtrate at concentrating step gives

$$\frac{dS_i}{dV_l} = \frac{(S_{i_0} - S_i)(1-R_i)}{V_{l_0} - V_l} \tag{9.10.3}$$

where

S_i = amount of component i in the filtrate (kg or mol),
S_{i_0} = initial amount of component i in the solution (kg or mol),
V_l = total volume of filtrate (m^3),
V_{l_0} = initial volume of solution (m^3).

Integration of equation (9.10.3) gives

$$S_i = S_{i_0} \left[1 - \left(\frac{V_l}{V_{l_0}} \right)^{1-R_i} \right] \tag{9.10.4}$$

Then, the concentration of component i in the solution is given as follows,

$$c_i = \frac{S_{i_0} - S_i}{V_{l_0} - V_l} = \frac{S_{i_0}(1 - V_l/V_{l_0})^{1-R_i}}{V_{t_0} - V_l} \tag{9.10.5}$$

The ratio of the final concentration to the initial concentration at concentrating step is obtained using equation (9.10.5),

$$\frac{c_i}{c_{i_0}} = \frac{V_{l_0}}{S_{i_0}} \cdot \frac{S_{i_0}(1 - V_l/V_{l_0})^{1-R_i}}{V_{l_0} - V_l} = \left(\frac{V_{l_0}}{V_{l_0} - V_l} \right)^{R_i} \tag{9.10.6}$$

The total volume of filtrate for 10-fold concentration of protein is obtained from equation (9.10.6)

$$V_l = V_{l_0}\left\{1-\left(\frac{c_i}{c_{i_0}}\right)^{-1/R_i}\right\}$$

$$= (10^{-2})\{1-(10.0)^{-1/0.99}\} = 9.02 \times 10^{-3}\,\text{m}^3$$

$$S_{i_0} = (1.0 \times 10^{-3}) - (4.5 \times 10^{-5}) = 9.55 \times 10^{-4}\,\text{kg}$$

Substituting the above values into equation (9.10.4), the loss of protein at concentrating step using PM-10 is given by

$$S_i = (9.55 \times 10^{-4})\left[1-\left(1-\frac{9.02 \times 10^{-3}}{10^{-2}}\right)^{1-0.99}\right]$$

$$= 2.19 \times 10^{-5}\,\text{kg}$$

The total loss of protein is

$$c_{PL} + S_i = (4.5 \times 10^{-5}) + (21.9 \times 10^{-5}) = 6.69 \times 10^{-5}\,\text{kg}$$

On the other hand, when PM-30 is used, the loss of protein at desalting step is

$$c_{PL} = (0.1)(10^{-2})\left[1-\exp\left\{-\frac{4.6 \times 10^{-2}}{1.0 \times 10^{-2}}(1-0.8)\right\}\right]$$

$$= 6.01 \times 10^{-4}\,\text{kg}$$

and the loss at concentrating step is,

$$S_i = (1.0 \times 10^{-3} - 6.01 \times 10^{-4})\left[1-\left(1-\frac{9.0 \times 10^{-3}}{1.0 \times 10^{-2}}\right)^{1-0.8}\right]$$

$$= 1.47 \times 10^{-4}\,\text{kg}$$

The total loss of protein is

$$c_{PL} + S_i = (6.01 + 1.47) \times 10^{-4}$$

$$= \underline{7.48 \times 10^{-4}\,\text{kg}}$$

PROBLEM 9.11

An enzyme is extracted from a crude solution in an aqueous polyethylene glycol (PEG)/Dextran two-phase system. A tubular bowl centrifuge is used for the separation step.

The system is 9% (w/w) PEG 4000 (upper phase), 2% (w/w) Dextran T 5000 (lower phase). The upper phase has a density of $1.046 \times 10^3\,\text{kg m}^{-3}$ and viscosity of $8.0 \times 10^{-3}\,\text{kg m}^{-1}\text{s}^{-1}$, and the lower phase has a density of $1.144 \times 10^3\,\text{kg m}^{-3}$ and viscosity of $8.0 \times 10^{-3}\,\text{kg m}^{-1}\text{s}^{-1}$ at 20°C. These phases are separated in a tubular bowl centrifuge 0.9 m long and with a 0.05 m radius rotating at 12,000 rpm. The radius of the dam over which the light phase flows is 0.02 m, whereas that over which the heavy phase flows is 0.022 m.

(a) Determine the location of the liquid–liquid interface within the centrifuge.
(b) If the centrifuge is fed at a rate of $1.0 \times 10^{-4} m^3 s^{-1}$ with the ratio of upper volume to lower volume = 3.50, what is the critical droplet diameter of lower phase (Dextran) held in the upper phase (PEG)?

Two diagrams, namely (a) representing a tubular bowl centrifuge and (b) explaining the principle of separation are shown in Fig. 9.6. The feed enters from a stationary nozzle inserted through an opening in the bottom of the bowl. Under rotation, it separates into two concentric layers of liquid inside the bowl. The inner (lighter) layer spills over a weir at the top of the bowl while the heavy liquid flows over another weir into a separate discharge spout.

References

Kroner, K. H., Hustedt, H., Granda, S. & Kula, M. R. (1978), *Biotechnol. Bioeng.*, **20**, 1967.
McCabe, W. K., Smith, J. C. & Hariot, P. (1985) *Unit Operations of Chemical Engineering*, McGraw-Hill, New York, p. 904.

Solution

(a) The location of the interface is fixed by a balance of forces arising from the centrifugal pressures $P(\mathrm{N/m^2})$ of the liquid phase layers. Therefore,

$$P = \frac{\rho_u \omega^2}{2}(r_s^2 - r_u^2) = \frac{\rho_l \omega^2}{2}(r_s^2 - r_l^2) \tag{9.11.1}$$

where

ρ_u = density of upper phase ($\mathrm{kg\,m^{-3}}$),
ρ_l = density of lower phase ($\mathrm{kg\,m^{-3}}$),
ω = angular velocity ($\mathrm{rad\,s^{-1}}$),
r_s = radius to liquid–liquid interface (m),
r_u = radius to outlet of upper liquid layer (m),
r_l = radius to outlet of lower liquid layer (m).

The interface location can be found directly from equation (9.11.1)

$$\frac{\rho_u}{\rho_l} = \frac{r_s^2 - r_l^2}{r_s^2 - r_u^2} \tag{9.11.2}$$

$$\frac{1.046 \times 10^3}{1.144 \times 10^3} = \frac{r_s^2 - (2.2 \times 10^{-2})^2}{r_s^2 - (2.0 \times 10^{-2})^2}$$

$$r_s = \underline{3.7 \times 10^{-2}\,\mathrm{m}}$$

(b) The volumetric feed rate F_u of upper phase is given

$$F_u = L\pi(r_s^2 - r_u^2)V_g\omega^2/g/\ln(r_s/r_u) \tag{9.11.3}$$

$$u_g = \frac{g(\rho_l - \rho_u)d_p^2}{18\mu_u} \tag{9.11.4}$$

where

L = bowl length (m),
u_g = settling velocity of droplet of lower phase in a gravitational field ($\mathrm{m\,s^{-1}}$),

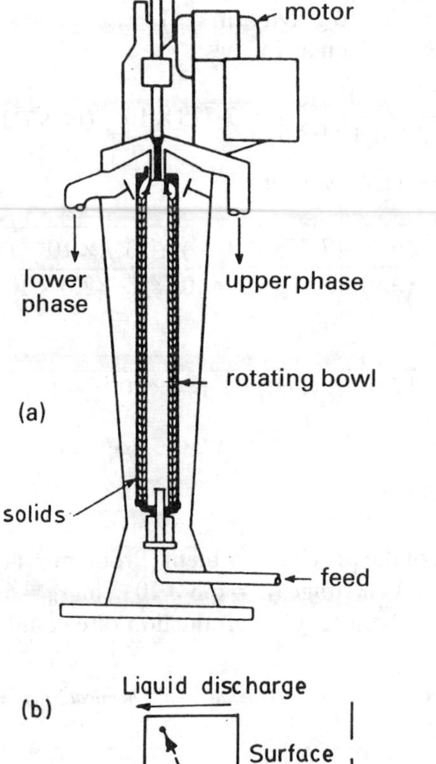

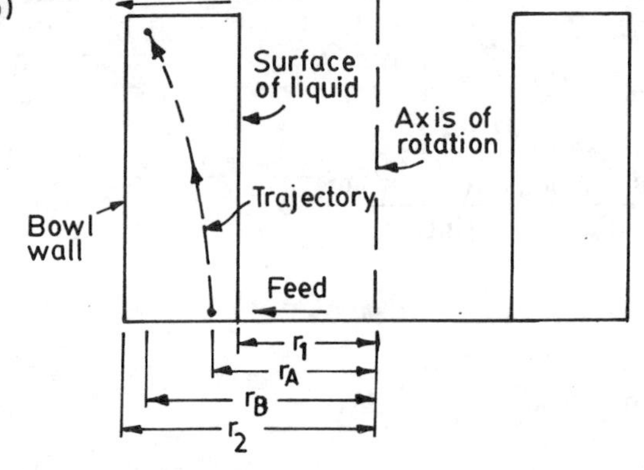

r_1 = radius of rotating bowl upto liquid surface
r_2 = radius of rotating bowl
r_A = distance between the settling particle and the axis of rotation
r_B = distance between the settled particle at the end and the axis of rotation

Fig. 9.6 — (a) Tubular bowl centrifuge, (b) principles of separation in a bowl centrifuge.

d_p = diameter of droplet (m),
μ_u = viscosity of upper phase ($kg\,m^{-1}s^{-1}$),
g = acceleration due to gravity ($m\,s^{-2}$).

On the other hand, F_u is given as follows:

$$F_u = \frac{1.0\times10^{-4}}{1+(V_l/V_u)} = \frac{1.0\times10^{-4}}{1+(1/3.5)} = 7.778\times10^{-5}(m^3s^{-1})$$

From Eq. (9.11.3) and (9.11.4), we get

$$d_p = \sqrt{\frac{18(8.0\times10^{-3})(7.778\times10^{-5})\ln\{(3.7\times10^{-2})/(2.0\times10^{-2})\}}{(0.9)(3.14)\{(3.7\times10^{-2})^2-(2.0\times10^{-2})^2\}}}$$

$$\sqrt{\frac{1}{\{(1.2\times10^4)(2)(3.14)/60\}^2\{1.144\times10^{-3}-1.046\times10^{-3}\}}}$$

$$= \underline{4.03\times10^{-6}\,m} \quad \text{or} \quad \underline{4.03\,\mu m}$$

PROBLEM 9.12

The liquid–liquid separation of the previous problem (Problem 9.11) is to be done in the plant using the same type of centrifuge ($r_l = 6.6\times10^{-2}$ m, $r_u = 6.0\times10^{-2}$ m, L = 1.50 m, rotational speed = 8000 rpm). What production rate could be expected?

Reference

McCabe, W. L., Smith, J. C. & Hariot, P. (1985) *Unit Operations of Chemical Engineering*, McGraw-Hill, New York, p. 904.

Solution

$$F = \frac{g(\rho_l-\rho_u)d_p^2}{18\mu_u}\frac{L\pi(r_s^2-r_u^2)\omega^2}{g\ln(r_s/r_u)} \tag{9.12.1}$$

$$= u_g\Sigma$$

where

$$u_g = \frac{g(\rho_l-\rho_u)d_p^2}{18\mu_u} \tag{9.12.2}$$

$$\Sigma = \frac{L\pi(r_s^2-r_u^2)\omega^2}{g\ln(r_s/r_u)} \tag{9.12.3}$$

Σ is a characteristic of the centrifuge itself. The Σ factor can then be used as a means of comparing centrifuges. For the laboratory centrifuge,

$$\Sigma_1 = \frac{L\pi(r_s^2-r_u^2)\omega^2}{g\ln(r_s/r_u)}$$

$$= \{(1.2\times10^4)(2)(3.14)/60\}^2\times$$

$$\times\frac{(0.9)(3.14)\{(3.7\times10^{-2})^2-(2.0\times10^{-2})^2\}}{(9.80)\ln\{(3.7\times10^{-2})/(2.0\times10^{-2})\}}$$

$$= 716\,\text{m}^2$$

For the large centrifuge, the interface location can be calculated

$$r_s = \sqrt{(\rho_l r_l^2 - \rho_u r_u^2)/(\rho_l - \rho_u)}$$

$$r_s = \sqrt{\frac{(1.144\times10^{-3})(6.6\times10^{-2})^2 - (1.046\times10^{-3})(6.0\times10^{-2})^2}{(1.144\times10^{-3}) - (1.046\times10^{-3})}}$$

$$= 0.111\,\text{m}$$

$$\Sigma_2 = \frac{L\pi(r_s^2 - r_u^2)\omega^2}{g\ln(r_s/r_u)}$$

$$= \frac{(1.50)(3.14)\{0.111^2 - (6.0\times10^{-2})^2\}\{(8.0\times10^3)(2)(3.14/60)\}^2}{(9.80)\ln\{0.111/(6.0\times10^{-2})\}}$$

$$= 4777\,\text{m}^2$$

$$F_2 = F_1\frac{\Sigma_2}{\Sigma_1}$$

$$= (1.0\times10^{-4})\frac{4777}{716} = \underline{6.67\times10^{-4}} \qquad (\text{m}^3\,\text{s}^{-1})$$

PROBLEM 9.13

On solid–liquid extraction, a liquid solvent is used to dissolve a soluble solid from an insoluble solid.

Soybean oil in 1 kg of soybean flake containing 20% oil is to be multistage cocurrently extracted three times with pure hexan, using 1 kg of solvent in each stage.

Determine the compositions and quantities of the stream from each stage and the content of oil remaining in the soybean after repetition of this procedure.

Equilibrium data for extraction are given in Table 9.5. The relationship between

Table 9.5 — Equilibrium data for extraction

Composition of raffinate			
$(\text{Wt\%}) = \dfrac{\text{kg (soybean oil)}}{\text{kg (soybean oil + hexan)}}$	0	20	30
Liquid adhering to the solid			
$\dfrac{\text{kg liquid (soybean oil + hexan)}}{\text{kg insoluble solid}}$	0.60	0.67	0.71
$\mathcal{N} = \dfrac{\text{kg insoluble solid}}{\text{kg liquid (soybean oil + hexan)}}$	1.67	1.49	1.41

the weight fraction of solute in the extract on an insoluble solid free basis, y and that in the raffinate, x can be represented by,

$$x = 1.1y \tag{9.13.1}$$

The insoluble solid in the extract is assumed to be negligible.

References
Yoshimura, K. (1963) In: *Extraction and Distillation in Food Engineering*, Korin Shoin, Tokyo, p. 31.
Treybal, R. E. (1955) *Mass Transfer Operations*, McGraw-Hill, New York, p. 584.

Solution

This is an extension of single-stage extraction, wherein the raffinate is successively contacted with fresh solvent. Refer to Fig. 9.7 which shows the flow sheet for three-stage extraction.

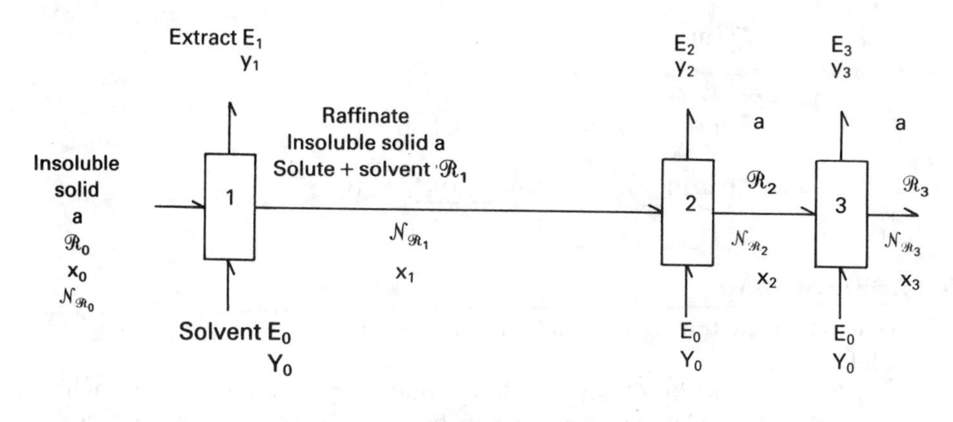

Fig. 9.7 — Multistage cocurrent extraction.

The following notation scheme and definition will be used to describe the fraction and the amount of mixture streams.

- E the mass of extract or solvent stream (kg)
- $\mathcal{R}$ the mass of raffinate stream on an insoluble solid free basis (kg)
- $\mathcal{N}$ the weight ratio of insoluble solid to the resultant liquid adhering to the insoluble solid (—)

$$\mathcal{N}_{\mathcal{R}} = \frac{a}{b+c} = \frac{a\ (\text{kg})}{\mathcal{R}\ (\text{kg})}\ \text{(in the raffinate)} \tag{9.13.2}$$

$$\mathcal{N}_{\mathrm{E}} = \frac{a}{b+c} = \frac{a\ (\text{kg})}{\mathcal{R}\ (\text{kg})}\ \text{(in the extract)} \tag{9.13.3}$$

- a the mass of insoluble solid A (kg)
- b the mass of solvent B (kg)

$$= 716\,\mathrm{m}^2$$

c the mass of solute C (kg)

x the mass fraction of solute C in the raffinate on an A-free basis

$$x = \frac{c}{b+c} = \frac{c \text{ (kg)}}{\mathcal{R} \text{ (kg)}} \tag{9.13.4}$$

y the mass fraction of solute C in the extract on an A-free basis

$$y = \frac{c}{b+c} = \frac{c \text{ (kg)}}{E \text{ (kg)}} \tag{9.13.5}$$

Material balance of solute C and solvent B give

$$\mathcal{R}_0 x_0 + E_0 y_0 = \mathcal{R}_1 x_1 = E_1 y_1 \tag{9.13.6}$$

$$\mathcal{R}_0(1-x_0) + E_0(1-y_0) = \mathcal{R}_1(1-x_1) + E_1(1-y_1) \tag{9.13.7}$$

From the definition of $\mathcal{N}_{\mathcal{R}}$, we get

$$a = \mathcal{N}_{\mathcal{R}0}\mathcal{R}_0 = \mathcal{N}_{\mathcal{R}1}\mathcal{R}_1 \tag{9.13.8}$$

From the statement of the problem,

$$b_0 = 0.0 \text{ (kg)} \qquad c_0 = 0.2 \text{ (kg)}$$

$$\mathcal{N}_{\mathcal{R}0} = \frac{0.80}{0.20} = 4.0 \text{ (—)}$$

$$x_0 = \frac{0.2}{0.0+0.2} = 1.0 \text{ (—)}$$

$$y_0 = \frac{0.0}{1.0+0.2} = 0.0 \text{ (—)}$$

$$E_0 = 1.0 \text{ (kg)}$$

$$\mathcal{R}_0 = 0.2 \text{ (kg)}$$

Stage 1:

In the first stage $\mathcal{R}_0$ and E_0 are mixed, and $\mathcal{R}_1$ and E_1 leave the stage in equilibrium.

$\mathcal{N}_{M1}$, x_{M1} in the mixture of the first stage are given from equation (9.13.2)

$$\mathcal{N}_{M1} = \frac{a}{\mathcal{R}_0 + E_0} = \frac{\mathcal{N}_{\mathcal{R}0}(b_0 + c_0)}{\mathcal{R}_0 + E_0} = \frac{(4.0)(0.2)}{0.2+1.0} = 0.667$$

$$x_{M1} = \frac{\mathcal{R}_0 x_0 + E_0 y_0}{\mathcal{R}_0 + E_0} = \frac{(0.2)(1.0) + (1.0)(0.0)}{1.0+0.2} = \frac{0.2}{1.2} = 0.167$$

$\mathcal{N}_{\mathcal{R}}$ versus x curve is plotted on the rectangular coordinates of Fig. 9.8, using equilibrium data listed in Table 9.5. The straight line between point $\mathcal{R}_0(x_0 = 1.0, \mathcal{N}_{\mathcal{R}0} = 4.0)$ and E_0 $(y_0 = 0.0, \mathcal{N}_{E0} = 0.0)$ is drawn on this figure. Point $\text{M}_1(x_{M1} = 0.167, \mathcal{N}_{M1} = 0.667)$ is plotted on the line $\mathcal{R}_0E_0$ in this figure. With the help of a distribution curve,

$$x = 1.1y$$

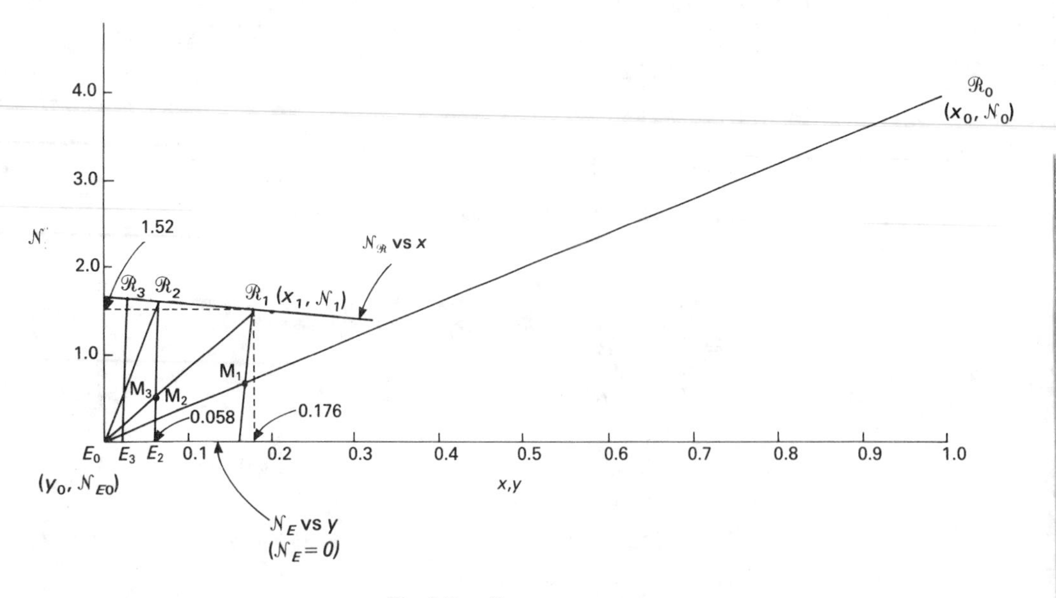

Fig. 9.8 — $\mathcal{N}$ versus x or y.

the tie line passing through M_1 is located so that the ratio of x_1 distance at the intersection of the tie line and the curve $\mathcal{N}_{\mathcal{R}}x$ to y_1 distance at the intersection of the tie line and the abscissa may be satisfied to be 1.1 as shown in Fig. 9.8.

From the coordinates $(x_1, \mathcal{N}_{\mathcal{R}_1})$ at point $\mathcal{R}_1$, we get $\mathcal{N}_{\mathcal{R}_1} = 1.52$. Then the weight of raffinate (A — free) can be estimated from equation (9.13.2).

$$\mathcal{R}_1 = \frac{a}{\mathcal{N}_{\mathcal{R}_1}} = \frac{0.80}{1.52} = 0.527\,\text{kg}$$

Balance of solution (B + C) in the first stage gives

$$\mathcal{R}_0 + E_0 = \mathcal{R}_1 + E_1$$

$$E_1 = (\mathcal{R}_0 + E_0) - \mathcal{R}_1 = (0.2 + 1.0) - 0.527 = \underline{0.673\,\text{kg}}$$

Stage 2:

$$x_{M2} = \frac{\mathcal{R}_1 x_1 + E_0 y_0}{E_0 + \mathcal{R}_1} = \frac{(0.527)(0.176) + 0}{1.0 + 0.527}$$
$$= \frac{0.093}{1.527} = 0.0608$$

$$\mathcal{N}_{M2} = \frac{a}{E_0 + \mathcal{R}_1} = \frac{0.80}{1.527} = 0.524$$

Point M_2 is located on line $\mathcal{R}_1E_0$, and the line $\mathcal{R}_2E_2$ through M_2 is determined similarly to the way in stage 1.

$$\left.\begin{aligned}\mathcal{N}_{\mathcal{R}_2} &= 1.61\\ x_2 &= 0.064\end{aligned}\right\} \text{at point } \mathcal{R}_2$$

$$\left.\begin{aligned}\mathcal{N}_{E_0} &= 0\\ y_2 &= 0.058\end{aligned}\right\} \text{at point } E_2$$

$$\mathcal{R}_2 = \frac{a}{\mathcal{N}_{\mathcal{R}_2}} = \frac{0.80}{1.61} = 0.497\,\text{kg}$$

$$E_2 = (E_0 + \mathcal{R}_1) - \mathcal{R}_2 = (1.0 + 0.527) - 0.497 = \underline{1.03\,\text{kg}}$$

Stage 3:

$$x_{M_3} = \frac{\mathcal{R}_2x_2 + E_0y_0}{E_0 + \mathcal{R}_2} = \frac{(0.497)(0.064)}{1.0 + 0.497} = \frac{0.0318}{1.497} = 0.0212$$

$$\mathcal{N}_{M_3} = \frac{a}{E_0 + \mathcal{R}_2} = \frac{0.8}{1 + 0.497} = 0.534$$

In a similar manner,

$$\left.\begin{aligned}\mathcal{N}_{\mathcal{R}_3} &= 1.65\\ x_3 &= 0.022\end{aligned}\right\} \text{at point } \mathcal{R}_3$$

$$\left.\begin{aligned}\mathcal{N}_{E_0} &= 0\\ y_3 &= 0.020\end{aligned}\right\} \text{at point } E_3$$

$$\mathcal{R}_3 = \frac{a}{\mathcal{N}_{\mathcal{R}_3}} = \frac{0.80}{1.65} = 0.485\,\text{kg}$$

$$E_3 = (E_0 + \mathcal{R}_2) - \mathcal{R}_3 = (1.0 + 0.497) - 0.485 = \underline{1.012\,\text{kg}}$$

Consequently, the composition of final raffinate is calculated

A: $a = 0.8\,\text{kg}$

B: $b_3 = \mathcal{R}_3(1 - x_3) = 0.485(1 - 0.022) = 0.474\,\text{kg}$

C: $c_3 = \mathcal{R}_3x_3 = (0.485)(0.022) = 0.011\,\text{kg}$

And the soybean oil content of soybean in the final raffinate after removing the solvent is

$$(100)\frac{c_3}{a + c_3} = (100)\frac{0.011}{0.8 + 0.011} = \underline{1.35\%}$$

PROBLEM 9.14

Soybean oil is extracted from 1 kg h^{-1} of soybean flake solid containing 20% oil by countercurrent extraction with 1 kg h^{-1} of pure hexan. The oil content of soybean is reduced to 0.6% in the exit raffinate. How many equilibrium stages are required?

Equilibrium data and the relationship between the weight fraction of solute in the extract, y, and that in the raffinate, x, in Problem 9.13 can be used.

Solution

Total net flow of solution is as shown in Fig. 9.9.

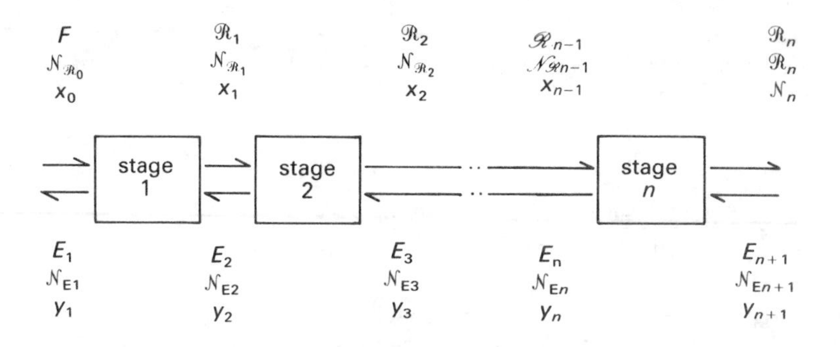

Fig. 9.9 — Multistage countercurrent extraction.

The following notation scheme and definition will be used to describe the fraction and the amount of mixture streams.

E the mass of extract or solvent stream (kg)

$\mathcal{R}$ the mass of raffinate stream on an insoluble solid free basis (kg)

$\mathcal{N}$ the weight ratio of insoluble solid to the resultant liquid adhering to the insoluble solid (—)

$$\mathcal{N}_{\mathcal{R}} = \frac{a}{b+c} = \frac{a\ (\mathrm{kg})}{\mathcal{R}\ (\mathrm{kg})} \text{ (in the raffinate)} \tag{9.14.1}$$

$$\mathcal{N}_{\mathrm{E}} = \frac{a}{b+c} = \frac{a\ (\mathrm{kg})}{\mathcal{R}\ (\mathrm{kg})} \text{ (in the extract)} \tag{9.14.2}$$

a the mass of insoluble solid A (kg)

b the mass of solvent B (kg)

c the mass of solute C (kg)

x the mass fraction of solute C in the raffinate on an A-free basis

$$x = \frac{c}{b+c} = \frac{c\ (\mathrm{kg}}{\mathcal{R}(\mathrm{kg})} \tag{9.14.3}$$

y the mass fraction of solute C in the extract on an A-free basis

$$y = \frac{c}{b+c} = \frac{c\ (\mathrm{kg})}{E\ (\mathrm{kg})} \tag{9.14.4}$$

A balance for solute C gives

$$Fx_0 + E_{n+1}Y_{n+1} = E_1Y_1 + \mathcal{R}_nx_n \tag{9.14.5}$$

Material balance for each stage is given as:

$$F - E_1 = \mathcal{R}_1 - E_2 = \mathcal{R}_2 - E_3 = \ldots = \mathcal{R}_{n-1} - E_n = \mathcal{R}_n - E_{n+1} = \Delta \tag{9.14.6}$$

$$Fx_0 - E_1y_1 = \mathcal{R}_1x_1 - E_2y_2 = \ldots = \mathcal{R}_nx_n - E_{n+1}y_{n+1} = \Delta x_\Delta \tag{9.14.7}$$

$$F\mathcal{N}_{\mathcal{R}0} - E_1\mathcal{N}_{E1} = \mathcal{R}_1\mathcal{N}_{\mathcal{R}1} - E_2\mathcal{N}_{E2} = \ldots = \mathcal{R}_n\mathcal{N}_{\mathcal{R}n} - E_{n+1}\mathcal{N}_{En+1} = \Delta\mathcal{N}_{\mathcal{R}\Delta} \tag{9.14.8}$$

Net flow to the right in the cascade shown in Fig. 9.9 is defined as the difference between the flow to the right and the flow to the left and it is constant.

Therefore, line $E_{m+1}\mathcal{R}_m$ extended must pass through Δ, as in Fig. 9.10. From

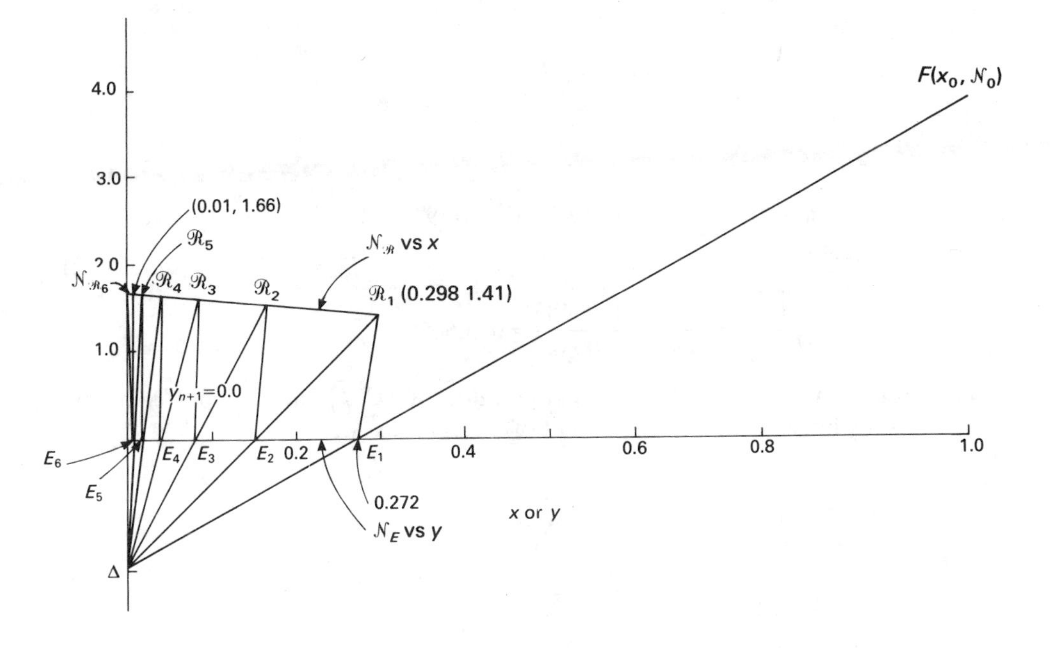

Fig. 9.10 — $\mathcal{N}$ versus x or y (multiistage countercurrent extraction).

equations (9.14.7) and (9.14.8), the coordinate of point Δ gives

$$x_\Delta = \frac{Fx_0 - E_1y_1}{F - E_1} = \frac{Fx_0 - E_1y_1}{\Delta} \tag{9.14.9}$$

$$\mathcal{N}_{\mathcal{R}\Delta} = \frac{F\mathcal{N}_{\mathcal{R}0}}{F - E_1} = \frac{F\mathcal{N}_{\mathcal{R}0}}{\Delta} \tag{9.14.10}$$

The graphical construction is as follows. After location of points F, E_{n+1}, $\mathcal{R}_n$ and E_1, the extended lines $\mathcal{R}_nE_{n+1}$ and FE_1 must intersect at Δ as shown in Fig. 9.10. A tie line through E_1 provides $\mathcal{R}_1$ using the relationship between x and y. A line from Δ to $\mathcal{R}_1$ gives E_2, a tie line through E_2 provides $\mathcal{R}_2$, etc.

F = feed stream on a insoluble solid free basis ($\mathrm{kg\,h^{-1}}$)

Feed

$F = 1\ \mathrm{kg\,h^{-1}}$

$a = 0.8\,kg$

$\mathcal{N}_{\mathcal{R}0} = 0.8/(0+0.20) = 4.0$

$x_0 = 0.20/(0+0.20) = 1.0$

Solvent

$E_{n+1} = 1.0$

$y_{n+1} = 0$

Raffinate

$a = 0.8$

The oil content of insoluble solid from the n-stage x_n from equation (9.13.3 of the previous problem) gives

$$\frac{c}{0.8+c} = 0.006$$

Then,

$$c = 0.00483\ \mathrm{kg} \tag{9.14.11}$$

On the other hand, from equations (9.14.1) and (9.14.2)

$$c = \mathcal{R}_n x_n \tag{9.14.12}$$

$$\frac{x_n}{\mathcal{N}_{\mathcal{R}n}} = \frac{\mathcal{R}_n x_n}{\mathcal{R}_n \mathcal{N}_{\mathcal{R}n}} = \frac{\mathcal{R}_n x_n}{a} = \frac{0.00483}{0.80} = 0.00605$$

Using the data shown in Table 9.5, the value of $x/\mathcal{N}_{\mathcal{R}n}$ can be approximately evaluated by interpolation from the plot of $\mathcal{N}_{\mathcal{R}}$ versus $x/\mathcal{N}_{\mathcal{R}}$.

$$\mathcal{N}_{\mathcal{R}n} = 1.66$$

$$\mathcal{R}_n = \frac{a}{\mathcal{N}_{\mathcal{R}n}} = \frac{0.80}{1.66}$$

$$= 0.482\ \mathrm{kg}$$

From equation (9.14.11) we obtain

$$x_n = \frac{0.00483}{0.482} = 0.010$$

From equation (9.14.6) and (9.14.7), E_1 and y_1 are calculated as

$$E_1 = F + E_{n+1} - \mathcal{R}_n = 0.2 + 1.0 - 0.482 = 0.718\ \mathrm{kg}$$

$$y_1 = \frac{Fx_0 - \mathcal{R}_n x_n}{E_1} = \frac{0.200 - 0.00483}{0.718} = 0.272$$

The intersection of extended lines $\mathcal{R}_n E_{n+1}$ and FE_1 provides point Δ. On the other hand, the tie line through E_1 gives $\mathcal{R}_1$ using the relation

$$x = 1.1y$$

By similar methods to those shown in Problem 9.13, the intersection of the lines $\mathcal{R}_1\Delta$ and $\mathcal{N}_E = \Delta$ (in extract) gives E_2. The graphical construction is continued up to

$$y < x_n/1.1 = 0.01/1.1 = 0.009$$

There are 5.6 stages. Therefore, the required stage numbers are 6.

PROBLEM 9.15

High-performance liquid chromatography (HPLC) is used for the separation of proteins in the solution.

The capacity factor of component i, k_i is defined by

$$k_i = \frac{t_{Ri} - t_0}{t_0} \tag{9.15.1}$$

where t_0 = dead time and t_{Ri} = retention time of component i.

The ratio of two capacity factors, k_2 and k_1, is designated as the relative retention δ

$$\delta = \frac{k_2}{k_1} \tag{9.15.2}$$

The number of theoretical plates N_p is defined by

$$N_{P2} = 16(t_{R2}/w_2)^2 \tag{9.15.3}$$

The resolution R_N of two component bands is defined in terms of the distance between the two peak and the arithmetic mean of the two band widths w.

$$R_N = \frac{2(t_{R2} - t_{R1})}{w_1 + w_2} \tag{9.15.4}$$

Express the resolution R_N related to the chromatographic parameters, k_i, δ and N_{P2}.

Reference
Engelhardt, H. (1979), *High Performance Liquid Chromatography*, p. 6, Springer-Verlag, New York.

Solution

If we can assume $w_1 = w_2$, then equation (9.15.4) becomes

$$R_N = \frac{t_{R2} - t_{R1}}{w_2} \tag{9.15.5}$$

Eliminating w_2 from equations (9.15.3) and (9.15.5), the relationship between R and N_{P2} is given by

$$R_N = \frac{1}{4}\sqrt{N_{P2}}\left(\frac{t_{R2} - t_{R1}}{t_{R2}}\right) \tag{9.15.6}$$

From equation (9.15.1), we obtain

$$k_1 = \frac{t_{R1} - t_0}{t_0} \tag{9.15.1i}$$

$$k_2 = \frac{t_{R2} - t_0}{t_0} \tag{9.15.1ii}$$

By rearranging equation (9.15.1ii), we obtain

$$t_{R2} = t_0(k_2 + 1) \tag{9.15.7}$$

From equations (9.151i) and (9.15.1ii), we get

$$t_{R2} - t_{R1} = t_0(k_2 - k_1) \tag{9.15.8}$$

Substitution of equations (9.15.7) and (9.15.8) into (9.15.6), gives

$$R_N = \frac{1}{4}\sqrt{N_{P2}}\left(\frac{k_2 - k_1}{k_2 - 1}\right) \tag{9.15.9}$$

The resolution R_N can be related to the chromatographic parameters, δ, k_2 and N_P, using the relation of equations (9.15.2) and (9.15.9)

$$R_N = \frac{1}{4}\sqrt{N_{P2}}\left(\frac{\delta - 1}{\delta}\right)\left(\frac{k_2}{k_2 + 1}\right) \tag{9.15.10}$$

PROBLEM 9.16

High-performance liquid chromatography (HPLC) is used for the separation of two proteins (bovine serum albumin (BSA) and myoglobin (Mb)) in the solution. Pilot runs in a laboratory type of HPLC indicates that the capacity factors k_1 and k_2 of two protein are 0.8 and 1.1, respectively.

Where the capacity factor, k_i of component i is defined by

$$k_i = \frac{t_{R_i} - t_0}{t_0}$$

t_0 = dead time (s)
t_{R_i} = retention time of component i (s).

The relationship between the heights equivalent to a theoretical plate (HETP) and the flow velocity is obtained as follows,

$$\text{HETP (m)} = A + Bu \tag{9.16.1}$$

$$A = 6.2 \times 10^{-5}\,(\text{m}) \qquad B = 1.13\,(\text{s})$$

where u = the flow velocity (m s^{-1}).

What resolution R_N of two enzymes may be expected under the following operating conditions?

Column length:	0.6 (m)
Flow velocity	2.17×10^{-4} (m s^{-1})

Column inside diameter: 7.5×10^{-3} (m)
Sample volume 0.1×10^{-6} (m^3)

Reference
Yamamoto, S. Nomura, M. & Sano, Y. (1987) *J. Chromatogr*, **394**, 363.

Solution

The ratio of two capacity factors is designated as the relative retention δ,

$$\delta = \frac{k_2}{k_1} = \frac{1.1}{0.8} = 1.37 \tag{9.16.2}$$

The number of theoretical plates N_P can be calculated using the column length L and HETP from the following equation:

$$\text{HETP} = 6.2 \times 10^{-5} + (1.13)(2.17 \times 10^{-4}) = 3.07 \times 10^{-4}\,(\text{m})$$

$$N_{P2} = L/\text{HETP} = 0.6/(3.07 \times 10^{-4}) = 1954 \tag{9.16.3}$$

The resolution R_N can be estimated as follows

$$R_N = \frac{1}{4}\sqrt{N_P}\left(\frac{\delta - 1}{\delta}\right)\left(\frac{k_2}{k_2 + 1}\right)$$

$$R_N = \frac{1}{4}\sqrt{1954}\left(\frac{1.37 - 1}{1.37}\right)\left(\frac{1.1}{1.1 + 1}\right)$$

$$= \underline{1.56}$$

Therefore, two enzymes can be almost completely separated.

(*Note*. Since the number of theoretical plates could also be a function of the packing materials and their diffusivities through their pores an alternative approach to solving the problem is not overruled.)

PROBLEM 9.17

It is required to calculate the flow rate and sample volume of separating two enzymes of the previous problem (Problem 9.16) on the large scale.

The characteristics of packing materials for a large HPLC column is same as that used in Problem 9.16. The column length and inside diameter of a large-scale HPLC are 0.5 m and 0.108 m, respectively.

The relationship between the heights equivalent to a theoretical plate (HETP) and the flow velocity is obtained as follows,

$$\text{HETP}\,(\text{m}) = A + Bu \tag{9.17.1}$$

$$A = 6.2 \times 10^{-5}\,(\text{m}) \qquad B = 1.13\ (\text{s})$$

where u = the flow velocity (m s^{-1}).

Solution

In order to get the same resolution obtained by the large-scale column as that obtained by the laboratory-scale column, their number of theoretical plates should be equal.

The number of theoretical plates for a laboratory scale column is $N_P = 1954$, from Problem 9.16.

Then, HETP for a large-scale column can be determined

$$\text{HETP} = L/N_P = 0.5/1954 = 2.56 \times 10^{-4}\,\text{m}$$

The flow velocity for the large-scale column can be estimated from equation (9.17.1)

$$u = (2.56 \times 10^{-4} - 6.2 \times 10^{-5})/1.13$$
$$= 1.72 \times 10^{-4}\ (\text{m s}^{-1})$$

The flow rate F_L ($\text{m}^3\,\text{s}^{-1}$) is

$$F_L = (1.72 \times 10^{-4})(0.108/2)^2\ (3.14)$$
$$= 1.57 \times 10^{-6}\ (\text{m}^3\,\text{s}^{-1})$$

The sample size can be determined using the previous data (Problem 9.16) by the following equation,

$$V_L/V_S = F_L/F_S$$

$$V_L = \frac{(1.57 \times 10^{-6})(0.1 \times 10^{-6})}{(2.17 \times 10^{-4})(7.5 \times 10^{-3}/2)^2(3.14)}$$
$$= \underline{1.64 \times 10^{-5}}\ (\text{m}^3)$$

PROBLEM 9.18

Separation and concentration of particles such as virus, plant cell and phage from cultures of large volume have been done by aqueous two-phase partitioning.

Consider c_0 (cells m^{-3}) and V_0 (m^3) are the initial concentration of particle and volume of solution respectively.

The total volume, V, of solution consisting of two kinds of polymers is added to the solution containing the particles to form the two-phase system. The volume of the top phase, V_t, is given as follows.

$$V_t = V_0 + V - V_b \tag{9.18.1}$$

where V_b is the volume of the bottom phase.

The partition coefficient, K of virus is defined as

$$K = \frac{c_t}{c_b} \tag{9.18.2}$$

where c_t and c_b are the concentrations of virus in the top and the bottom phases, respectively.

The degree of concentration, δ_c and the particle yield, y are used for the index of concentration effect as follows,

$$\delta_c = \frac{c_b}{c_0}(c_b > c_t) \tag{9.18.3}$$

or

$$= \frac{c_t}{c_0}(c_t > c_b)$$

$$Y = \frac{c_b V_b}{c_0 V_0}(c_b > c_t)$$

or

$$= \frac{c_t V_t}{c_0 V_0}(c_t > c_b) \tag{9.18.4}$$

Express the δ_c and Y related to the V_0, V_b and K, respectively.

Reference

Albertsson, P. A. (1971) *Partition of Cell Particles and Macromolecules*, 2nd edn, Almqvist Wiksell, Stockholm; John Wiley, New York.

Solution

Material balance of particles gives,

$$V_t c_t + V_b c_b = V_0 c_0 \tag{9.18.5}$$

δ_c is given from equations (9.18.2) and (9.18.3)

$$\delta_c = \frac{V_0}{V_b\left(1 + \frac{V_t}{V_b}K\right)}(c_b > c_t) \tag{9.18.6}$$

or

$$= \frac{V_0}{V_t\left(1 + \frac{V_b}{V_t}\frac{1}{K}\right)}(c_t > c_b)$$

$$Y = \delta_c \frac{V_b}{V_0}(c_b > c_t) \tag{9.18.7}$$

or

$$= \delta_c \frac{V_t}{V_0}(c_t > c_b)$$

PROBLEM 9.19

An aqueous two-phase system is used for the separation and concentration of virus from the culture solution.

Initial volume of culture solution containing virus:

$$V_0 = 3.0 \times 10^{-3}\,(\text{m}^3)$$

Volume of polymer solution added to the culture:

$$V = 1.0 \times 10^{-4}\,(\text{m}^3)$$

Partition coefficient of virus in the two-phase system:

$$K = 4.0 \times 10^{-4}\,(\text{—})$$

Volume of the bottom phase:

$$V_b = 5.0 \times 10^{-5}\,(\text{m}^3)$$

Estimate the degree of concentration δ_c for virus and virus yield.

Reference

Albertsson, P. A. (1971) *Partition of Cell particles and Macromolecules*, 2nd edn, Almqvist Wiksell, Stockholm; John Wiley, New York.

Solution

$$\delta_c = \frac{V_0}{V_b\left(1 + \dfrac{V_t}{V_b}K\right)}$$

$$V_t = V_0 + V - V_b$$

$$= 3.0 \times 10^{-3} + 1.0 \times 10^{-4} - 5.0 \times 10^{-5}$$

$$= 3.05 \times 10^{-3}\,(\text{m}^3)$$

$$\delta_c = \frac{3.0 \times 10^{-3}}{5.0 \times 10^{-5}\left(1 + \dfrac{3.05 \times 10^{-3}}{5.0 \times 10^{-5}} \times 4 \times 10^{-4}\right)}$$

$$= 59.5$$

$$Y = 59.5 \times \frac{5.0 \times 10^{-5}}{3.0 \times 10^{-3}}$$

$$= 0.975$$

$$= 97.5\,(\%)$$

PROBLEM 9.20

A rotary-drum vacuum filter with string discharge is used to separate starch from a slurry.

The filtration rate is $8.33 \times 10^{-3}\,\text{m}^3\,\text{s}^{-1}$ at constant vacuum with 6.93×10^4 $\text{N}\,\text{m}^{-2}$.

The filter medium resistance is assumed to be negligible.

(a) If the rotating rate is increased from 0.3 rpm to 0.5 rpm, determine the increase of the filtration ability.
(b) If the drum surface area which is partly submerged in an open tank of slurry is increased from 30% to 40%, determine the increase of the filtration ability.

Solution

(a) The filtration equation for a rotary-drum vaccum filter is expressed as

$$V^2 + 2V_0V = K\frac{R_{\text{Area}}}{n} \tag{9.20.1}$$

where

V = volume of filtrate per revolution (m^3),
n = rotational rate (s^{-1}),
R_{Area} = ratio of the area which is submerged in the liquid to that of the total drum surface (—),
K = filtration parameter, defined as

$$K = \frac{2(-\Delta P)A^2}{\mu\alpha W_1} \qquad (\text{m}^6\text{s}^{-1})$$

μ = viscosity of solution (kg m^{-1} s^{-1}),
$(-\Delta P)$ = pressure drop through the cake (Nm^{-2}),
A = filtration area (m^2),
α = specific cake resistance (m kg^{-1}),
W_1 = weight of the dry cake per volume of filtrate (kg m^{-3}),
V_0 = filtration parameter, defined as

$$V_0 = \frac{AR_{\text{M}}}{\alpha W_1} \qquad (\text{m}^3)$$

R_{M} = filter medium resistance (m^{-1}).

If the filter medium resistance, R_{M} is negligible, then the equation (9.20.1) is simplified as:

$$V^2 = K\frac{R_{\text{Area}}}{n} \tag{9.20.2}$$

V is given from equation (9.20.2)

$$V = \sqrt{\frac{KR_{\text{Area}}}{n}} \tag{9.20.3}$$

The ratio of the volume of filtrate per minute by rotating filter operating at n' rpm to that at n rpm gives

$$\frac{V'}{V} = \frac{V'n'}{Vn} = \sqrt{\frac{kR_{\text{Area}}/n'}{kR_{\text{Area}}/n}}\;\frac{n'}{n} = \sqrt{\frac{n'}{n}} \tag{9.20.4}$$

where

V' = volume of filtrate per a revolution by filter operating at n' rpm ($m^3 min^{-1}$),
V = volume of filtrate per a revolution by filter operating at n rpm ($m^3 min^{-1}$).

Substituting $n = 0.3$ rpm and $n' = 0.5$ rpm into equation (9.20.4) gives

$$\frac{V'}{V} = \sqrt{\frac{0.5}{0.3}} = 1.29$$

Then, the filtration ability increase by 29%.

The ratio of the volume of filtrate per minute between two filters is calculated from equation (9.20.4):

$$\frac{V'}{V} = \frac{V'n}{Vn} = \sqrt{\frac{KR'_{\text{Area}}}{n}} \Big/ \sqrt{\frac{KR_{\text{Area}}}{n}} = \sqrt{\frac{R'_{\text{Area}}}{R_{\text{Area}}}}$$

$$= \sqrt{\frac{0.4}{0.3}}$$

$$= \underline{1.15}$$

An increase in the filtration performance by 15% takes place by increasing the submerged area from 30% to 40%.

10

Bioproject engineering

Contributors: Purnendu Ghosh, K. B. Ramachandran

PROBLEM 10.1

Show by computer simulation that for alcohol production two CSTR in series will give more productivity than a single CSTR for the same conversion. Also show that to achieve optimum productivities, the second reactor should be smaller than the first reactor. Assume the following kinetics and parameter values as given by Lee *et al*. (1983).

Substrate consumption is given by the expression:

$$\frac{dc_S}{dt} = -\frac{1}{Y_{X/S}}\frac{dx}{dt} \tag{10.1.1}$$

$$\frac{dc_P}{dt} = \frac{1}{Y_{X/P}}\frac{dx}{dt} \tag{10.1.2}$$

and growth is given by

$$\mu = \mu_0\left(1 - \frac{c_P}{c_{P_m}}\right)^n\left(1 - \frac{x}{x_m}\right)^m \frac{c_S}{c_S + K_S} \tag{10.1.3}$$

Parameter values are:
Saturation constant K_S 1.6 kg m^{-3}
Maximum specific growth rate

$$\mu_0 = 0.249\ \mathrm{h}^{-1}$$

Yield

$$Y_{X/P} = 0.16$$
$$Y_{X/S} = 0.06$$

Product concentration above which growth rate becomes zero, $c_{P_m} = 90$ kg m^{-3}
Exponents in the equations (10.1.1), (10.1.2) and (10.1.3), $m = n = 1$

Cell concentration above which growth rate becomes zero, $x_m = 100\ \text{kg m}^{-3}$

Assume inlet substrate is $100\ \text{kg m}^{-3}$ and conversion is 98%. Compare the productivity with a single CSTR.

Reference

Lee, J. M., Pollard, J. F. & Coulman, G. A. (1983) *Biotechnol. Bioeng.*, **25**, 497.

Solution

The system can be schematically shown as in Fig. 10.1, where F = flow rate of substrate, x = cell mass concentration, c_s = substrate concentration, c_p = product concentration, and V_1, V_2 = reactor volumes.

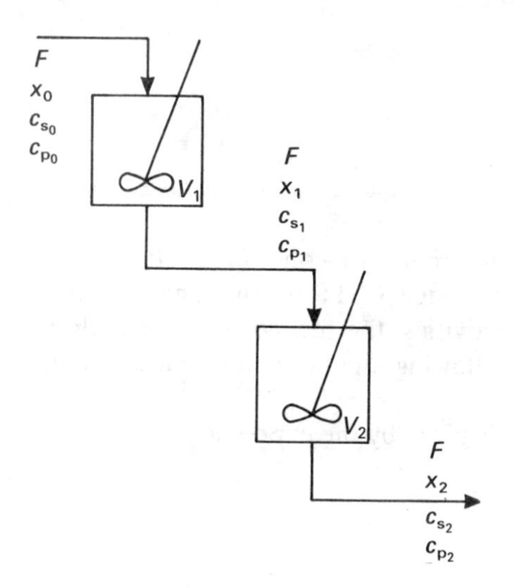

Fig. 10.1.

Material balance for cell, substrate and product over the reactor gives.

$$\frac{d}{dt}(Vx) = Fx_0 - Fx + Vq_x \tag{10.1.4}$$

$$\frac{d}{dt}(Vc_s) = Fc_{s_0} - Fc_s + Vq_s \tag{10.1.5}$$

$$\frac{d}{dt}(Vc_p) = Fc_{p_0} - Fc_p + Vq_p \tag{10.1.6}$$

At steady state,

$$x = Y_{X/S}(c_{s_0} - c_s) + x_0 \tag{10.1.7}$$

$$c_{\mathrm{p}} = (Y_{\mathrm{X/S}})/(Y_{\mathrm{X/P}})(c_{\mathrm{s_0}} - c_{\mathrm{s}}) + c_{\mathrm{p0}} \tag{10.1.8}$$

and dilution rate, $D = F/V$

$$D = \frac{\mu_0(1 - c_{\mathrm{p}}/c_{\mathrm{p_m}})^n \, (1 - x/x_{\mathrm{m}})^m \, c_{\mathrm{s}}x}{(x - x_0)\,(K_{\mathrm{s}} + c_{\mathrm{s}})} \tag{10.1.9}$$

The above equations can be solved by the computer program given here for two reactors in series. From Fig. 10.2 it can be seen that the productivity (Dc_{p}) of single reactor is only 2.7287 kg $\mathrm{m^{-3}\,h^{-1}}$ whereas the productivity for the combined system is 3.989 kg $\mathrm{m^{-3}h^{-1}}$ at a volume fraction $\alpha = (V_1/(V_1 + V_2)) = 0.86$. This means the first reactor should be 0.86/0.14 = 6.14 times larger than the second reactor. The relative productivity in the program is defined as the productivity of the combined reactor divided by the productivity of a single reactor of same volume.

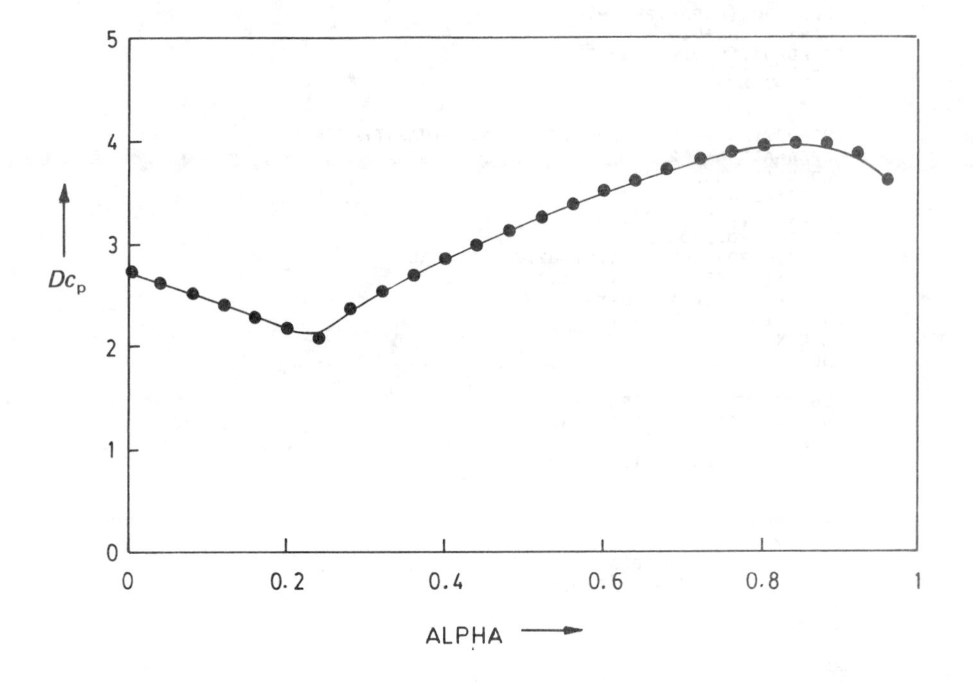

Fig. 10.2.

```
-----------------------------------------------------------------------
C         COMPUTER SIMULATION OF ETHANOL PRODUCTION IN A TWO STAGE
C         C S T R  CASCADE
C-----------------------------------------------------------------------
          REAL MUO,KS
          READ(5,*) MUO,PM,XM,KS,YXS,YXP,SO,FC,XO,PO
          M=2
          S2=SO*(1.0-FC)
          X2=YXS*(SO-S2)
          P2=YXS*(SO-S2)/YXP
          D1MAX=MUO*SO/(KS+SC)
```

```
          AC=MUO/PM/XM
          A1=AO*(PM-P2)*(XM-X2)*S2*X2
          A=A1/YXS/(KS+S2)
          S1=SO
          ALPHA=0.00000
          D2=A/(S1-S2)
          PROSIN=D2*(1-ALPHA)*P2
          PRO=PROSIN
          RELPRO=PRO/PROSIN
          WRITE(6,20)ALPHA,S1,D1,D2,PRO,RELPRO
   20     FORMAT(5X,F5.2,4X,F7.3,4(4X,F7.4))
          CALL PLOT(ALPHA,PRO,M)
          CALL WAZOUT(ALPHA,S1,D2,P2,D1MAX,PROSIN,M)
   10     S1=S1-0.01000
          X1=XO+YXS*(SO-S1)
          P1=PO+YXS*(SO-S1)/YXP
          D1I=AO*(PM-P1)*(XM-X1)*S1*X1
          D1=D1I/(X1-XO)/(KS+S1)
          D2=A/(S1-S2)
          DIFF=D2-D1*ALPHA/(1-ALPHA)
          IF(ABS(DIFF).GT.1.0E-2) GO TO 10
          PRO=D1*ALPHA*P2
          RELPRO=PRO/PROSIN
          WRITE(6,20) ALPHA,S1,D1,D2,PRO,RELPRO
          CALL PLOT(ALPHA,PRO,M)
          ALPHA=ALPHA+0.02
          IF(ALPHA.LE.0.98) GO TO 10
          STOP
          END

          SUBROUTINE WAZOUT(ALPHA,S1,D2,P2,D1MAX,PROSIN,M)
   10     ALPHA=ALPHA+.02
          PRO=D2*(1-ALPHA)*P2
          D1=D2*(1-ALPHA)/ALPHA
          IF(D1.LT.D1MAX) RETURN
          RELPRO=PRO/PROSIN
          WRITE(6,20) ALPHA,S1,D1,D2,PRO,RELPRO
   20     FORMAT(5X,F5.2,5(4X,F7.4))
          CALL PLOT(ALPHA,PRO,M)
          GO TO 10
   30     RETURN
          END

          SUBROUTINE PLOT(ALPHA,PRO,M)
          INTEGER IROW(80),B,D,Y,AX
          DATA B,D,AX/' ','.','|'/
          IF(M.EQ.1) GO TO 15
          DO 5 I=1,80
   5      IROW(I)=B
          Y=20*PRO+.5
          IROW(Y)=D
          WRITE(7,10)PRO,ALPHA,AX,IROW
   10     FORMAT(F7.2,2X,F5.2,T20,81A1)
          M=1
          RETURN
   15     M=2
          RETURN
          END
```

	ALPHA	S1	D1	D2	PRO	RELPRO
1	0.00	100.000	0.0000	0.0743	2.7287	1.0000
2	0.02	100.000	3.6383	0.0743	2.6741	0.9800
3	0.04	100.000	1.7820	0.0743	2.6195	0.9600
4	0.06	100.000	1.1633	0.0743	2.5650	0.9400
5	0.08	100.000	0.8539	0.0743	2.5104	0.9200
6	0.10	100.000	0.6683	0.0743	2.4558	0.9000
7	0.12	100.000	0.5445	0.0743	2.4012	0.8800
8	0.14	100.000	0.4561	0.0743	2.3467	0.8600

9	0.16	100.000	0.3898	0.0743	2.2921	0.8400
10	0.18	100.000	0.3383	0.0743	2.2375	0.8200
11	0.20	100.000	0.2970	0.0743	2.1836	0.8000
12	0.22	100.000	0.2633	0.0743	2.1284	0.7800
13	0.24	99.990	0.2362	0.0743	2.0834	0.7635
14	0.26	99.980	0.2362	0.0743	2.2569	0.8271
15	0.28	93.944	0.2292	0.0791	2.3584	0.8643
16	0.30	87.548	0.2218	0.0851	2.4453	0.8961
17	0.32	81.752	0.2151	0.0912	2.5297	0.9271
18	0.34	76.477	0.2090	0.0977	2.6128	0.9573
19	0.36	71.642	0.2035	0.1045	2.6923	0.9866
20	0.38	67.198	0.1984	0.1116	2.7708	1.0154
21	0.40	63.084	0.1937	0.1191	2.8470	1.0434
22	0.42	59.260	0.1893	0.1271	2.9215	1.0707
23	0.44	55.697	0.1852	0.1355	2.9943	1.0973
24	0.46	52.364	0.1813	0.1445	3.0653	1.1234
25	0.48	49.230	0.1777	0.1541	3.1345	1.1487
26	0.50	46.288	0.1743	0.1643	3.2020	1.1735
27	0.52	43.515	0.1710	0.1753	3.2679	1.1976
28	0.54	40.882	0.1679	0.1871	3.3319	1.2211
29	0.56	38.390	0.1649	0.2000	3.3941	1.2439
30	0.58	36.028	0.1621	0.2138	3.4546	1.2660
31	0.60	33.775	0.1593	0.2290	3.5131	1.2875
32	0.62	31.623	0.1567	0.2456	3.5694	1.3081
33	0.64	29.571	0.1541	0.2639	3.6237	1.3280
34	0.66	27.599	0.1515	0.2842	3.6154	1.3470
35	0.68	25.718	0.1491	0.3068	3.7250	1.3651
36	0.70	23.906	0.1466	0.3322	3.7715	1.3822
37	0.72	22.164	0.1442	0.3609	3.8151	1.3981
38	0.74	20.482	0.1418	0.3937	3.8558	1.4127
39	0.76	18.871	0.1393	0.4313	3.8913	1.4261
40	0.78	17.309	0.1369	0.4753	3.9229	1.4377
41	0.80	15.798	0.1343	0.5274	3.9491	1.4472
42	0.82	14.328	0.1317	0.5902	3.9686	1.4543
43	0.84	12.908	0.1289	0.6671	3.9802	1.4587
44	0.86	11.518	0.1260	0.7645	3.9814	1.4591
45	0.88	10.168	0.1228	0.8909	3.9702	1.4549
46	0.90	8.848	0.1192	1.0626	3.9412	1.4443
47	0.92	7.523	0.1150	1.3140	3.8869	1.4243
48	0.94	6.248	0.1099	1.7130	3.7959	1.3911
49	0.96	4.938	0.1032	2.4769	3.6418	1.3346
50	0.98	3.588	0.0936	4.5828	3.3695	1.2348
	1.00				2.7287	1.0000

PROBLEM 10.2

Fluid flow behaviour in an anaerobic sludge blanket reactor (UASB) has been studied by Heertjes and Van der Meer for waste-water treatment by means of stimulus–response experiment with Li^+ tracer. The 30 m^3 reactor consists of three compartments: sludge bed (25%), sludge blanket (70%) and the settler (5%). The fluid flow in the settler, being more or less laminar, is described as plug flow. The sludge blanket is described as a perfectly mixed region. Nearly 20% of the sludge bed is considered as dead zone and the remaining portion as well-mixed region. Bypassing (30%) and return flow occurs in the sludge bed. Tracer (concentration = 10 kg m^{-3}) is introduced in the reactor at a flow rate of 10 $m^3\ h^{-1}$. For this system, find the concentration — time tracer profile at different sections of the reactor.

Reference

Heertjes, P. M. and Van der Meer, R. R. (1978) *Biotechnol. Bioeng.* **20**, 1577.

Solution

Based on the description of flow behaviour, a flow diagram as shown in Fig. 10.3 can be proposed. A mass balance of the tracer can be written as

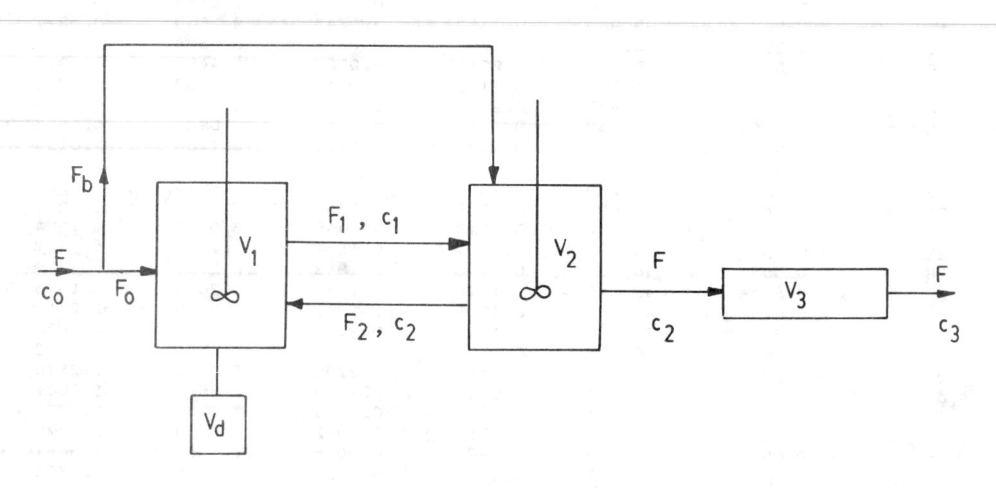

Fig. 10.3.

$$V_1\frac{dc_1}{dt} = F_0c_0 + F_2c_2 - F_1c_1 \tag{10.2.1}$$

$$V_2\frac{dc_2}{dt} = F_1c_1 + F_bc_0 - F_2x_2 - Fc_2 \tag{10.2.2}$$

Rewriting (10.2.1) and (10.2.2) as

$$(D-m)c_1 - nc_2 = l \tag{10.2.3}$$

$$-qc_1 + (D-r)c_2 = p \tag{10.2.4}$$

Using Cramer's rule and simplifying, we can write

$$D^2c_1 - (m+r)Dc_1 + (mr-qn)c_1 = pn - rl \tag{10.2.5}$$

$$D^2c_2 - (m+r)Dc_2 + (mr-qn)c_2 = ql - mp \tag{10.2.6}$$

where $D = d/dt$; $l = F_0c_0/V$; $m = -F_1/V_1$; $n = F_2/V_1$; $p = F_0c_0/V_2$; $q = F_1/V_2$; $r = -(F_2+F)/V_2$

Putting $F = 10\ m^3\ h^{-1}$; $c_0 = 10\ kg\ m^{-3}$; $V = 30\ m^3$,

$V_1 = 0.2V = 6\ m^3$ $\quad$ $F_b = 0.3F = 3\ m^3\ h^{-1}$
$V_2 = 0.7V = 21\ m^3$ $\quad$ $F_1 = 2.3F = 23\ m^3\ h^{-1}$
$V_3 = 0.05V = 1.5\ m^3$ $\quad$ $F_2 = 1.6F = 16\ m^3\ h^{-1}$
$V_d = 0.05V = 1.5\ m^3$

Using the above values, equations (10.2.5) and (10.2.6) can be written as

$$D^2c_1 + 5.1\ Dc_1 + 1.83\ c_1 = 18.3 \tag{10.2.7}$$

$$D^2c_2 + 5.1\ Dc_1 + 1.83\ c_2 = 18.3 \qquad (10.2.8)$$

Solution for c_1 and c_2 will be dependent on the boundary conditions. Solving for c_1 with boundary conditions

$$t = 0;\ c_1 = 0$$

$$t = \infty;\ \frac{dc_1}{dt} = 0$$

gives

$$c_1 = -10.9\ \exp(-0.4t) + 0.9\ \exp(-4.7t) + 10 \qquad (10.2.9)$$

solution for c_2 in

$$c_2 = A\ \exp(-0.4t) + B\ \exp(-4.7t) + 10 \qquad (10.2.10)$$

Using boundary condition

$$t = \infty;\ \frac{dc_2}{dt} = 0$$

$$A = -11.75B$$

To find the time up to which c_2 is zero, write equation (10. 2.3) with $c_2 = 0$ as

$$Dc_1 = l + mc_1 \qquad (10.2.11)$$

Substituting equation (10.2.9) in (10.2.10) and simplifying,

$$37.4\ \exp(-0.4t) + 0.8\ \exp(-4.7t) - 26.6 = 0 \qquad (10.2.12)$$

By trial and error $t \simeq 0.85$ h; i.e. up to this time c_2 is zero and at $t > 0.85$ h, c_2 increases.

Using the boundary condition

$$t = 0.85\ c_2 = 0$$

the solution of c_2 is

$$c_2 = -14.1\ \exp(-0.4t) + 1.2\ \exp(-4.7t) + 10 \text{ for } t \geqslant 0.85.$$

The stimulus–response curve of the tracer is shown in Fig. 10.4.

In the present example, a tracer profile is drawn for one set of parameters while taking different flow behaviour in different parts of the reactor. The possibility of bypass, return flow and dead volume are considered. If the experimental tracer output matches with that of the predicted one, then the proposed assumptions are correct. However, if the experimental and predicted values do not match, another set of assumptions (based on experimental observations) will be needed for a better correlation between experiment and theory.

PROBLEM 10.3

A molasses-based ethanol plant discharges liquid effluent containing 90 kg m^{-3} COD. Each cubic metre of alcohol produced results in 12 m^3 of effluent. Three

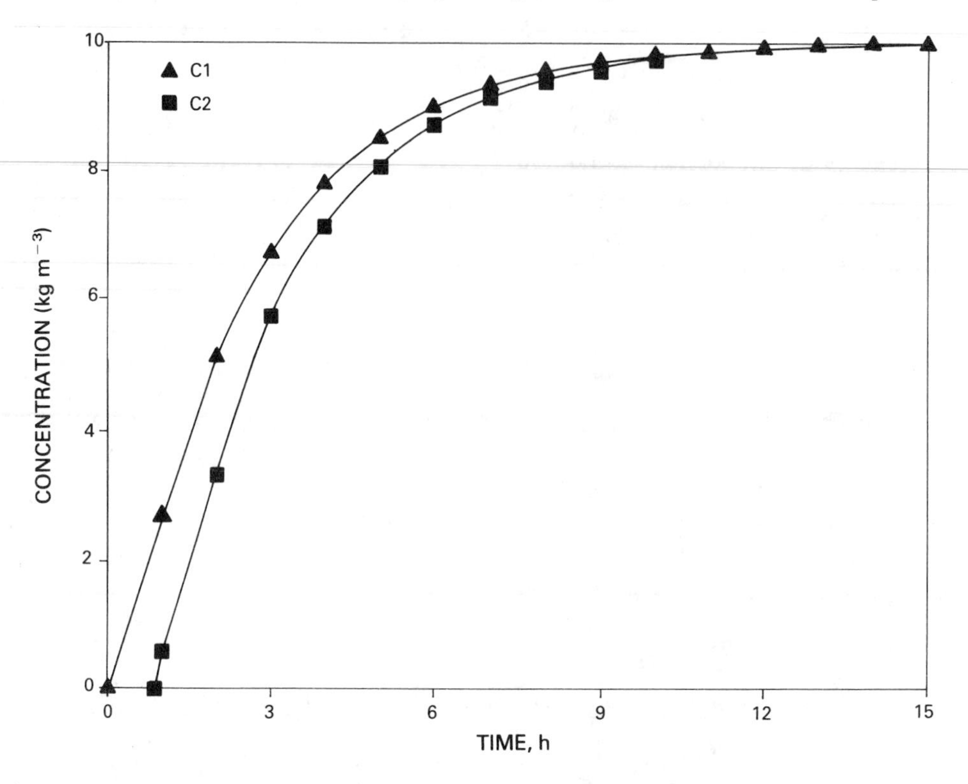

Fig. 10.4.

options are available for the treatment of distillery effluent: anaerobic digestion, incineration/evaporation, and concentrate as animal feed. In the anaerobic digestion process, the methane yield is 0.3 m^3 per kg of COD removed and COD reduction is 75%. The biogas contains 65% methane. In the incineration process, the effluent containing 9% total solids is concentrated to 60% in a multiple-effect evaporator. The solids when incinerated generate sufficient heat so that after meeting evaporation steam requirement, a net surplus steam is available. The resultant ash containing 30–40% potash can be used as high-value potassium fertilizer. In the third option, the concentrated effluent is used as animal feed.

Suggest a process option which would give the minimum pay-off period. What would be the impact of these process options on ethanol cost? Consider the following data for analysis:

1. Capacity of ethanol plant $= 100\ m^3\ d^{-1}$
2. Working period $= 300$ days $(\text{year})^{-1}$
3. Capital investment
   - Anaerobic digestion $= \$3.5 \times 10^6$
   - Incineration $= \$6.4 \times 10^6$
   - Animal feed $= \$3.0 \times 10^6$
4. Annual operating cost
   - Anaerobic digestion $= \$0.85 \times 10^6$

Incineration	$= \$1.6 \times 10^6$
Animal feed	$= \$1.3 \times 10^6$
5. Depreciation	= 10% per year
6. Yield of animal feed	= 90 kg per m^3 effluent
7. Yield of potash fertilizer (ash)	= 0.025 tonnes per m^3 effluent
8. Surplus steam generated	= 0.15 tonnes per m^3 effluent
9. Cost of production of ethanol	$= \$0.30\ (\text{litre})^{-1}$
10. Product cost	
Animal feed concentrate	$= \$35\ (\text{tonne})^{-1}$
Ash from incinerator	$= \$90\ (\text{tonne})^{-1}$

Heating value of biogas containing 65% methane = 25 MJ m^{-3}. Steam equivalent 3.4 MJ kg^{-1} steam

Steam cost $= \$15\ (\text{tonne})^{-1}$

Assume 80% efficiency for conversion of biogas to steam. Therefore credit from biogas $= (25 \times 0.8/3.4) \times 15 \times 10^{-3} = \$0.088$ per m^3 biogas.

Solution

Effluent generated per day = 1200 m^3.

(a) Anaerobic digestion

Biogas generated per day $= 1200 \times 90 \times 0.75 \times \frac{1}{0.65} \times 0.3$
$= \underline{37{,}384.6\ m^3}$

Revenue per year $= 37{,}384.6 \times 0.088 \times 300$
$= \underline{\$0.98 \times 10^6}$

(b) Incineration

Surplus steam available per day $= 0.15 \times 1200$
= 180 tonnes

Potash fertilizer (ash) per day $= 0.025 \times 1200$
= 30 tonnes

Revenue generated per year $= \$[(180 \times 15) + (30 \times 90)]300$
$= \underline{\$1.62 \times 10^6}$

(c) Animal feed

Animal feed concentrate available per day $= 0.09 \times 1200$
= 108 tonnes

Revenue from animal feed per year $= 108 \times 300 \times 35$
$= \$1.13 \times 10^6$

$$\text{Pay-off period} = \frac{\text{Investment on treatment plant}}{\begin{matrix}\text{Revenue generated} \\ \text{by effluent treatment plant}\end{matrix} - \begin{matrix}\text{Undepreciated} \\ \text{operating cost}\end{matrix}}$$

Pay-off period

(a) Anaerobic digestion $\dfrac{3.5 \times 10^6}{0.98 \times 10^6 - (0.85 - 0.35)10^6}$

$= \underline{7.30\ \text{years.}}$

(b) Incineration

$$= \frac{6.4 \times 10^6}{1.62 \times 10^6 - (1.6 - 0.64)10^6}$$

$$= \underline{9.7 \text{ years}}$$

(c) Animal feed

$$= \frac{3 \times 10^6}{1.13 \times 10^6 - (1.3 - 0.3)10^6}$$

$$= \underline{23.0 \text{ years.}}$$

The treatment process will have impact on the cost of ethanol production. Additional investment on effluent treatment plant in an alcohol distillery will increase the cost of ethanol production. However, revenue generated from the treatment plant may offset this additional investment. Impact of treatment facility on ethanol cost is given in Table 10.1. The table indicates that additional investment for the installation of treatment facilities does not change ethanol cost. However, the anaerobic digestion process has minimum pay-off period and this indicates better economic incentive for this process.

Table 10.1

Process	Percentage change in ethanol cost
Anaerobic digestion	−1.4
Incineration	−0.2
Animal feed	+1.9

PROBLEM 10.4

By computer simulation for the two-stage CSTR with cell recycle system schematically shown in Fig. 10.5:

(1) Find out the effect of recycle on the productivity as a function of the ratio of the volumes of two CSTRs in series for a constant bleed rate ($B/F = 0.4$).
(2) Find out the effect of bleed stream on productivity for constant recycle rate ($R/F = 1.0$).

Use the data given in problem 10.1 for computer simulation.

A cell balance equation around the separation unit gives

$$(L + R)x = Rx_R$$

Material balance around the entire system gives

$$\frac{d(Vx)}{dt} = F_0 x_0 - Bx + Vq_x$$

$$\frac{d(Vc_s)}{dt} = F_0 c_{s0} - (B + L)c_s + Vq_s$$

$$\frac{d(Vc_p)}{dt} = F_0 c_{p0} - (B + L)c_p + Vq_p$$

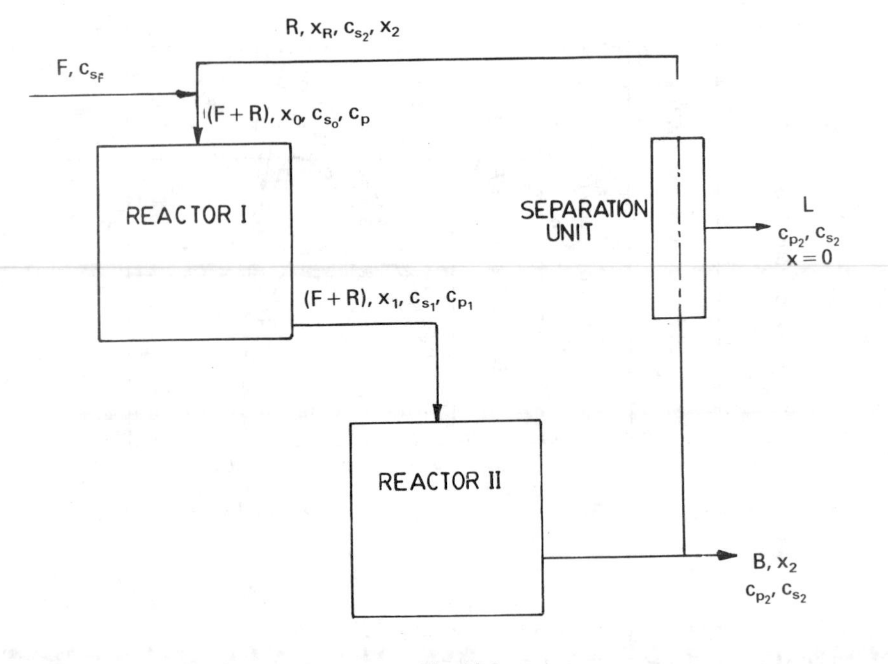

Fig. 10.5.

At steady state, the above equations after substituting for kinetic equations can be written as

$$x = (F/B)\ [Y_{X/S}(c_{s_0} - c_s) + x_0]$$

$$c_p = (Y_{x/s}/Y_{x/p})(c_{s_0} - c_s) + c_{p0}$$

$$D = \frac{\mu(1 - c_p/c_{pm})^n(1 - x/x_m)^m c_s x}{[(B/F)x - x_0]\ (K_s + c_s)}$$

Productivity $D_1 \alpha c_{p2} = Dp_2$, where $\alpha = V_1/V_1 + V_2$

For $c_{s_f} = 100\ \mathrm{kg\ m}^3$, 98% conversion means $x_{s_2} = 2\ \mathrm{kg\ m}^{-3}$. The program and the computer simulation are shown (Fig. 10.6). It can be seen for all values of α productivity in decreasing with increase in recycle ratio owing to feedback inhibition by alcohol. Also for a given R/F, productivity showed optimum at a single value of α. Again with cell recycle system also, a larger reactor followed by a smaller reactor is optimal.

Fig. 10.7 shows the effect of B/F at a fixed value of $R/F = 1.0$. Reducing the bleed stream increased the productivity. Reducing the bleed stream B/F from 0.4 to 0.2 increased the productivity by more than 65%.

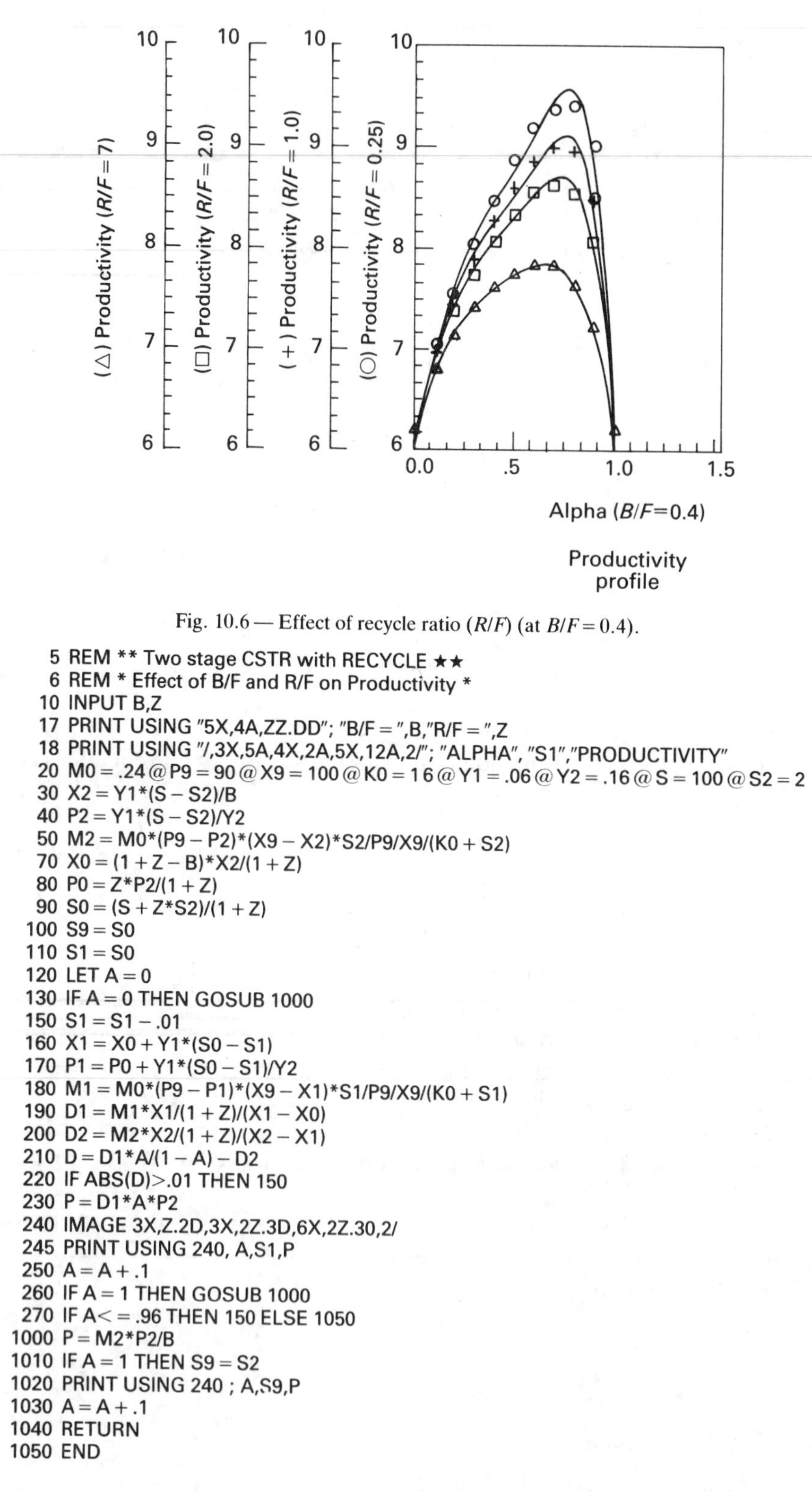

Fig. 10.6 — Effect of recycle ratio (R/F) (at $B/F = 0.4$).

```
5 REM ** Two stage CSTR with RECYCLE ★★
6 REM * Effect of B/F and R/F on Productivity *
10 INPUT B,Z
17 PRINT USING "5X,4A,ZZ.DD"; "B/F = ",B,"R/F = ",Z
18 PRINT USING "/,3X,5A,4X,2A,5X,12A,2/"; "ALPHA", "S1","PRODUCTIVITY"
20 M0 = .24 @ P9 = 90 @ X9 = 100 @ K0 = 16 @ Y1 = .06 @ Y2 = .16 @ S = 100 @ S2 = 2
30 X2 = Y1*(S - S2)/B
40 P2 = Y1*(S - S2)/Y2
50 M2 = M0*(P9 - P2)*(X9 - X2)*S2/P9/X9/(K0 + S2)
70 X0 = (1 + Z - B)*X2/(1 + Z)
80 P0 = Z*P2/(1 + Z)
90 S0 = (S + Z*S2)/(1 + Z)
100 S9 = S0
110 S1 = S0
120 LET A = 0
130 IF A = 0 THEN GOSUB 1000
150 S1 = S1 - .01
160 X1 = X0 + Y1*(S0 - S1)
170 P1 = P0 + Y1*(S0 - S1)/Y2
180 M1 = M0*(P9 - P1)*(X9 - X1)*S1/P9/X9/(K0 + S1)
190 D1 = M1*X1/(1 + Z)/(X1 - X0)
200 D2 = M2*X2/(1 + Z)/(X2 - X1)
210 D = D1*A/(1 - A) - D2
220 IF ABS(D)>.01 THEN 150
230 P = D1*A*P2
240 IMAGE 3X,Z.2D,3X,2Z.3D,6X,2Z.30,2/
245 PRINT USING 240, A,S1,P
250 A = A + .1
260 IF A = 1 THEN GOSUB 1000
270 IF A< = .96 THEN 150 ELSE 1050
1000 P = M2*P2/B
1010 IF A = 1 THEN S9 = S2
1020 PRINT USING 240 ; A,S9,P
1030 A = A + .1
1040 RETURN
1050 END
```

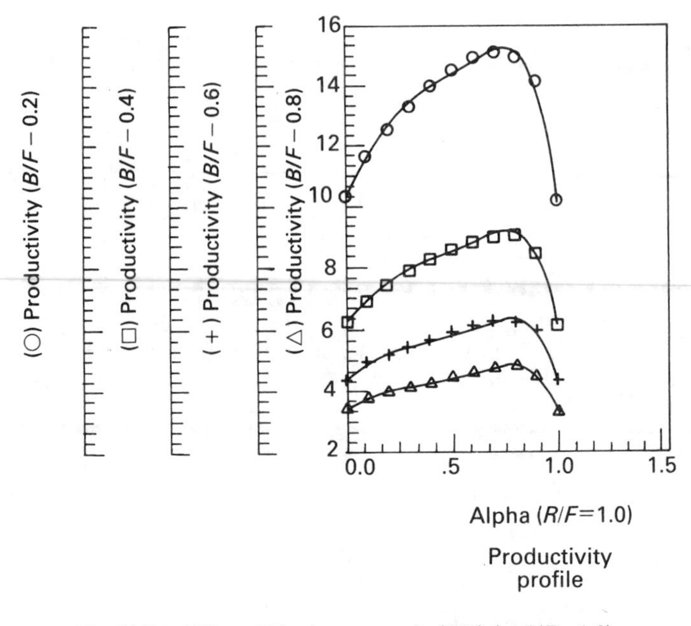

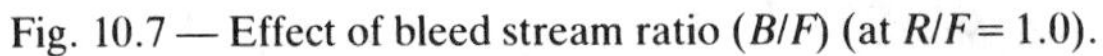

Fig. 10.7 — Effect of bleed stream ratio (B/F) (at $R/F = 1.0$).

11

Unsolved problems on various topics

Contributors: Tarun Ghose, John Villadsen, K. Suga, V. S. Bisaria, Subhash Chand, K. B. Ramachandran, S. N. Mukhopadhyay

PROBLEM 11.1

Free energies of hydrolysis of a few phosphorylated compounds (which transfer phosphate groups in many biological reactions) are as given in Table 11.1. What information do these data provide about the relative rates of hydrolysis?

Table 11.1

Compound	Standard free energy change $\Delta G^{\circ\prime}$ (J mol^{-1})
Phosphoenol pyruvate	−61,923.2
Acetyl phosphate	−45,187.2
Pyrophosphate	−33,472.0
ATP (to ADP)	−30,543.2

PROBLEM 11.2

Glutamate dehydrogenase catalyses the following reaction:

$$\alpha\text{-ketoglutarate} + \text{NADPH} + \text{NH}_4^+ \rightarrow \text{glutamate} + \text{NADP}^+ + \text{H}_2\text{O}$$

If the intracellular ratios of glutamate/α-ketoglutarate and NADPH/NADP$^+$ are each 10 and K_{eq} for the reaction is 9.55×10^5, calculate the minimum intracellular concentration of NH_4^+ required to drive the reaction towards glutamate formation.

PROBLEM 11.3

The value of standard redox potential, E_0' of succinate to fumarate is:

$$E_o' = 0.0331 \text{ V at pH 7 and 25°C} \; .$$

Compute the redox potential of the system at the concentration of 70% of fumarate and 30% of succinate in the reaction mixture.

PROBLEM 11.4

Determine the value of free energy change $\Delta G'$ in the undernoted reactions and also assess the efficiency of ATP formation. The value of free energy change of ATP hydrolysis can be taken as 50.2 kJ mol^{-1}.

(a) Ethanol formation by yeasts

$$C_6H_{12}O_6 \rightarrow 2CH_3.CH_2.OH + 2CO_2$$

with net production of 2 ATP/mol glucose.

(b) Pyruvate oxidation

$$CH_3.CO.COOH + 2.5O_2 \rightarrow 3CO_2 + 2H_2O$$

with net production of 15 ATP/mol pyruvic acid.

The standard free energy of formation of each molecular species involved are as given in Table 11.2.

Table 11.2

Molecular species	Standard free energy of formation (kJ mol^{-1})
$C_6H_{12}O_6$	−908.01
C_2H_5OH	−181.96
CO_2	−394.38
H_2O	−237.19
$CH_3COCOOH$	−474.13

PROBLEM 11.5

The value of equilibrium constant, K_{eq} for the following reaction at 20°C, pH 7.8 is assumed as 3×10^3.

$$\text{3-phosphoglyceraldehyde} + H_3PO_4 + ADP + NAD^+ \rightleftharpoons \text{3-phosphoglyceric acid} + ATP + NADH + H^+ \qquad (11.1.5)$$

Calculate

(a) The value of standard free energy change $\Delta G°$ of the above reaction.

(b) By using other relevant data given below, assess the value of ΔG_1° for hydrolysis of ATP.

$$NAD^+ + 2H \rightleftharpoons NADH + H^+, \ \Delta G_2^\circ = 56{,}484 \text{ J mol}^{-1}$$

$$\text{3-phosphoglyceraldehyde} + H_2O \rightleftharpoons \text{3-phosphoglyceric acid} + 2H, \ \Delta G_3^\circ = -125{,}938.4 \text{ J mol}^{-1}$$

PROBLEM 11.6

An aerobic culture of *Aerobacter aerogenes* in a minimal medium (glycerine as the sole energy source) results:

$$Y_{X/S} = 41.3 \text{ cell mol}^{-1} \text{ substrate}$$

and

$$Y_{X/O} = 31.0 \text{ g cell mol}^{-1} \text{ O}_2$$

calculate, Y_Δ of cell (kJ^{-1}).

Given, $\Delta H_a = 22.175$ kJ g^{-1} cell

and

$$\Delta H_{ave^-} = 110.876 \text{ kJ (electron)}^{-1}$$

PROBLEM 11.7

Sulphur is oxidized by *Thiobacillus thiooxidans* as per following stoichiometry:

$$S + \tfrac{3}{2}\, O_2 + H_2O \rightarrow H_2SO_4$$

$$\Delta G^\circ = -493.7 \text{ kJ/g-atm S} \ .$$

Supposing that the oxidation of sulphur (35 g) was attained upon harvest of 2.4 g (dry) cells, what is the efficiency of free energy conserved in the bacterial growth?
Heat of combustion of cells is taken as 19.246 kJ g^{-1} cell.

PROBLEM 11.8

Hydrogenomonas assimilates ambient carbon dioxide, utilizing free energy of the following reaction.

$$H_2 + \tfrac{1}{2}O_2 \rightarrow H_2O;\ \Delta G^{\circ\prime} = -240.16 \text{ kJ mol}^{-1} \text{ H}_2$$

Based on the observation that 7.5 moles of hydrogen are oxidized in the assimilation of carbon dioxide to yield 24 g cells, calculate the efficiency of free energy change in the bacterial growth. Heat of combustion of dry cells and carbon content of the cells are 20.92 kJ g^{-1} cell and 50% respectively.

PROBLEM 11.9

When animal cells are exposed to a hypotonic medium, they swell and their cytoplasm leaks out, leaving an empty membrane or ghost. If these ghosts are subsequently placed in an isotonic medium, they shrink to their normal size and regain their usual impermeability. In this process a sample of the isotonic solution is trapped inside. The directionality of the linked $Na^+ - K^+$ pump has been demonstrated by analysing for ATP hydrolysis after resealing these ghosts in one medium and transferring them to another. From the results of the experiments in Table 11.3, predict which of the experiments will lead to hydrolysis of ATP.

Table 11.3 — Conditions of experiments on ATP hydolysis by resealed ghosts

Experiment[a]	ATP	Na^+	K^+
1	In	In	In
2	In	In	Out
3	In	Out	In
4	In	Out	Out
5	Out	In	In
6	Out	In	Out
7	Out	In	0
8	Out	Out	Out

[a]The ghosts have been resealed in one medium and then placed in another so that ATP, Na^+ and K^+ each are present inside (In) or outside (Out), *or* are absent (0) as indicated.

Reference

Wood, W. B., Wilson, J. H., Benbow, R. M. & Hood, L. E. (1981), *Biochemistry: A Problems Approach*, 2nd edn., The Bejamin Cummings Publishing Company, Menlo Park, California, p. 267.

PROBLEM 11.10

What is the maximum number of Na^+ ions that can be transported out of a cell per ATP hydrolysed by a simple Na^+ pump that operates independently of K^+. Given are the following data:

$$\text{external } [Na^+] = 170 \text{ mM}$$
$$\text{internal } [Na^+) = 20 \text{ mM}$$

The free energy change for hydrolysis of ATP to ADP and P_i is -52.3 kJ mol^{-1} under intracellular conditions and the temperature is 25°C.

PROBLEM 11.11

The value of denaturation rate constant, k of a spore suspension was found to be 0.25 min^{-1} at 109°C. What is the value of k at 117°C? The activation energy, E of the denaturation reaction can be taken as 2.88×10^5 J mol^{-1} in the spore suspension.

PROBLEM 11.12

If the following two oxidation — reduction systems are mixed and come into equilibrium, show the relation between the change of redox potential of the coupled reaction and the equilibrium constant

$$A \rightleftharpoons B + H_2$$

$$C + H_2 \rightleftharpoons CH_2$$

PROBLEM 11.13

An anaerobic culture of *Streptococcus agalactiae* by using a complex medium showed the scheme of intermediate metabolites given in Fig. 11.1 when pyruvic acid was used as energy source. Assess Y_{ATP}; given: $Y_{X/S} = 7.4$ g dry cell mol^{-1} pyruvic acid.

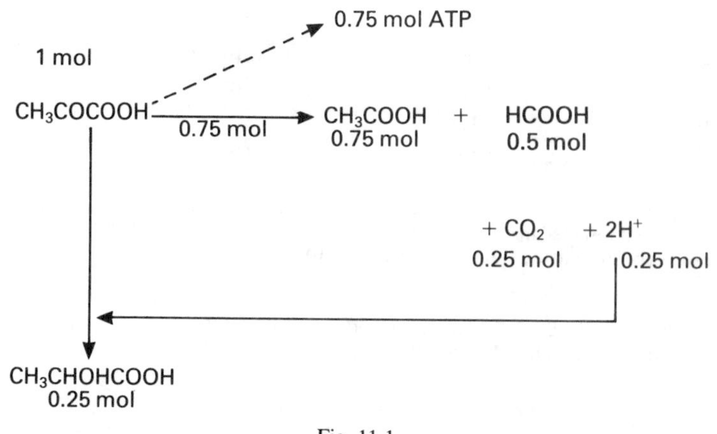

Fig. 11.1.

PROBLEM 11.14

A yeast species can grow on either a hydrocarbon (CH_2) or carbohydrate (CH_20) as substrate. The cell mass composition was determined as C, 47%; H, 6.5%; N, 7.5%; ash, 8.0%; and O, 31.0% (by difference). If the cell yield coefficient ($Y_{X/S}$) based on carbohydrate and hydrocarbon substrates are 15 g mol^{-1} and 14 g mol^{-1} respectively, compute the relative quantities of the oxygen required for cell growth using the two substrates separately.

Reference

Darlington, W. A. (1964), *Biotechnol. Bioeng.*, **6**, 241.

PROBLEM 11.15

The data on Figs 11.2(a) and (b) were used by Bijkerk and Hall (1977) in an attempt to set up a structured model for aerobic fermentation of glucose by *Saccharomyces cerevisiae*. They did have some trouble fitting their model to the data for $D > 0.25$ h^{-1} where considerable amounts of ethanol is produced besides biomass. We shall examine the data for consistency. The biomass formula is $CH_{1.83}O_{0.56}N_{0.17}$. It contains 8 wt% ash. Consequently 'cell dry matter' on the figure is higher than corresponding to the formula. The nitrogen source is NH_3.

(a) First consider $0.1 < D < 0.22$ h^{-1} where $x \simeq 14$ g l^{-1} and $s = p \sim 0$ in the exit from the chemostat.

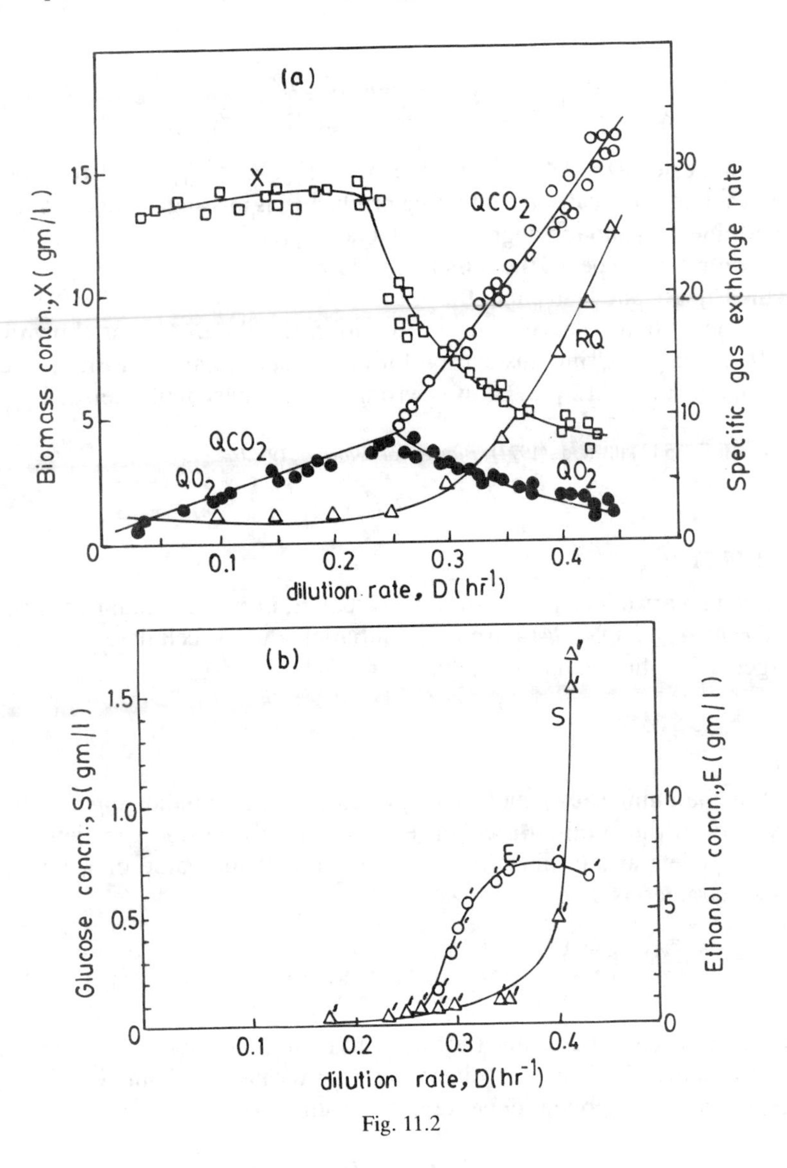

Fig. 11.2

Determine $Y_{X/S}$, $Y_{P/S}$, $Y_{O/S}$ and RQ and check that RQ fits the experimental data reasonably well.
Draw a straight line through the Q_{O_2} points and (0,0).
Check that the slope of the line corresponds reasonably well with the $Y_{O/S}$ determined above.
Find Q (kJ h^{-1} l^{-1} fermenter broth) for $D = 0.15$ h^{-1}.

(b) For $D > 0.25$ h^{-1} Q_{CO_2} and Q_{O_2} can be approximated by the straight lines:

$$Q_{CO_2} = 116\ \mathrm{D} - 20.4\ \frac{\text{mmol CO}_2}{\text{g dry matter h}}$$

$$Q_{O_2} = -29.1\,D - 15.3\ \frac{\text{mmol } O_2}{\text{g dry matter h}}$$

Determine for $D = 0.4\ h^{-1}$ and $0.3\ h^{-1}$, $Y_{X/S}$ and $Y_{P/S}$, and for both cases confirm that some carbon containing product P_1 is missing. For both values of D determine $Y_{P/S}$ and the degree of reduction γ'_{P1}.
Determine the 'type' of the missing product.

(c) In the Bijkerk and Hall paper the authors claim that some '12–13 wt% is missing from the carbon balance', i.e. not found in the three streams shown on Figs 11.1(a) and (b). Using this extra information determine the formula weight of the missing product P_1. What was wrong in the experimental set-up?

Reference

Bijkerk, A. H. E. and Hall, R. J. (1977), *Biotechnol. Bioeng.*, **19**, 267.

PROBLEM 11.16

Considering binary adhesive collision of mycelium, increase in number of pellets of *Trichoderma reesei* QM 9414 from its initiation during cellulase production by submerged cultivation in liquid medium is given by

$$\frac{dN}{dt} = \eta N^2 \qquad (11.16.1)$$

where N is the number of pellets and η denotes a factor indicating the degree of homogeneity in the broth. Based on equation (11.16.1) one can determine the number of pellets at any time from its initiation. If the value of η is related to fermenter parameters as

$$\eta = \left(\frac{\pi}{4K}\right)^3 \left(\frac{D_i}{D_t}\right)^{1+3\alpha} \frac{1}{N_p} \qquad (11.16.2)$$

determine the pellet radius at any time in terms of mass and density of the pellets and η. [K = constant, D_i = impeller diameter; D_t = fermenter diameter, N_p = power number and α is an exponent depending on broth rheology.]

Reference

Mukhopadyay, S. N. & Ghose, T. K. (1977), *Proc. Symp. Bioconversion of Cellulosic Substances into Energy, Chemicals and Microbial Protein*, (T. K. Ghose, ed.), Indian Institute of Technology-Delhi, New Delhi, pp. 97.
Pirt, S. J. (1975), *Principles of Microbes and Cell Cultivation*, Blackwell Sci. Pub., Oxford.

PROBLEM 11.17

The rate of interfacial subsidence, i_u of a yeast suspension in a cylinder under gravity was $i_u = 2.8 \times 10^{-3}$ m h^{-1}. Psychometric determinations gave the following data:

$\rho_c = 1.007 \times 10^3$ kg m^{-3} (yeast suspension density)
$\rho_m = 1.000 \times 10^3$ kg m^{-3} (liquid density)
$\rho_{cell} = 0.0780$ (volume fraction of yeast cells in the suspension)

Taking liquid viscosity, μ as 1.13×10^{-3} N sec m^{-2}, calculate the equivalent diameter, d_e of the yeast cells. The equivalent diameter is synonymous with the Stokes' resistance in the settling of suspension. Note that the law of additivity is applicable among the densities of ρ_c, ρ_m and ρ_y (yeast density) as follows:

$$\rho_c = f_{cell}\ \rho_y + (1 - f_{cell})\ \rho_m$$

PROBLEM 11.18

Citric acid

$$\begin{array}{c} H_2C-COOH \\ | \\ HO-C-COOH \\ | \\ H_2C-COOH \end{array}$$

is produced from glucose by fermentation with *Aspergillus niger* ($CH_{1.8}O_{0.5}N_{0.2}$) in an aerated tank. pH is between 1.8 and 2. The nitrogen source is NH_4NO_3. No CO_2 is formed.

(a) A typical yield at these conditions is 68 g citric acid per 100 g glucose. Determine the corresponding oxygen consumption assuming that nothing but biomass and citric acid is formed.

(b) What is the heat generation Q (kJ mol^{-1} glucose C)? Given: Δh°_c (NH_4NO_3) = 216 kJ mol^{-1}.

(c) Accidentally pH increases to 4. Analysis shows that the biomass yield $Y_{x/s}$ remains the same as before, but besides citric acid another organic acid with $Y_{p/s_1} = 0.1$ is formed. The oxygen consumption increases to 0.11875 mol O_2 mol^{-1} glucose C.
Which other organic acid P_1 is formed?

PROBLEM 11.19

Penicillin G has the formula

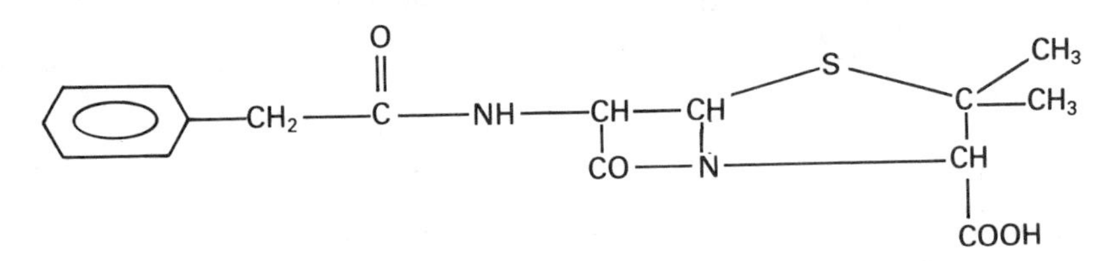

Fig. 11.3.

or ($C_{16}\ H_{18}\ O_4\ N_2S$).

It is produced as a secondary metabolite by *Penicillum crysogenum* $CH_{1.64}\ O_{0.52}\ N_{0.16}\ S_{0.0046}\ P_{0.0054}$ which grows on glucose with SO_4^{-2}, PO_4^{-3} and NH_3 as sources for

S, P and N respectively. To produce penicillin-G another substrate (S_2), phenylacetic acid must be available

$$\underset{(S_1)}{\text{Glucose}} + \underset{(S_2)}{\varphi - CH_2 - COOH} \rightarrow X + \underset{(P)}{\text{Pen} - G}$$

$$+ \text{ other substrates} \qquad + CO_2 \text{ and } H_2O$$

For a fed-batch fermentation in the laboratory the following average results were obtained for the first 120 hours of the production phase.

$$Y_{x/s_1} = 0.127 \text{ and } Y_{p/s_1} = 0.194$$
$$Y_{p/s_2} = 2 \text{ (exactly)}$$

All yield coefficients are in units of C-mole/C-mole.

1. Why can Y_{p/s_2} be set to 2? Any assumptions?

1a. Determine $Y_{o/s_1}\left(\dfrac{\text{mol } O_2}{\text{C-mol glucose}}\right)$

and the respiratory quotient RQ (the values determined are approximately averaged values for the first 120 h).

1b. Write down the complete stoichiometric equation per C-mol glucose.

2a. Make the assumption that the rate r_x of penicillin production r_p is proportional to the rate of biomass production. Discuss if this assumption is reasonable.
The assumption means assuming $Y_{p/x}$ constant.
If the assumption holds, examine the effect on productivity Y_{p/s_1} of changing the operating conditions so that $Y_{o/x}$ increases (more oxygen used per C-mole biomass formed).

2b. The on-line RQ measurements for the first 120 h show that (with considerable data scatter) RQ is in the range 0.95–1.15. After 150 h when the growth has virtually stopped, RQ seems to increase rapidly above 1.2. Discuss the probable causes of this effect. Is it due to break down of the assumption made in 2a? Are other products formed? Could there be some straight forward experimental error?

3. When growth stops the rate of penicillin formation is a C-moles h^{-1} and the rate of CO_2-formation is b moles $CO_2 \, h^{-1}$ (this is an assumption completely different from that of question 2a). Determine Y_{p/s_1} in terms of the constants a and b.

4. Calculate the heat of reaction per kg penicillin formed. Use the mass balance of question 1.

PROBLEM 11.20

Bacterial spores are inactivated by heat following first order reaction. The effectiveness of the inactivation conditions is usually expressed as decimal reduction time (θ_d), which is the time required to reduce the viable population to one-tenth of original. Derive the relationship between the inactivation rate constant and the decimal reduction time.

PROBLEM 11.21

The probability, $P(t)$ deals with a life-span distribution between t and $t + \mathrm{d}t$ in a specific bacterial clump of n single spores, written as

$$P(t) = \int_0^t f(t)_n \, \mathrm{d}t$$

or,

$$f(t)_n = \frac{\mathrm{d}P(t)}{\mathrm{d}t}$$

Again, $P(t) = (1 - \mathrm{e}^{-k't})^n$ where k' = specific rate of denaturation of spores with heat.

Show that the average life span, $\bar{t}_n$, for a clump of n spores becomes larger than $\bar{t}_f$ for single spores with the increase of n.

PROBLEM 11.22

Calculate the overall efficiency, $\bar{\eta}$ of collecting airborne microbes, using the following data:

$$N = 1.5\ s^{-1}$$

$$\overline{\Delta t}_{n'=\infty} = 432\ s$$

where N = number of airborne microbes blown into the filter per second;

$\overline{\Delta t}_{n'=\infty}$: time interval between successive appearance of microbes from air filter (statistically averaged).

PROBLEM 11.23

Derive the inertial parameter as the ratio of stopping distance of a microbial spore (d_p in diameter, ρ in density and u_0 in linear velocity) to a fibre diameter, d_{fi}. The slip effect of the spore is assumed to be negligible. It is also assumed that air-resistance to the particle's motion is governed by the Stoke's Law. Air viscosity is designated as μ.

PROBLEM 11.24

Sterilized air is sparged into a fermenter at a rate of 10 $m^3\,h^{-1}$. The original number of airborne microbes that are hazardous to the fermentation is $10^3\ m^{-3}$. The air is sterilized by passing it through a Gibsons filter of appropriate design.

If the time-interval between successive passages of the undesirable microbes from the filter exit to the fermenter inlet is 10 h in a statistical average of the term, what is the overall collection efficiency, $\bar{\eta}$, of this specific filter?

PROBLEM 11.25

A fermentation medium, 0.01 m^3 in volume and having 10^7 spores m^{-3} is exposed to thermal sterilization at 121°C for 18.5 min. The degree of sterilization obtained was: $1-P=10^{-2}$. If the contamination level is 10^8 spores m^{-3}, how long should the medium be exposed to heat at 121°C in order to expect the same degree of sterility, $(1-P=10^{-2})$?

PROBLEM 11.26

The data listed in Table 11.4 show the effect of temperature, t on the value of denaturation rate constant, k, of a bacterial spore suspension. Compute the activation energy, E (kJ mol^{-1}) of this reaction.

Table 11.4

t, °C(K)	k, min^{-1}
117 (390)	2.0
115 (388)	1.0
112 (385)	0.5
109 (382)	0.25

PROBLEM 11.27

The value of denaturation rate constant, k, min^{-1}, of a spore suspension is 0.25 min^{-1} at 109°C. What is the value of k at 120°C? The activation energy, E, of the denaturation reaction is given as 287.44 kJ mol^{-1}.

PROBLEM 11.28

(i) If the value of void fraction, ε, of a specific mycelial cake obtained after filtration is reduced from 0.6 to 0.5 at constant pressure, what is the percentage reduction in the filtration rate? It can be assumed that all factors excepting ε remain unchanged.

(ii) In the same filter working at constant pressure ($\Delta P=1.96\times10^5$ N m^{-2}), the volume of filtrate obtained from filtering a mycelial slurry during the initial stage of operation for 2 h was 8 m^3. Calculate the time required for obtaining 10 m^3 of filtrate when the applied pressure is increased to $\Delta P=2.94\times10^5$ N m^{-2}. It is assumed that the cake is incompressible and the resistance of the filter medium is negligible.

PROBLEM 11.29

HPLC is used for purification of two proteins in a mixture. The capacity factors, k_1 and k_2 of the proteins were determined as 0.9 and 1.1 respectively.

The relationship between HETP and the flow velocity, u, is expressed as follows,

$$\text{HETP(m)} = A + Bu$$

where $A = 6.2 \times 10^{-5}$(m), $B = 1.13$(s), and u = the flow velocity (m s^{-1}).

If the flow velocity is 1.7×10^{-4} (m s^{-1}), what column length is required for getting the resolution value of more than 1.5?

PROBLEM 11.30

The parameters for the filtration of a fermenter broth were determined by a laboratory vacuum filter with 0.05 m^2 of filtration area under 3.33×10^4 N m^{-2} of reduced pressure as follows.

$$K = 10^{-8}\ m^6\ s^{-1}$$

$$V_0 = 1.0 \times 10^{-4}\ m^3$$

If the same broth is filtered in the Oliver-type rotary-drum vacuum filter with 1.5 m long and 1.0 m diameter at 4.67×10^4 N m^{-2}, what filtration rate could be expected?

The rotating rate is 5 rpm and one-third of the rotary drum surface is submerged in the broth. The weight of dry solid per unit weight of broth, W is 0.05 and the ratio of the weight of wet cake to that of dry cake is 2.0. The density of filtrate ρ and that of solid ρ_P are 1.0×10^3 kg m^{-3}, and 1.10×10^3 kg m^{-3}, respectively. When the cake is scraped off the drum by a knife, the cake with thickness 1.0×10^{-3} m is left on the drum. The cake is assumed to be incompressible.

PROBLEM 11.31

Alcohol dehydrogenase (ADH) is desired to be purified from baker's yeast cell-free extract by affinity chromatography. An affinity adsorption column containing cibachron blue coupled sepharose is used for this purpose. The concentration of the ligand (cibachron blue) in the column is 0.08 M. If phosphate buffer of 0.1 M ionic strength is used to elute the desired enzyme and the dissociation constant for ADH/ cibachron blue affinity under the eluting buffer conditions is 4.2×10^{-3} M, determine by theoretical analysis the amount of eluting buffer to be passed for collection of pure ADH, if the void volume of the column is 80 ml.

PROBLEM 11.32

Show that the interfacial area, a, of the air bubbles per unit volume of a fermentation broth under aeration is given by:

$$a = \frac{6\theta}{d_B}$$

where d_B = mean diameter of air bubbles that are assumed spherical and θ = hold-up of air bubbles.

PROBLEM 11.33

An aerobic fermentation using *Aspergillus niger* tends to form pellets. The average mycelial density of the pellets is 2.0 kg m^{-3} of pellet volume. If the average dissolved oxygen concentrations in the bulk broth is 6×10^{-3} kg m^{-3} and if the specific oxygen uptake rate by the mould is 1.6×10^{-5} kg O_2 s^{-1} kg^{-1} cells, the diffusivity of O_2 is 1.8×10^{-9} m^2 s^{-1} in the fermentation broth, estimate the size of the mould pellet which will just cause O_2 deficiency within the pellet.

PROBLEM 11.34

A fermenter (nominal volume = 60 m^3, working volume = 40 m^3, inside (absolute) pressure (averaged) = 151,987.5 N m^{-2}, temperature 30°C) is aerated at a rate of 30 m^3 min^{-1} (standard temperature and pressure). Assuming that the average values of oxygen partial pressures in air bubbles and in the liquid are 31,410.75 N m^2 and 30,397.5 N m^{-2} respectively, calculate the efficiency of oxygen transfer rate of the system. Power required for liquid-agitation under aeration is taken as 1,103.25 W m^{-3}.

PROBLEM 11.35

An enzymatic reaction takes place at the surface of a pellet in which specific substrate uptake rate is given as

$$q_s = \frac{ks}{as^2 + s + K_m} = K_s(s_b - s) \tag{11.35.1}$$

The following dimensionless variable is introduced:

$$y = \frac{s}{s_b}$$

and parameters $A = as_b$, $f = K_m/s_b$ and $Da = k/s_b k_s$. s_b is the bulk concentration of substrate.

(a) Show that the steady state solution(s) of (11.35.1) can be obtained by the solution of

$$F(y) = \frac{(1-y)(Ay^2 + y + f)}{y} = Da \tag{11.32.2}$$

Plot $F(y)$ versus y and show that there is at least one solution $y \in (0,1)$ for any set of positive (A, f, Da).

(b) Show that necessary and sufficient conditions for three solutions to (a) are

$$\frac{f}{A} < \frac{1}{27} \wedge A > \left(1 - 3\left(\frac{f}{A}\right)^{1/3}\right)^{-1} \tag{11.35.3}$$

Compute and plot in an A versus f diagram the curve $A(f)$ which separates three from one steady state solutions to (a). Show that for values (A, f) along this curve (the 'separatrix')

$$Da = (A-1)\left(\frac{f}{A}\right)^{1/3} + 1 - f \tag{11.35.4}$$

(c) Choose $f = 0.05$ and $A = 5$.
Plot $F(y) = Da$ versus y and obtain [by differentiation of (11.35.2)] the upper and the lower limit of Da for which three solutions exist.

(d) For $f = 0.05$ and $A = 5$, plot

$$\eta = \frac{q_s(s)}{q_s(s_b)} \text{ versus } Da$$

It is easiest to use y as a parameter along the curve. Choose y and compute the corresponding value of (η, Da).
Calculate the y value for which η is maximum.

(e) Consider the steady state solution ($f = 0.005$, $A = 5$) for which $y = 0.3 = y_2$. Calculate the corresponding Da and the two other steady states y_1 and y_3, which belong to this particular value of Da.
Show by a perturbation analysis that y_1 and y_3 are stable steady states while y_2 is an unstable steady state.
Show that all steady states with $0.1381 < y < 0.3610$ (i.e. steady states between the two extrema of $F(y)$ in (c) are unstable steady states.

(f) The preceding analysis has shown that transport resistance may lead to an effectiveness factor η considerably greater than 1 for the substrate-inhibited enzyme kinetics of (a). Reconsider the whole problem — now in terms of homogeneous enzymatic reaction where the substrate has to be absorbed from a gas phase. Here k_s is an overall mass transfer coefficient from the bulk gas phase to the bulk liquid — or perhaps a pure liquid phase transfer coefficient if s_b is taken to be the interface concentration of substrate (in units of liquid phase concentration).
Reflect on the influence of stirrer speed, gas distribution device and slow deactivation of the enzyme on the reactor performance.

PROBLEM 11.36

Which type of flow, complete mixing or plug flow, is recommendable to minimize nominal holding time of liquid in a reactor, where cell concentration is increased from $X = X_1$ to $X = X_2$? The reactor (working volume $= V$) is operated continuously with an equal flow rate, F for both the inflow and the outflow.

PROBLEM 11.37

The rotational speed N_{c_1} of an impeller to realize uniformity of mycelial suspension in a vessel (diameter, $D_{t_1} = 0.3$ m) without baffle plates was 120 rpm, the ratio of impeller span to vessel diameter being 0.8. Physical properties of the suspension are: $\rho_1 = 1.1 \times 10^3$ kg m^{-3}; viscosity $\mu_1 = 4.6$ N sec m^{-2}. It is desired to conduct the same operation with a larger vessel diameter ($D_{t_2} = 0.9$ m). What are the kinematic viscosity (μ_2/ρ_2) of another mycelial suspension and rotation speed N_{c_2} of impeller that could secure the same kind of uniformity as obtained in the similar vessel? Note that the dynamic similarity of liquid flow in geometrically similar vessels without baffle plates require the following two kinds of dimensionless numbers to be constant, respectively:

$$N_{Re} = \frac{N\,D_i^2\rho}{\mu} = a \;; \qquad N_{Fr} = \frac{N^2\,D_i}{g} = b$$

where:

N	= rotation speed of impeller
D_1	= impeller diameter
g	= acceleration due to gravity
a, b	= constants

PROBLEM 11.38

An allosteric enzyme which catalyzes the conversion of substrate S to product P exhibits the phenomenon of positive cooperativity with respect to its substate. The dependence of its reaction velocity, v on concentration of the substrate S exhibits sigmoidal kinetics as shown in Fig. 11.4a. A compound S′ acts as a competitive inhibitor of the enzyme. The dependence of reaction velocity on concentration of S′ in the presence of a constant low level of substrate S (indicated by an arrow in Fig. 11.4a) is shown in Fig. 11.4b. S′ does not participate as a substrate in the reaction. Explain the nature of velocity profile shown in Fig. 11.4b.

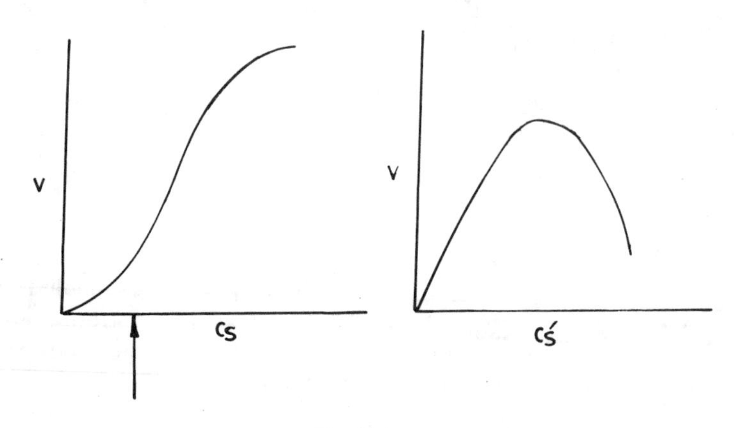

Fig. 11.4

PROBLEM 11.39

In many enzymatic reactions such as the one catalysed by lysozyme, the amino acids constituting the active site must be present in ionized forms. These ionized groups are required for the binding of the substrate as well as for the reaction to take place. Consider that the active site of an enzyme contains a single ionizable group that must be in the negative form before the substrate can bind and catalysis can occur. The pK of this group is 3.0. The substrate is a positively charged compound and remains completely ionized over the pH range (4.0–6.0) studied.

(a) Write the reactions showing the effect of pH on distribution of enzyme species.
(b) What type of velocity equation (competitive/noncompetitive/uncompetitive) is applicable for this system?
(c) What would be a plot of v versus pH at saturating concentrations of substrate?

PROBLEM 11.40

A number of enzyme-catalysed reactions are inhibited by compounds which alter their reaction kinetics. A reversible inhibitor is termed competitive if it increases K_m (Michaelis–Menten constant) without affecting V_m (maximum rate velocity). A noncompetitive inhibitor, on the other hand, lowers V_m without affecting K_m. Common noncompetitive inhibitors are heavy metal ions, which combine reversibly with the sulfhydryl (–SH) group of cysteine residues. The rates of enzyme catalysed reaction in the absence and the presence of a noncompetitive inhibitor are given by equations (11.40.1) and (11.40.2) respectively.

$$v_0 = \frac{V_m\,[c_s]}{K_m + [c_s]} \tag{11.40.1}$$

and,

$$v_i = \frac{V_m\,[c_s]}{(K_m + [c_s])(1 + c_i/K_i)} \tag{11.40.2}$$

where,

v_0 = velocity of the reaction in absence of an inhibitor,
v_i = velocity of the reaction in presence of the noncompetitive inhibitor,
V_m = maximum reaction velocity,
K_m = Michaelis–Menten constant,
$[c_s]$ = substrate concentration,
$[c_i]$ = inhibitor concentration, and
K_i = inhibition constant.

Derive an expression for the relative activity and calculate the degree of inhibition caused by a noncompetitive inhibitor when $[c_s] = K_m$ and $[c_i] = K_i$.

12

Symbols, terms and units

Symbol	Definition of terms	SI units	Customary units
A	Area	m^2	cm^2
A_{O_2}	Ratio of oxygen consumption to glucose uptake	dimensionless	dimensionless
ATP_{pool}	Intracellular ATP per gram cell	kg ATP $(kg\ cell)^{-1}$	mg ATP $(g\ cell)^{-1}$
a	Interfacial area per unit volume	m^{-1}	cm^{-1}
a'	Percentage water content of floc	dimensionless	dimensionless
a_m	Surface area per unit weight	$m^2\ kg^{-1}$	$cm^2\ g^{-1}$
B	Batch number	dimensionless	dimensionless
b	Membrane thickness	m	microns
C_{fi}	Capacity factor of component i	dimensionless	dimensionless
C_{PL}	Amount of protein lost	kg kmol	g gmol
C_T	Total enzyme	kg kmol	g gmol
c	Concentration	$kg\ m^{-3}$	$g\ l^{-1}$
c^*	Dissolved oxygen concentration at equilibrium with gas phase	$kg\ m^{-3}$	$g\ l^{-1}$
$\overline{c}$	Average dissolved oxygen concentration in the liquid phase	$kg\ m^{-3}$	$g\ l^{-1}$
c_{ATP}	Intracellular concentration of ATP	kg ATP $(kg\ cell^{-1})$	mg ATP $(g\ cell^{-1})$
c_e	Enzyme concentration	$kg\ m^{-3}$, $kmol\ m^{-3}$	$g\ l^{-1}$, $mol\ l^{-1}$
c_{e_0}	Initial enzyme concentration	$kg\ m^{-3}$, $kmol\ m^{-3}$	$g\ l^{-1}$, $mol\ l^{-1}$
c_i	Inhibitor concentration	$kg\ m^{-3}$	$g\ l^{-1}$
c_L	Dissolved oxygen concentration in the liquid phase	$kg\ m^{-3}$	$g\ l^{-1}$

c_l	Concentration of ligand bound per unit concentration of enzyme	$kg\ m^{-3}$	$g\ l^{-1}$
c_p	Product concentration	$kg\ m^{-3}$	$g\ l^{-1}$
c_s	Substrate concentration	$kg\ m^{-3}$	$g\ l^{-1}$
c_{sb}	Bulk substrate concentration	$kg\ m^{-3}$	$g\ l^{-1}$
c_{ss}	Substrate concentration at the surface	$kg\ m^{-3}$	$g\ l^{-1}$
c_{sf}	Substrate concentration in feed	$kg\ m^{-3}$	$g\ l^{-1}$
D	Dilution rate	s^{-1}	h^{-1}, d^{-1}
$\mathcal{D}$	Diffusivity	$m^2\ s^{-1}$	$cm^2\ s^{-1}$
D_i	Diameter of impeller	m	m, cm
D_r	Diameter of reactor	m	m, cm
D_t	Diameter of tank	m	m, cm
d_B	Bubble diameter	m	cm
d_f	Diameter of floc	m	cm, micron
d_{fi}	Fibre diameter	m	cm, micron
d_p	Particle diameter	m	cm, micron
E	Activation energy	$J\ (mol^{-1})$	$cal\ mol^{-1}$, $kcal\ mol^{-1}$
$\bar{E}$	Actual electrochemical potential	V	mV
E^*	Standard electrochemical potential	V	mV
E_h	Redox potential	V	mV
E_m	Measured electrical potential difference	V	mV
E^N	Nerst potential	V	mV
E'_0	Standard redox potential	V	mV
F	Faraday constant	$J\ mol^{-1}\ V^{-1}$	—
F	Flow rate	$m^3\ s^{-1}l$	min^{-1}, $l\ h^{-1}$
F_F	Filtration rate	$m^3\ s^{-1}$	$l\ min^{-1}$, $l\ h^{-1}$
f_b	Fraction of solution withdrawn in each batch	dimensionless	dimensionless
f_c	Correction factor	dimensionless	dimensionless
f_{cell}	Volume fraction of cell	dimensionless	dimensionless
f_f	Volume fraction of fibrous material	dimensionless	dimensionless
f_i	Volume fraction of oxygen in inlet air	dimensionless	dimensionless
f_o	Volume fraction of oxygen in outlet air	dimensionless	dimensionless
G	Mass velocity	$kg\ m^{-2}\ s^{-1}$	$g\ cm^{-2}\ s^{-1}$
$G(p)$	Transfer function with p-Laplacian	dimensionless	dimensionless
$\Delta G'$	Actual free energy change	$J\ mol^{-1}$	$cal\ mol^{-1}$
$\Delta G°$	Standard free energy change	$J\ mol^{-1}$	$cal\ mol^{-1}$
g	Acceleration due to gravity	$m\ s^{-2}$	$cm\ s^{-2}$
H	Henry's law constant	$Pa\ kmol^{-1}\ m^{-3}$	$atm\ mol^{-1}\ l^{-1}$
HETP	Height equivalent of a theoretical plate	m	cm, m

H_L	Liquid depth	m	m, cm
H_0	Holdup of bubbles	dimensionless	dimensionless
ΔH	Enthalpy change	J (kg cell)$^{-1}$	cal (g cell)$^{-1}$
$\Delta H°$	Standard heat of reaction	J kmol^{-1}	cal mol^{-1}
ΔH_a	Enthalpy change of dry cell	j (kg cell)$^{-1}$	cal (g cel)$^{-1}$
ΔH_{ave}	Enthalpy change of available electron	J (electron)$^{-1}$	cal (electron)$^{-1}$
ΔH_c	Heat of combustion of cell	J kg^{-1}	cal g^{-1}
ΔH_f	Heat of formation	J kg^{-1}	cal g^{-1}
ΔH_r	Heat of reaction	J kg^{-1}	cal g^{-1}
ΔH_s	Heat of combustion of substrate	J kg^{-1}	cal g^{-1}
I_p	Output current of probe	A	mA, amp
i_t	Unsteady state current	A	amp, mA
i_u	Rate of interfacial subsidence	m h^{-1}	m h^{-1}, cm h^{-1}
i_∞	Steady state current	A	amp, mA
J_s	Mass flux	kg m^{-2}	g cm^{-2}
K	Consistency index	kg m^{-1} s^{-1}	g cm^{-1} s^{-1}
K_2	Turnover number	s^{-1}	min^{-1}, h^{-1}
K_a, K_b	Ionization constants	dimensionless	dimensionless
K_d	Dissociation constant	dimensionless	dimensionless
K_{eq}	Equilibrium constant	dimensionless	dimensionless
K_i	Inhibition constant for growth	kg m^{-3}	g l^{-1}
K_{ip}	Inhibition constant for product formation	kg m^{-3}	g l^{-1}
K_m	Michaelis–Menten constant	kg m^{-3}	g l^{-1}
K_p	Partition coefficient	dimensionless	dimensionless
K_s	Saturation constant for growth	kg m^{-3}	g l^{-1}
K_s'	Saturation constant for product formation	kg m^{-3}	g l$_{-1}$
K_{si}	Substrate inhibition constant	kg m^{-3}	g l^{-1}
k	Rate constant (first order)	s^{-1}	s^{-1}, min^{-1}, h^{-1}
k'	Specific rate of denaturation	s^{-1}	s^{-1}, min^{-1}, h^{-1}
k_{av}	Average rate constant	s^{-1}	s^{-1}, min^{-1}, h^{-1}
k_d	Deactivation rate constant (first order)	s^{-1}	s^{-1}, min^{-1}, h^{-1}
k_e	Maintenance constant	s^{-1}	s^{-1}, min^{-1}, h^{-1}
k_i	Inhibition rate constant	s^{-1}	s^{-1}, h^{-1}, min^{-1}
k_i'	First order inactivation rate constant	s^{-1}	s^{-1}, h^{-1}, min^{-1}
k_L	Liquid side mass transfer coefficient	s^{-1}	s^{-1}, h^{-1}, min^{-1}
k_La	Volumetric mass transfer coefficient	s^{-1}	s^{-1}, h^{-1}, min^{-1}
k_v	Mass transfer coefficient	kg mol m^{-3}s^{-1}	mol l^{-1} h^{-1} atm^{-1}

L	Length, thickness	m	m, cm
l	Length of fibre	m	m, cm,
M	Mass	kg	kg, g
m	Maintenance coefficient	dimensionless	dimensionless
N	Number of cells or pellets	dimensionless	dimensionless
$\mathcal{N}$	Weight ratio of insoluble solids to the resultant liquid adhering to the insoluble solid	dimensionless	dimensionless
N_a	Aeration number	dimensionless	dimensionless
N_{abs}	Absorption number	dimensionless	dimensionless
N_b	Number of ligand binding sites	dimensionless	dimensionless
N_c	Number of cells at any time, t	dimensionless	dimensionless
N_g	Number of generations	dimensionless	dimensionless
N_0	Initial number of cells	dimensionless	dimensionless
N_{O_2}	Oxygen transfer rate	$kmol\ m^{-3}\ s^{-1}$	$mol\ l^{-1}\ h^{-1}$
N'_P	Number of theoretical plates	dimensionless	dimensionless
N_P	Power number	dimensionless	dimensionless
N_{Pe}	Peclet number	dimensionless	dimensionless
N_{Re}	Reynolds number	dimensionless	dimensionless
N_r	Number of reactors	dimensionless	dimensionless
N_{Sc}	Schmidt number	dimensionless	dimensionless
N_{Sh}	Sherwood number	dimensionless	dimensionless
n	Impeller speed	s^{-1}	s^{-1}, min^{-1}, h^{-1}
n'	Flow behaviour index	dimensionless	dimensionless
n_e	Number of electrons per mole	mol^{-1}	—
P	Total pressure	$N\ m^{-2}$	atm, bar, $kg\ cm^{-2}$
P	Ungassed power requirement	$W\ m^{-3}$	$W\ l^{-1}$
ΔP	Pressure drop	$N\ m^{-2}$	atm, bar, $kg\ cm^{-2}$
P_g	Gassed power requirement	$W\ m^{-3}$	$W\ l^{-1}$
P_m	Product concentration at which cell growth ceases	$kg\ m^{-3}$	$g\ l^{-1}$
P'_m	Product concentration at which product formation ceases	$kg\ m^{-3}$	$g\ l^{-1}$
P/O	ATP yield based on oxygen	kg mol ATP $(kg\ mol\ O_2)^{-1}$	mol ATP $(mol\ O_2)^{-1}$
P_s	Probability of successful sterilization	dimensionless	dimensionless
P_t	Total product formed	kg	kg, g
P_V	Agitation power per unit volume	$W\ m^{-3}$	$W\ l^{-1}$
P_W	Agitation power per unit mass	$W\ kg^{-1}$	$W\ kg^{-1}$, $W\ g^{-1}$
p	Partial pressure	$N\ m^{-2}$	atm, bar, $kg\ cm^{-2}$
p_{gm}	Mean oxygen partial pressure	$N\ m^{-2}$	atm, bar, $kg\ cm^{-2}$
Q	Rate of heat generation	$J\ h^{-1}$	$cal\ h^{-1}$

Q	Productivity	$kg\ m^{-3}\ s^{-1}$	$g\ l^{-1}\ h^{-1}$
Q_{CO_2}	Specific CO_2 production rate	$kg\ m^{-3}$	$g\ l^{-1}$
Q_m	Maximum rate of reaction	$kg\ m^{-3}\ s^{-1}$	$g\ l^{-1}\ h^{-1}$
Q_O	Heat of reaction per electron transferred to oxygen	J (kg equivalent)$^{-1}$	cal (g equivalent)$^{-1}$
Q_{O_2}	Specific oxygen uptake	$kg\ m^{-3}$	$g\ l^{-1}$
Q_p	Product productivity	$kg\ m^{-3}\ s^{-1}$	$g\ l^{-1}\ h^{-1}$
Q_r	Rate of reaction	$kg\ m^{-3}\ s^{-1}$	$g\ l^{-1}\ h^{-1}$
Q_{res}	Respiration rate	$kg\ O_2\ m^{-3}\ s^{-1}$	$g\ O_2\ l^{-1}\ h^{-1}$
Q_s	Specific substrate utilization rate	$kg\ mol\ kg^{-1}\ s^{-1}$	$mol^{-1}\ g^{-1}\ s^{-1}$
Q_x	Cell productivity	$kg\ m^{-3}\ s^{-1}$	$g\ l^{-1}\ h^{-1}$
q	Reaction velocity	$kmol\ kg^{-1}\ s^{-1}$	$mol\ g^{-1}\ s^{-1}$
q_{mp}	Maximum rate of product formation	$kg\ m^{-3}\ s^{-1}$	$g\ l^{-1}\ h^{-1}$
q_{O_2}	Specific oxygen uptake rate	$kg\ m^{-3}\ s^{-1}$ kmol (kg cell)$^{-1}$ s^{-1}	$g\ l^{-1}\ h^{-1}$ mol (g cell)$^{-1}$ h^{-1}
q_s	Specific substrate uptake rate	$kg\ m^{-3}\ s^{-1}$	$g\ l^{-1}\ h^{-1}$
q_x	Cell production rate	$kg\ cells\ m^{-3}\ s^{-1}$	$g\ l^{-1}\ h^{-1}$
R	Recycle ratio	dimensionless	dimensionless
R	Gas constant	$J\ K^{-1}\ mol^{-1}$	$cal\ mol^{-1}\ K^{-1}$ $l\ atm\ K^{-1}\ mol^{-1}$
$\mathcal{R}$	Mass of raffinate stream	kg kmol	g gmol
R_p	Weight of protein released per unit weight of packed yeast	$lg\ kg^{-1}$	$g\ g^{-1}$
R_A	Reaction rate	$kg\ s^{-1}$, $kmol\ s^{-1}$	$g\ s^{-1}$, $gmol\ s^{-1}$
R_{Area}	Ratio of area submerged in liquid to that of the total drum	dimensionless	dimensionless
R_E	Ratio of enzyme present in solution to that on carrier	dimensionless	dimensionless
R_i	Rejection of component i	dimensionless	dimensionless
R_m	Filter medium resistance	m^{-1}	cm^{-1}
R_p^m	Total protein per unit weight of packed yeast	$kg\ kg^{-1}$	$g\ g^{-1}$
R_N	Resolution of two components	dimensionless	dimensionless
RQ	Respiratory quotient	dimensionless	dimensionless
$R_{W/D}$	Weight ratio of wet cake to that of dry cake	dimensionless	dimensionless
r	Radius	m	m, cm
$\Delta S°$	Standard entropy change	$J\ kmol^{-1}\ K^{-1}$	$cal\ mol^{-1}\ K^{-1}$
T	Temperature	K, °C	°C, °F
$\mathcal{T}$	Turbidity	dimensionless	dimensionless
t	Time	s	s, min, h
$\bar{t}$	Mean life span	s	s, min, h
Δt	Time interval	s	s, min, h
$t_{1/2}$	Half-life	s	s, min, h
t_b	Batch processing time	s	s, min, h

t_f	Batch culture time	s	s, min, h
t_g	Generation time	s	s, min, h
t_{growth}	Time when cell growth ceases	s	s, min, h
t_p	Processing time	s	s, min, h
u	Velocity	$m\ s^{-1}$	$cm\ s^{-1}$
u_B	Bubble velocity	$m\ s^{-1}$	$cm\ s^{-1}$
u_{c0}	Velocity of single particle in centrifugal field	$m\ s^{-1}$	$cm\ s^{-1}$
u_g	Settling velocity of droplet	$m\ s^{-1}$	$cm\ s^{-1}$
u_{gs}	Superficial gas velocity	$m\ s^{-1}$	$cm\ s^{-1}$
u_0	Relative velocity between fluid and single particle	$m\ s^{-1}$	$cm\ s^{-1}$
u_s	Superficial linear velocity	$m\ s^{-1}$	$cm\ s^{-1}$
V	Volume	m^3	l
V_e	Elution volume	m^3	l
V_G	volume of gaseous phase	m^3	l
V_i	Inner gel volume	m^3	l
V_l	Volume of filtrate	m^3	l
V_0	Void volume	m^3	l
W_B	Width of impeller blade	m	m, cm
W_g	Mass of gel	kg	kg, g
W_l	Mass of dry cake per unit volume of filtrate	$kg\ m^{-3}$	$g\ l^{-1}$
W_m	Weight of dry cake per unit weight of slurry	$kg\ kg^{-1}$	$g\ g^{-1}$
W_r	Water regain value	$m^3\ kg^{-1}$	$l\ kg^{-1}, l\ g^{-1}$
X	Fractional conversion	dimensionless	dimensionless
x	Cell mass concentration	$kg\ m^{-3}$	$g\ t^{-1}$
x	Weight fraction of solute in the raffinate	dimensionless	dimensionless
$\bar{x}^*$	Steady state value of x^*	dimensionless	dimensionless
x_0	Initial cell concentration	$kg\ m^{-3}$	$g\ l^{-1}$
$\bar{Y}^*$	Steady state value of Y^*	dimensionless	dimensionless
Y_Δ	Yield based on heat generated	$kg\ J^{-1}$	$g\ cal^{-1}$
Y_{ATP}	Yield based on ATP	$kg\ kmol^{-1}$	$g\ mol^{-1}$
Y_C	Biomass carbon yield	dimensionless	dimensionless
Y_G	True growth yield with respect to glucose taking maintenance into account	dimensionless	dimensionless
$Y_{\Delta H}$	Yield based on enthalpy	$kg\ cell\ J^{-1}$	$g\ cell\ cal^{-1}$
$Y_{p/s}$	Yield of product based on substrate	dimensionless	dimensionless
$Y_{x/O}$	Yield based on oxygen	dimensionless	dimensionless
$Y_{x/p}$	Cell yield based on product	dimensionless	dimensionless
$Y_{x/s}$	Cell yield based on substrate	dimensionless	dimensionless
y	Weight fraction of solute in the extract	dimensionless	dimensionless
z	Valency	dimensionless	dimensionless

Symbol	Term	SI unit	Common unit
z_c	Centrifugal effect ($= r\omega^2/g$)	dimensionless	dimensionless
z_{O_2}	Specific oxygen requirement	$kg\ O_2\ (kg\ cell)^{-1}$	
α	Specific cake resistance	$m\ kg^{-1}$	$cm\ g^{-1}$
γ	Shear rate	s^{-1}	s^{-1}, min^{-1}, h^{-1}
γ_s, γ_b, γ_p	Number of equivalents of electron per gram atom carbon of substrate, cell mass and product	dimensionless	dimensionless
δ	Relative retention	dimensionless	dimensionless
δ_c	Degree of concentration	dimensionless	dimensionless
ε	Void fraction	dimensionless	dimensionless
ε'	Fraction of available electron transferred to oxygen	dimensionless	dimensionless
η	Energetic yield coefficient (fraction of available electron transferred to cell mass)	dimensionless	dimensionless
$\bar{\eta}$	Overall filter collection efficiency	dimensionless	dimensionless
η_0	Single fibre efficiency	dimensionless	dimensionless
θ_d	Decimal reduction time	s	s, min, h
κ	Electrostatic double layer	Å	Å
κ_m	Oxygen permeability of membrane	$m^3\ m^{-1}\ s^{-1}$	$ml\ cm^{-1}\ h^{-1}$
μ	Specific growth rate	s^{-1}	s^{-1}, h^{-1}
μ	Viscosity	$Pa\ s\ (N\ s\ m^{-2})^{-1}$	poise
μ_a	Apparent viscosity	$Pa\ s\ (N\ s\ m^{-2})^{-1}$	poise
μ_{air}	Viscosity of air	$Pa\ s\ (N\ s\ m^{-2})^{-1}$	poise
μ_m	Maximum specific growth rate	s^{-1}	s^{-1}, h^{-1}
ν	Kinematic viscosity	$m^2\ s^{-1}$	$cm^2\ s^{-1}$
ν	Specific productivity (product)	$kg\ m^{-3}$	$g\ l^{-1}$
ξ_p	Fraction of available electron transferred to product	dimensionless	dimensionless
P	Production rate	$kg\ s^{-1}$	$g\ s^{-1}$, $kg\ h^{-1}$
ρ	Density	$kg\ m^{-3}$	$g\ l^{-1}$
ρ_a	Density of air	$kg\ m^{-3}$	$g\ l^{-1}$
ρ_c	Density of wet cake	$kg\ m^{-3}$	$g\ l^{-1}$
ρ_f	Density of filtrate	$kg\ m^{-3}$	$g\ l^{-1}$
ρ_m	Density of medium	$kg\ m^{-3}$	$g\ l^{-1}$
ρ_y	Density of cell	$kg\ m^{-3}$	$g\ l^{-1}$
Σ	Surface area of centrifuge	m^2	m^2, cm^2
σ_b, σ_p, σ_s	Weight fraction of carbon in cell mass, product and substrate	dimensionless	dimensionless
τ	Shear stress	$N\ m^{-2}$	$kg\ cm^{-2}$
τ	Reactor space time	s	s, min, h
ϕ	Hold up of air bubbles	dimensionless	dimensionless
ψ	Electric potential	V	mV
ω	Angular velocity	$rad\ s^{-1}$	—
Δ	Degree of sterilization	dimensionless	dimensionless

13

Biotechnology glossary

Abundance of mRNA The average number of mRNA molecules per cell.

Abundant mRNA A small number of individual species present in a large number of copies per cell.

Acceptor splicing site The junction between the right end of an intron and the left end of the exon.

Acetoin Acetyl methyl carbinol, an intermediate in the 2,3-butanediol fermentation. It is used to detect the pathway.

Acetone–butanol fermentation A fermentation pathway in which the major waste products are carbon dioxide, acetone, and butanol. Species of *Clostridium* carry out this type of fermentation.

Acclimatization A biological process whereby an organism adapts to a new environment. It applies to process of developing microorganisms the degrade toxic wastes in the environment.

Acentric fragment Part of a chromosome generated by breakage, and lacking a centromere. It is lost at cell division.

Activated sludge A composite mixture of materials containing a large active microbial population used in the purification of waste water.

Activation energy The minimum collision energy required for a chemical reaction to occur.

Active immunity The disease resistance in a person or animal due to antibody production after exposure to a microbial antigen following disease, inapparent infection or inoculation. Active immunity is usually long-lasting. (*See also* Passive immunity.)

Adaptation The ability of a cell to exist in a different environment.

Adsorption The process of uptake of molecules of gases, dissolved substances, or liquids by the surfaces of solids or liquids with which they are in contact.

Aerial mycelium A mycelium composed of fungal hyphae that project above the surface of the growth medium and produce asexual spores.

Aerobe The kind of organism that needs molecular oxygen for growth and metabolism.

Aerobic The condition of living or acting only in the presence of molecular oxygen.

Aerobic reactor A reaction vessel operating under aerobic conditions with sterile air being used as the source of oxygen in dissolved state.

Affinity The degree of binding of a substrate to an active site.

Affinity chromatography A powerful method of separation and purification based on a highly specific interaction between two compounds, one of which is bound to an immobile matrix followed by elution of the bound substance by change of pH, salt concentration addition of cofactors, etc. (e.g. antigen binding to antibody or enzyme binding to coenzymes).

Aflatoxin $C_{17}H_{10}O_6$ a carcinogenic toxin produced by *Aspergillus flavus*.

Agglutination The process of aggregation of polymeric molecules such as antigens or cells like bacteria, erythrocytes, etc., induced by other substances in a highly specific manner (e.g. agglutination of blood group substances by specific antisera).

Alcoholic (ethanolic) fermentation A fermentation in which the major products are carbon dioxide and ethanol (alcohol).

Algae Heterogeneous groups of eukaryotic photosynthetic unicellular or multicellular organisms, generally aquatic plants containing characteristic pigments.

Allosteric control The ability of an interaction at one site of a protein to influence the activity at another site.

Allosteric site The site on an enzyme at which an inhibitor binds.

Allosterism The effect of binding of molecules to an enzyme at a site or sites distinct from the active site resulting in changes in the catalytic properties of the active site.

Alpha-amino acid An amino acid with –COOH and $-NH_2$ attached to the same carbon atom.

Amber codon The nucleotide triplet UAG, one of three nonsense codons that causes termination of protein synthesis.

Amber mutation A change in DNA that creates an amber codon at a site previously occupied by a codon representing an amino acid in some protein.

Amino acids The building blocks of proteins. There are 20 common amino acids.

Amino acids sequence The linear order of amino acids in a protein.

Ammonification Removal of amino groups from amino acid to form ammonia.

AMP Adenosine monosphosphate.

Amplification The production of additional copies of a chromosomal sequence found as either intra- or extrachromosomal DNA.

Anabolism The phase of metabolism that encompasses the synthesis of low-molecular-weight precursors and from that the synthesis of macromolecular cell component.

Anerobic The condition of living or biological activity in the absence of oxygen. (*See also* Aerobic.)

Anaerobic reactor Bioreactor in which no dissolved oxygen or nitrate is present and microbial activity is due to anaerobic bacteria.

Anaerobic respiration Respiration in which the final electron acceptor is an inorganic molecule (sulphate or nitrate) other than molecular oxygen.

Angstron (Å) A unit of measurement equal to 10^{-10}m, 10^{-4}fm and 10^{-1}nm. It is no longer an official unit.

Anoxic reactor Bioreactor in which no dissolved oxygen is present, but biochemical oxidation takes place by (*See also* Aerobic reactor) using oxygen from the nitrate ion.

Antibiotic A specific type of chemical substance that is administered to fight infections caused usually by bacteria, in humans or animals. Many antibiotics are produced by microorganisms; some are chemically synthesized.

Antibody A protein (immunoglobulin) produced by humans or higher animals in response to exposure to a specific antigen and characterized by specific reactivity with its complementaty antigen. (*See also* Monoclonal antibodies.)

Anticoding strand of duplex DNA Used as template to direct the synthesis of mRNA and complementary to it.

Anticodon A triplet of nucleotides in the structure of tRNA that is complimentary to the codon(s) in mRNA to which the tRNA responds.

Antigen A substance, usually a protein or carbohydrates, which, when introduced into the body of a human or higher animal, stimulates the production of an antibody that will react specifically with it.

Apoenzyme An enzyme that requires activation by a coenzyme.

Apparent viscosity Ratio of shear stress to shear rate. It is dependent on the rate of shear, μ_a $[ML^{-1}T^{-1}]$.

Archaebacteria A class of microorganism which can be subdivided into three groups (methanogenic, halophilic, thermoacidophilic) characterized by special constitutents such as ether-bound phytane containing lipids and special coenzymes.

Asepsis The absence of contamination by unwanted organisms.

Aseptic technique Procedures that reduce the risk of contaminating materials or infecting patients.

Asexual reproduction Reproduction without opposite mating strains.

Assay A technique that measures a specific biological response.

Assimilation Absorption by and formation of simple nutrients into the complex constituents of the organism.

ATEE *N*-Acetyl-1-tyrosine ethyl ester.

ATP Adenosine triphosphate.

Attenuated vaccine Whole, pathogenic organisms that are treated with radioactive chemicals or other means to render them incapable of producing infection. Attenuated vaccines are injected into the body which then produces protective antibodies against the pathogen so as to be able to provide protection against disease.

Attenuation Lessening of virulence of a microorganism. Also, a regulatory mechanism for protein synthesis.

Autotroph An organism that uses carbon dioxide as its principal carbon source.

Auxotroph A mutant microorganism with a nutritional requirement not possessed by the parent.

Auxotrophy The inability of microorganisms to synthesize organic compounds required for their own growth.

Available electrons (av e^-) Electrons in a substrate that are not involved in orbitals with oxygen, but can be considered as 'available' for (a) transfer of oxygen, or (b) reductive reactions. The number of available electrons per mole of substrate is 4 × TOD (*see also* total oxygen demand) where the number of electrodes required to reduce one molecule of oxygen is 4.

***Bacillus subtilis* (*B. subtilis*)** An aerobic bacterium used as a host in rDNA experiments.

Bacteria Any of a large group of microscopic organisms having round, rod-like, spiral or filamentous, unicellular or non-cellular bodies that are often aggregated into colonies. These are enclosed by a wall or membrane and lack fully differentiated nuclei. Bacteria may exist as free-living organisms in soil, water or organic matter, or as parasites on the bodies of plants and animals.

Bacteriophage (phage bacterial virus) A virus that invades bacteria and multiplies in them. Bacteriophage lambda is commonly used as a vector in rDNA experiments. Bacteriophage are often abbreviated as phages.

Base pairing The specific association of the nucleotide bases of double stranded DNA (A=T, G=C) and RNA (A=U, G=C) responsible for the stability of the helical structure formed from the two complementary strands. (*See also* Bases.)

Base pairs The arrangement of nitorgenous bases in nucleic acid based on hydrogen bonding; in DNA, base pairs are A-T and G-C; in RNA, base pairs are A-U and G-C.

Bases The heterocyclic compounds which constitute all nucleic acids like adenine (A), guanine (G) and cytosine (C) are found in both DNA amd RNA. Thymine (T) is found only in DNA and uracil (U) only in RNA.

Batch processing A processing technique in which a bioreactor is supplied with substrate and essential nutrients, sterilized and inoculated with microorganisms and the process is run to completion followed by removal of products. (*See also* continuous processing.)

Binary fission Bacterial reproduction by division into two daughter cells.

Bioassay The procedure for determining the concentration or biological activity of biomolecules (e.g. vitamins, hormones, plant growth substances, etc.) by measuring its effect on an organism compared to a standard preparation.

Biocatalyst An enzyme that plays a fundamental role in living organisms or in industry by activating or accelerating a bioprocess.

Biochemical oxygen demand (BOD) The molecular oxygen used in meeting the metabolic needs of aerobic organisms in aqueous medium containing oxidizable organic molecule.

Biochemical pathway A sequence of enzymatically catalysed reactions occurring in a cell.

Biochip An electronic device that uses biological molecules as the framework for substances that act as semiconductors. It functions as an integrated circuit.

Biocoenosis A naturally occurring community of (micro)organisms having interactions with themselves and with the surrounding environment.

Bioconversion Chemical conversion of naturally occurring biodegradable substance using a biocatalyst.

Biodegradation The breakdown of substances by microorganisms, mainly by aerobic bacteria.

Bioeletronics The application of properties of biomolecules to microelectronics in biosensor and biochips.

Biological response modifier Generic term for hormones, immunoactive and neuroactive compounds which act at the cellular level; many of these are possible production targets employing biotechnology.

Biological transmission Transmission of a pathogen from one host to another when the pathogen reproduces itself in the vector.

Biologics Vaccines, therapeutic serums, toxoids, antitoxins and analogous biological products used to induce immunity to infectious diseases or disorders caused by harmful substances of biological origin.

Biomass All organic matters including those belonging to aquatic environment that grow by the photosynthetic conversion of low energy carbon compounds employing solar energy.

Biomass hold-up Cell biomass contained in a bioreactor or associated with matrix-supporting enzymes or cells in an immobilized state.

Biooxidation Oxidation (loss of electrons) process accelerated by a biocatalyst.

Biopolymers Naturally occurring macromolecules that include proteins, nucleic acids and polysaccharides.

Bioprocess The process that uses complete living cells or their components (e.g. enzymes, organelles and chloroplasts) to effect desired chemical and/or physical changes.

Bioreactor The vessel of defined configuration in which a bioprocess is carried out.

Biosensor An electronic device that transforms signals produced by biological molecules to detect concentration or transport of specific compounds.

Biosurfactant A compound produced by living organisms that helps dissolve organic molecules such as oil and tar by reducing surface tension between the compound and the liquid.

Biosynthesis Production, by synthesis or degradation, of a chemical compound by living organism, plant or animal cells or enzymes elaborated by them.

Biotechnology Commercial processes that use living organisms, or substances from those organisms, to make or modify a product, and including techniques used for the improvement of the characteristics of economically important plants and animals and for the development of microorganisms to act on the environment. In some literature this term is used to mean 'new' biotechnology, which generally includes the use of some novel biological techniques namely recombinant DNA, cell fusion and hybridoma techniques for the production of monoclonal antibodies, and development of new bioprocesses including bioseparations used in commercial production.

Biotransformations Any chemical conversion of substances not necessarily natural substrates only, but structurally related and mediated by microorganisms or enzymes obtained therefrom. (*See also* Bioconversion.)

Blotting A technique used for transferring DNA, RNA or protein from gels to a suitable binding matrix.

Budding A form of a sexual reproduction.

Burst size The number of newly synthesized bacteriophage particles released from a single cell.

Burst time The time required from bacteriophage adsorption to release.

2,3-Butanediol fermentation A fermentation in which the major waste products are 2,3-butanediol and carbon dioxide. Bacteria such as *Enterobacter aerogenes* carry out such a fermentation.

Butyric acid bacteria Bacteria that release butyric acid during fermentation. An example is *Clostridium*.

Butyric acid–butanol fermentation A fermentation in which the major waste products are butyric and/or butanol, and carbon dioxide. Bacteria such as *Clostridium* carry out such a fermentation.

Callus An undifferentiated cluster of plant cells produced in higher plants. It is also formed *in vitro* as a first step in regeneration of plants from tissue culture.

Calvin–Benson cycle Conversion of CO_2 into reduced organic compounds.

Capsule A gelatinous layer surrounding the cell wall of many bacteria.

Carbon cycle The series of processes that coverts carbon dioxide (CO_2) to organic substances and back to carbon dioxide in nature.

Carboxylation Addition of an organic acid group (COOH) to a molecule.

Catabolism Reactions involving the breakdown of complex organic substrates to provide energy in the form of ATP and to generate metabolic intermediates used in subsequent anabolic reactions.

Catabolic repression A process of inhibition of enzyme synthesis pathways, whereby glucose or metabolites produced from glucose decrease the intracellular level of cAMP, thereby blocking an essential positive regulator protein for expression of glucose-sensitive operons.

Catalase An enzyme that catalyses the breakdown of hydrogen peroxide to water and oxygen.

Catalysis A process by which the rate of a chemical reaction is increased by a substance (namely enzymes in biochemical reactions) that remains chemically unchanged at the end of the reaction.

cDNA (complementary DNA) A single-stranded DNA that is formed from a mRNA template by the enzyme reverse transcriptase. cDNA is used for cloning or as a probe in DNA hybridization studies.

Cell The smallest structural unit of living matter capable of functioning independently; it is a microscopic mass of protoplasm surrounded by a semipermeable membrane, including one or more nuclei and various non-living substances which are capable, either alone or with other cells, of performing all the fundamental functions of life.

Cell culture The *in vitro* growth of cells isolated from multicellular organisms. These cells are usually of one type.

Cell cycle A cycle consisting of the four growth phases of a cell from one division to the next.

G1-phase A cell that has just undergone mitosis and has only one chromatid and half the normal DNA content.

S-phase Each chromatid is duplicated.

G2-phase Duplication is complete, but mitosis has not been initiated again.

Cell differentiation The process whereby descendants of a common parental cell achieve and maintain specialization of structure and function of the parent cells.

Cell fusion Formation of a single hybrid cell with nuclei and cytoplasm from different cells.

Cell line Cells that acquire the ability to multiply indefinitely *in vitro*.

Cell wall The outer covering of most bacterial, fungal, algal, and plant cells; in eubacteria, it consists of peptidoglycan.

Cellulase The enzyme that cuts the linear chain of cellulose, a glucose polymer at 1–4-β-linkages into cellodextrins and glucose.

Cellulose A polymer of six-carbon glucose sugars found in all plant matter; the most abundant biological substance on earth.

Centrifuge A machine for whirling fluids rapidly to separate substances of different densities by centrifugal force.

Chemical oxygen demand (COD) A measure of the amount of oxygen, expressed in milligrams per litre, required to oxidize organic matter present in a substance using a chemical oxidation method.

Chemically defined medium A culture medium in which the exact chemical composition is known.

Chemiosmosis A proton gradient across a cytoplasmic membrane; can be used to generate ATP.

Chemoautotroph An organism that uses an inorganic chemical as an energy source and carbon dioxide as a carbon source.

Chemoheterotroph An organism that uses organic moleules as a source of carbon and energy.

Chemostat A bioreactor in which steady-state growth of microorganisms is maintained over prolonged periods of time under sterile conditions by providing the cells with a constant input of nutrients and continuously removing effluent with cells as output. (*See also* Turbidostat.)

Chemostat selection Screening process used to identify microorganisms with desired characteristics like degrading toxic chemicals, increasing tolerance of substrates or products, to higher or lower temperatures and pH level, etc. (*See also* Acclimatization.)

Chemotaxis A response to the presence of a chemical.

Chlorophyll Green pigment of plants consisting of closely related pigmenting components, chlorophyll a and chlorophyll b, etc.

Chromosome A self-replicating structure in cells which contains the nuclear DNA, complexed with a large number of different proteins and storing the transmitting genetic information. Chromosomes are also the general physical structures that contain genes.

Eukaryotic cells have a characteristic number of chromosomes per cell (*see also* Ploidy); these chromosomes are linear DNA duplexes. The bacterial chromosome consists of double-stranded circular DNA molecules.

Chromatography The process by which components from a gaseous or liquid mixture are separated by passing them over an absorbent to which each compound has a different degree of interaction. The absorbent material usually exists in a packed column and separation is brought about by the differential rate of migration of the absorbed components.

Cistron *See* Gene (2).

Citric acid cycle (*See* TCA cycle).

Clone A group of genetically identical cells or organisms produced asexually from a common ancestor.

Cloning The amplification of segments of DNA, usually genes.

CM Carboxymethyl group.

CMC Carboxymethyl cellulose.

***Coccus* sp.** Spherical bacterium.

Coding sequence The region of the gene in a DNA molecule that encodes the amino acid sequence of a protein.

Codon The sequence of three adjacent nucleotides that occurs in messenger RNA (mRNA) and that functions as a coding unit for a specific amino acid in protein synthesis.

Coenzyme The dissociable, low-molecular-weight active group of enzyme which transfers chemical groups, hydrogen or electrons. A coenzyme binds with its associated protein (apoenzyme) to form the active enzyme (holoenzyme).

Coenzyme A: A coenzyme that functiuons in decarboxylation.

Cofactors Additional molecules needed for enzymatic functions.

Colibacillosis A bacterial disease that causes diarrhoea, dehydration, and death in calves and piglets.

Collagen The main structural protein of muscles.

Collision theory The principle that chemical reactions occur because energy is gained as reacting particles collide.

Colony A visible accumulation of cells that usually arises from the growth of a single microorganism.

Commensalism A system of interaction in which two organisms live in association and one of which is benefited while the other is neither benefited nor harmed.

Commodity chemicals Chemicals produced in large volumes that sell for less than $1 per pound (say 50c per kg). (*Compare* Speciality chemicals.)

Competitive inhibition The process by which a chemical competes with the normal substrate for the active site of an enzyme.

Condensation reaction A chemical reaction in which a molecule of water is released.

Conjugation The transfer of a part or all of the chromosomes from the cell of one bacteria (donor) to another (recipient) of opposite mating type which are associated side by side.

Continuous cell line Animal cells that can be maintained through an indefinite number of generations *in vitro*.

Consistency General term used to define property of a material by which it resists permanent change of shape.

Constitutive equation Equation relating to stress, strain, time and sometimes other variables, such as temperature; example: rheological equation of state.

Continuous culture Method of cultivation in which nutrients are supplied and products are removed continuously at volumetrically equal rates (*see* Chemostat), maintaining the cells in a condition of stable multiplication and growth.

Continuous processing A processing techniques in which raw materials are supplied and products are removed continuously at volumetrically equal rates (compare batch processing).

Copy number The number of molecules per genome of a plasmid or gene which a cell contains.

Cosmid A DNA cloning vector consisting of plasmid and phage sequences. These are plasmids into which phage lambda *cos* sites have been inserted; as a result, the plasmid DNA can be packaged *in vitro* in the phage coat.

Covalent bond A chemical bond in which the electron of one atom are shared with another atom.

Crossing over The reciprocal exchange of genetic material between chromosomes that occurs during meiosis and is responsible for genetic recombination.

Culture Population of microorganisms.

CTP Cytidine triphosphate.

Culture medium Any nutrient system for the cultivation of bacteria or other cells; usually a complex mixture or organic and inorganic substances.

Cyclic AMP (cAMP) 3′–5′ cyclic adenosine monophosphate. The phosphate group is connected to the 3′ and 5′ carbons of the ribose sugar, thus creating a cyclic structure.

Cytochromes Conjugated proteins, associated with electron transport and redox couples, containing heme as the prosthetic group. The terminal electron transport chain of oxidative respiration contains at least five different cytochromes.

Cytoplasm The 'liquid' portion (membrane) of a cell outside and surrounding the nucleus.

DCC *N,N*-dicyclohexylcarbodiimide.

Denaturation The process of partial or total alteration of the native structure of a macromolecule, e.g. by a change in conformation and a loss of secondary and tertiary structure. It occurs when proteins and nucleic acids are exposed to extreme temperatures, pH-conditions, non-physiological concentration of salt, organic solvents, or other chemical agents.

Denitrification The reduction of nitrates to nitrites and finally to nitrous oxide or even to molecular nitrogen catalysed by facultative aerobic soil bacteria working under anaerobic conditions.

Deoxyribonucleic acid (DNA) A linear polymer, made up of deoxyribonucleotide units, that is the carrier of genetic information, present in chromosomes and chromosomal material of cell orgenelles such as mitrochondria and chloroplasts, and also in some viruses. Every characteristic inherited trait has its origin in the code of each individual's DNA.

Dextran Polysaccharides in which most of the glucose is held together by alpha-(1,6), alpha-(1,3) or alpha-(1,4) linkages which are responsible for their branching. The dextrans form viscous and slimy solutions.

Dextrin A branched polysaccharide of glucose which remains after limited amylase hydrolysis of starch or glycogen, also called limit dextrins. Most of the glucose is held together by alpha-(1,4) linkages, whereas alpha-(1,6) linkages are responsible for branching.

Diacetyl A compound released by certain bacteria and yeast which gives an off-flavour to beer and a buttery aroma and taste to butter.

Dialysis The removal of low molecular weight ions and molecules by diffusion across a membrane.

Diatomaceous earth Earth consisting of diatom cell walls.

Diauxie The two-step growth curve that results when a population of organisms metabolizes one carbon source completely before beginning to metabolize another.

Dilution rate A measure of the rate at which the existing medium in continuous cultivation of organisms is replaced with fresh medium.

Differential medium A solid culture medium that makes it easier to distinguish colonies of the desired organism.

Diffusion The net movement of molecules or ions from an area of higher concentration to an area of lower concentration.

DNA base pair A pair of DNA nucleotide bases. All nucleotide bases pair across the double helix in a very specific way, namely, adenine can only pair with thymine and cytosine with guanine.

DNA probe A short sequence of DNA that is used to detect the presence of a particular nucleotide sequence.

DNA sequence The order of nucleotide bases in the DNA chain. A correct DNA sequence is essential for the storage of genetic information.

DNA synthesis The synthesis of DNA in the laboratory by the sequential addition of nucleotide base.

Donor cell A cell that gives DNA to a recipient cell in recombination.

Doubling time The time required for a cell population to double in the number of cells or in the amount of active cell mass.

Downstream processing Operations associated with purification and separation of products following bioconversion of substances in a bioreactor.

Dynamic viscosity *See also* Viscosity, $\mu_a \eta_a$ $[ML^{-1}T^{-1}]$.

Ecology Study of organisms in relation to their environment.

Effector A molecule which regulates the activity of an allosteric protein by binding itself to the protein at a site other than the substrate binding site.

Electron transport chain A series of compounds that transfer electrons from one compound to another, generating ATP by oxidative phosphorylation.

Electrophoresis The separation of substances (for example, serum proteins) by their relative rate of movement through an electric field.

Elongation viscosity Ratio of the tensile stress to the rate of extension, $\mu_E \eta_E$ $[ML^{-1}T^{-1}]$.

Elution The removal of adsorbed material from an adsorbent, such as the removal of a product from an enzyme bound on a column.

Embden–Meyerhof pathway A specific sequence of enzymatic reactions that convert glucose of pyruvic acid. (*See also* Glycolysis.)

Emulsification The process of making lipids soluble in water employing some chemical agents.

Endergonic Class of chemical reactions in which energy in the form of heat is absorbed. (*See also* Exergonic.)

Endonuclease An enzyme that leaves phosphodiester bonds within a nucleic acid chain, either randomly or at specific base sequences.

Entropy The molecular chaos of a system.

Enzyme A protein that functions as a biocatalyst in a chemical reaction. Enzymes are produced by living cells mediating and promoting the chemical processes of life without themselves being altered or destroyed. Enzymes operate on only one kind of substrate and at specific sites. (*See also* Biocatalyst.)

Enzyme induction The process whereby an inducible enzyme is synthesized in response to an inducer. The inducer combines with a repressor and thereby prevents the blocking of an operator by the repressor.

Epitope Any part of a molecule that acts as an antigenic determinant. A biological macromolecule can contain many different epitopes each giving rise to a different specific antibody.

***Escherichia coli* (*E. coli*)** A species of bacteria whose habitat is the intestinal tract of most vertebrates. Some strains are pathogenic to humans and animals. Many non-pathogenic strains are used experimentally as host of rDNA experiments.

Eukaryote An organism that carries its generic material physically constrained within a nuclear membrane and separated from the cytoplasm. Organisms include fungi, moulds and yeasts. (*See also* Prokaryotes.)

Exergonic Class of chemical reaction in which energy in the form of heat is released. (*See also* Endergonic.)

Exon A section of DNA which carries the coding sequence for a protein. It is found only in eukaryotes where most genes consist of a number of exons separated by non-coding sequences. (*See also* Intron.)

Exonuclease On class of enzymes that cleaves nucleotides, one by one, from either (or both) of the 5′- or the 3′- hydroxyl ends of the nucleic acid.

Exponential growth The phase of cell growth in which the number of cells or the cell mass increases exponentially.

Expression (1) A recombinant gene inserted into the host cell using a vector is expressed in the synthesis of the protein it codes for can be demonstrated. (2) The cellular production of proteins encoded by a particular gene.

Facilitated diffusion The transfer of a substance across a plasma membrane from an area of lower concentration mediated by carrier proteins (permeases).

Facultative Of the ability to live under more than one set of conditions, usually in relation to oxygen tension prevailing in the medium, i.e. aerobic or anaerobic; for example: facultative anaerobes.

FAD Flavin adenine dinucleotide.

Fatty acids Organic acids with long carbon chains. Fatty acids are abundant in cell membranes and are widely used as emulsifiers, as metallic soaps and for other industrial uses.

Fed-batch culture A cell cultivation technique in which one or more nutrients are supplied to the bioreactor in a given sequence during the growth or bioconversion process while the products remain in the vessel till the end of the run.

Feedback inhibition Inhibition of an enzyme in a particular pathway on account of accumulation of end-product from the pathway.

Feedstocks Raw materials used for the production of chemicals.

Fermentation An anaerobic bioprocess. Fermentation is used in various industrial processes for the manufacture of products such as alcohols, organic acids, solvents and cheese by the addition of yeasts, moulds and bacteria.

Fermenter Microbiological reactor in which addition of nutrients, removal of products and insertion of measuring sensors and control devices are maintained while accessories like heating, aeration, agitation and sterilization systems are provided.

Fibrinolytic agents Blood-borne compounds that activate fibrin in order to dissolve blood clots.

Flocculation Agglomeration of colloidal material by addition of chemical that causes the colloidal particles to coalesce.

Fluidized bed Slug of small particles freely suspended in upward flow of liquid and air. At the point of minimum fluidization, the pressure drop across the bed equals the total weight of particles in the bed (corrected for their buoyancy in the liquid). Increase in flow beyond the minimum fluidization velocity results in expansion of the bed with corresponding increase in the voidage.

FMN Flavin monophosphate or riboflavin 5′-phosphate.

Fractionation (of blood) Separation of blood components by centrifugation, resulting in fractions like plasma, serum albumin, antithermophilic factor and other products.

Free-living organism An organism that does not depend on other organisms for survival.

Fungi A major group of saprophytic, parasitic non-photosynthetic unicellular or multicellular eukaryotic microorganisms like moulds, mushrooms, nuts, rusts, etc., including the single-cell types generally termed yeasts. Fungi are used in the production of a wide range of substances of commercial importance, such as antibiotics, steroids, hormones, enzymes, SCP, solvents, organic acids, etc., and in the decomposition of a large variety of organic materials in the process of composting and waste treatment.

Fusion The combination of two distinct cells or macromolecules (genes) into a single integrated unit.

Gamma globulin (GG) A protein component of blood that contains antibodies and confers passive immunity.

Gene (1) The basic unit of heredity; an ordered sequence of nucleotide bases which comprises a segment of DNA. A gene contains the sequence of DNA that encodes one polypeptide chain via RNA. (2) (Also **cistron.**) The segment of DNA that is involved in producing a polypeptide chain; it includes segments preceding and following the coding region (leader and trailer) as well as intervening sequences (introns) between individual coding segments (exons).

Gene amplification An increase in gene number for a certain protein so that the same protein is produced in increased amounts.

Gene bank A collection of cloned DNA fragments from a single genome.

Gene expression The mechanism whereby the genetic directions in any particular cell are decoded and processed into the final functioning product (target) usually a protein. (*See also* Transcription and Translation.)

Generation time The set of rules which governs the relationship between the linear order of the four different nucleotides in a mRNA molecule and that of the twenty amino acids in the protein which it encodes. The genetic code does not overlap. A mutation which alters only a single nucleotide in a gene can only change one amino acid in the encoded protein.

Genetic code The time required by a cell for the completion of one growth cycle.

Genetic engineering *In vitro* techniques to produce DNA molecules containing novel combinations of genes of other sequences and to insert them into vectors so as to allow their incorporation into host organisms or cells in which they do not naturally occur but where they are capable of continued propagation.

Genetics The study of the nature and transfer of hereditable information, that controls the development of living organisms, and the distribution of this information during reproduction and growth.

Genome The genetic endowment of an organism or individual.

Genotype The genetic constitution of an organism as revealed by molecular biological analysis, i.e. the complete set of genes, both dominant and resessive, possessed by a particular cell or organism.

Gene transfer The use of genetic or physical manipulation to incorporate foreign genes into host cells to achieve desired characteristics in progeny.

Genus A taxonomic category that includes groups of closely related species.

Germ cell The male and female reproductive cells namely egg and sperm.

Germplasm The total genetic variability available to a species.

Glucose A 6-carbon sugar molecule, the building block of natural substances like cellulose, starch, dextrans, xanthan and some other biopolymers used as a basic energy source by the cells of most organisms.

Glycolysis Metabolic pathway involving the conversion of glucose to lactic acid or ethanol. (*See also* Embden–Meyerhof pathway.)

Glycoproteins Proteins with attached sugar groups.

Glycosylation The attachment of sugar groups to a molecule such as a protein.

Gram-negative cell wall A cell wall compound of a peptidoglycan layer surrounded by a lipopolysaccharide outer membrane.

Gram-positive cell wall A cell wall composed of peptidoglycan.

Gram stain A set of two stains (stain crystal violet and counter-stain iodine) used to stain bacteria of the following two groups.

(a) Gram-negative bacteria are completely decolourized after treating the stained bacteria with ethanol or acetone. They have a complex cell wall in which the peptidoglycan layer is covered by lipopolysaccharide.

(b) Gram-positive bacteria resist decolourization by ethanol or acetone; they retain the initial Gram stain and are not stained by counter-stain. They have a cell wall consisting predominantly of peptidoglycan not overlain by an outer membrane.

Growth curve Graphical representation of the growth of an organism in nutrient medium in a batch reactor under predetermined environmental conditions.

Growth factors Various organic molecules such as amino acids, nucleosides, and vitamins which an organism is unable to synthesize.

Growth hormone (GH) A group of peptides involved in regulating growth in higher animals.

GTP Guanosine triphosphate.

Growth rate The fingerprint of the growth of an organism of culture, expressed in terms of the increase of cell mass or number of cells per unit time.

Halophile An organism capable of tolerating relatively high sodium chloride concentration in its environment.

Haploid Describes cells having a single set of chromosomes.

Hapten A low-molecular-weight substance that contains an antigenic determinant but which itself is not antigenic unless complexed with a protein carrier. Examples: dinitrophenol, phosphorycholine and dextran.

Helminths Parasitic worms.

Hemicelluloses Polymers of D-xylose held together by (1,4)-β- linkages with side chains of arabinose and other sugars.

Herbicide A chemical agent used to destroy or inhibit plant growth: specifically, a selective weedkiller that is not injurious to crop plants.

Heterofermentation A fermentation in which a number of waste products are generated; it is used in connection with lactic acid bacteria that have major waste products in addition to lactic acid.

Heterotrophic A class of organisms which is unable to synthesize cell components from carbon dioxide. These organisms require preformed organic substrates as carbon and energy source.

Hexose Sugar with six carbon atoms.

High temperature short-time (HTST) pasteurization The process of pasteurization at 72°C for 15 seconds.

Histone A class of basic proteins associated with DNA in the chromosomes of eukaryotic cells. They contain an unusually large proportion of arginine and lysine.

Homofermentation A fermentation in which there is only one major waste. Used in connection with lactic acid bacteria that produce lactic acid and very small amounts of other acids.

Homologous chromosome A chromosome that has the same sequence as another. In a diploid cell, one of a pair of chromosomes.

Host A cell whose metabolism is used for growth and reproduction of a virus, plasmid or other form of foreign DNA introduced by a vector.

Host–vector system A compatible combination of host and vector that allows stable introduction of foreign DNA in cells.

Hot-air sterilization The use of an oven at 170°C for approximately 2 hours.

Hybridization (1) Formation of stable duplexes of both DNA and RNA (complementary) species via Watson–Crick base pairing used for locating or identifying nucleotide sequences and to establish the effective transfer of nucleic acid material to the new host. (2) Formation of a novel diploid organism either by sexual process or by protoplast fusion.

Hybridoma Product of fusion between myeloma cells (which divide continuously in culture and are 'immortal') and lymphocytes (antibody-producing cells). The resulting cells grow in culture and produce monoclonal antibodies. It is also the cell line produced by fusing a myeloma with a lymphocyte and it continues indefinitely to express the immunoglobulin to both parents.

Hydrolase An enzyme that catalyses a reaction involving addition of water to or removal of water from the substrate molecule. These enzymes act on ester bonds, glycosyl compounds, ether bonds, peptide bonds, C–N bonds, acid anhydride bonds, C–C bonds and halide bonds.

Hydrolysis Chemical reaction involving addition of water to break bonds.

Hydrophilic forces Forces that arise from chemical groups or molecules that arrange themselves in an aqueous environment in such as way as to be surrounded by water.

Hydrophobic forces Forces that arise from chemical groups or molecules that arrange themselves in an aqueous environment in such a way as to exclude water.

Hydroxylation Chemical reaction involving addition of hydroxyl (–OH) group to chemical compound.

Hypertonic A solution that has a higher concentration of solutes than an isotonic solution.

Hypotonic A solution that has a lower concentration of solutes than an isotonic solution.

IMP Inosine monophosphate.

Immobilization The techniques used for the fixation (physically attached or chemically bound) of enzymes, cells organelles or other proteins (e.g. monoclonal antibodies) onto a solid support, into a solid matrix or encapsulated by a membrane, in order to increase their stability, and in most cases turnover and to make possible their repeated or continued use. Such systems offer a number of process advantages.

Immune response The response of an organism against invasion by a foreign substance. Such responses are often complex and may involve the production of antibodies from special cells (lymphocytes) as well as the removal of the foreign substance by other cells.

Immunization The administration of an antigen to an animal organism to stimulate the production of antibodies by that organism. Also, the administration of antigens, antibodies or lymphocytes to an animal to produce the corresponding active, passive or adoptive immunity.

Immunoassay The use of antibodies to identify and quantify various biomolecules. In the binding of antibodies to antigen, the substance being measured is often followed by isotope tracers.

Immunogen A substance that elicits a cellular immune response and/or antibody production.

Immunotoxin A molecule, attached to an antibody, capable of killing cells that display the antigen to which the antibody binds.

Immunoglobulin (Ig) A protein of the globulin-type found in serum or other body fluids that possesses antibody activity. An individual Ig molecule is built up from light (L) and heavy (H) polypeptide chains linked together by disulphide bonds. Igs are classified into five groups and into sub-groups based on antigenic and structural differences in the H chains.

Incubation The introduction of microorganisms into a culture medium with a view to growing a large number of cells of the same all by propagation.

Inducer A small molecule that triggers gene transcription by binding to a regular protein.

Induction (1) An increase in the rate of synthesis of an inducible enzyme in response to the action of an inducer. (2) The experimental elucidation of phase development from a prophage by a lysogen.

Inhibition The decrease of the rate of an enzyme-catalysed reaction by a chemical compound including substrate analogues. Such inhibition may be competitive with the substrate (binding at the active site of the enzyme) or non-competitive (binding at an allosteric site).

Inhibitor An agent that prevents the normal action of an enzyme without destroying it.

Insert A sequence of foreign RNA introduced into an unique restriction site (insertion site, cloning site) of a vector with DNA without any loss of the latter.

Interferons (IFNS) A class of glycoproteins important in immune functions and believed to inhibit viral infections.

Intron An intervening section of DNA within a eukaryotic gene which does not code for an amino acid in the gene product, that is not expressed in the protein molecule or in mature RNA. (*See also* Exon.)

In vitro In a test tube outside the living cell.

In vivo Literally 'in life'. Pertaining to a biological reaction taking place in a living cell or organism. *In vivo* products are substances used within the cell.

Ionization Separation of a molecule into groups of atoms with electrical charges.

Ionization radiation High-energy radiation that causes ionization; examples: X-rays and gamma rays having wave-lengths less than 1 mm.

Isoelectric focusing Electrophoresis of a protein until it reaches its isoelectric point.

Isoenzyme One of the groups of related enzymes catalysing the same reaction but having different molecular structures and characterized by varying physical, biochemical and immunological properties.

Isomerase An enzyme that catalyses intramolecular rearrangements. Isomerases are classified into racimases and epimerases, cis–trans isomerases, intramolecular oxidoreductases, intramolecular transferases, intramolecular lyases and other isomerases.

Isomorph A class of organism that is very similar in appearance to another organism to which it is genetically unrelated.

Isotonic Referring to a solution in which osmotic pressure is equal across a membrane.

ITP Inosine triphosphate.

Kilobase An abbreviation for 1000 bases or base pairs in DNA, nucleic acids or oligonucleotides. The term is used to define the size of the genome of a specific organism or of various fractions of nucleic acids.

Kinases (1) Bacterial enzymes that break down fibrin (blood clots). (2) Enzymes that remove phosphate from ATP and attach it to another molecule.

Krebs cycle A pathway that converts two-carbon compounds to carbon dioxide (CO_2), transferring electrons to NAD + and other carriers.

Lactic acid bacteria A group of gram positive, catalase negative, exclusively fermentative, spherical and rod-shaped bacteria that carry out a lactic acid fermentation. Examples are *Streptococcus* and *lactobacillus*.

Lactobacilli One group of lactic acid bacteria, Gram-positive, rod-shaped, non-motile, aerotolerant, anaerobic and catalase negative.

Lag phase The growth phase of a cell, commonly after inoculation, that precedes the exponential phase and during which there is only little or no growth.

Leaching The removal of soluble compounds mixed with insoluble solids by washing or percolating.

Lectins Proteins isolated from plants (wheatgerm, lentil, pea, etc.) and animals (snail) that react with specific carbohydrate components of other molecules (e.g.

cell wall polysaccharides) in much the same way as an antibody does resulting cells to agglutinate. (*See also* Agglutination.)

Ligase An enzyme that catalyses the condensation of two molecules forming C–C, C–O, C–S, or C–N bonds. In recombinant DNA technique, ligase covalently joins together two sequences of DNA (e.g. host DNA and foreign DNA) by a phosphodiester bond.

Lignin A major component of all lignocellulosic biomass very difficult to biodegrade.

Lignocellulose The major building material of woody and agroresidue biomass which includes lignin, cellulose and hemicellulose.

Lignolytic Pertaining to the breakdown of lignin.

Lineweaver–Burk plot Method of analysing kinetic data (growth rates of enzyme catalysed reactions) in linear form using a double reciprocal plot of rate versus substrate concentration.

Linker A tiny fragment of synthetic DNA that has a restriction site useful for gene cloning. It is used for joining DNA strands together. Chemically a linker fragment is a short, synthetic duplex oligonucleotide containing the target site for some restriction enzyme. It may be added to the ends of a DNA fragment prepared by cleavage with some other enzyme during reconstruction of recombinant DNA.

Lipids A large, varied class of water-insoluble organic molecules, including steroids, fatty acids, prostaglandins, terpenes and waxes.

Liposome transfer The process of enclosing biomolecules inside a lipid membrane to form a complex (or complexes) and allowing the complex to be taken up by a cell.

Log phase Period of bacterial growth with logarithmic increase in cell number.

Lyase An enzyme that catalyses the addition of groups to multiple bonds or the formation of multiple bonds. The class of enzyme are classified into C–C, C–O, C–N, C–S, C-halide, and other lyases.

Lymphocytes Specialized white blood cells involved in the immune response; B-lymphocytes produce antibodies.

Lysis Rupture of the cell wall by means of biological, physical or chemical agents.

Mapping The plotting of the position of genes along a chromosome.

Mass transfer Irreversible and spontaneous transport of mass of a chemical component in a space with a non-homogeneous field of the chemical potential of the component. The driving force causing the transport can be the difference in concentration (in liquids) or partial pressure (in gases) of the component. In biological systems, mass transfer may result from diffusion, facilitated transport or active transport.

Maximum oxygen transfer rate Volumetric mass transfer coefficient times the oxygen solubility at constant temperature and pressure. The value, which is often obtained by the sulphite method, wrongly though, has been in use in the past for the evaluation of oxygen transfer rate of a chemical reaction as a rough approximation. For bioreactor systems gassing out, dynamic methods and oxygen sensors are employed.

Medium Mixture of nutrient substances required by cells for growth and metabolism.

Meiosis Form of cell division in which a diploid cell after division gives rise to haploid cells.

Memory cells B-lymphocytes of T-lymphocytes that have been sensitized to a specific antigen and which can mount an immune response more rapidly than unsensitized lymphocytes involved in a primary immune response.

Meristem A plant tissue consisting of cells which are undergoing mitotic division. Such tissues are capable of being cultured under sterile conditions to economically valuable plants.

Mesophile An organism that has an optimum growth rate at temperatures between 20 and 45°C.

Messenger RNA (mRNA) RNA that serves as the template for protein synthesis which carries the transcribed genetic code from the DNA to the protein synthesizing complex directing the synthesis.

Metabolism The physicochemical transformations through which foodstuffs are synthesized into complex elements, complex substances are rendered into simple one, and energy is made available for use by an organism.

Metabolite A product of metabolism resulting from one or more biochemical reactions that breaks down a compound such as a drug or pesticide in a biological system namely in human or animal bodies and plants.

Metallothionines Proteins found in higher organisms that have a higher affinity for heavy metals.

Methane-oxidizing bacteria Bacteria that utilize methane as a source of carbon and energy. Examples are *Methylobacter* and *Methylococcus*.

Methane-producing bacteria (methanogens) Bacteria that release methane. These bacteria belong to the family Methanobacteriaceae. Examples are *Methanobacteria*, *Methanosarcina* and *Methanococcus*.

Methylotroph An organism that grows on C-1 substrates which are in lower oxidative state than carbon dioxide.

Michaelis–Menten kinetics Kinetics of conversion of substrates in enzyme-catalysed reactions.

Microaerophile Bacterium that grows rapidly in the presence of small amounts of molecular oxygen.

Microbial film Thin layer of microbial cells adsorbed on a rigid surface.

Microbial floc Adherent mass of clustered microbial cells.

Microcarrier A small beaded material derived from silica, glass, dextran, natural/synthetic polysaccharides or similar substances including ceramic or clay bodies employed as a support matrix for the cultivation of anchorage-dependent animal cell lines.

Mitochondria Minute granular, rod-like structures contained inside the cytoplasm.

Microencapsulation The technique of enveloping cells with a permeable membrane.

Microorganisms Microscopic living entities; they can be viruses, prokaryotes (e.g. bacteria), or eukaryotes (e.g. fungi and algae).

Mitosis Cycle of repeating process by which a cell nucleus divides into two daughter nuclei resulting two new cells each having the same genetic complement as the parent cell. Nuclear division is usually followed by cell division.

Mixed culture Culture containing two or more types of microorganisms.

Monoclonal antibodies (MAbs) Homogeneous antibodies derived from a single clone of cells. They recognize only one antigen. MAbs are useful in a variety of industrial and medical capabilities as they are easily produced in large quantities and have remarkable specificity.

Morphology Structure, shape and form of an organism.

Monod kinetics Kinetics of microbial cell growth as a function of substrate concentration proposed by Jacques Monod and widely used to understand growth–substrate relationships.

Multigenic Of trait specified by several genes.

Mutagenesis The process by which permanent changes in the DNA sequences in the genome of an organism are introduced. Researchers may use physical or chemical means to cause such changes that improve the production capabilities of target molecules by an organism.

Mutant An organism which has undergone with one or more DNA mutations rendering its genetic function or structure different from that of a corresponding wild-type organism.

Mutation rate The frequency with which mutations occur in a given organism or gene. Mutation rates vary between one in 10^4 to one in 10^8 per genome per generation. This can be considerably increased by mutagens.

Myeloma Malignant lymphocyte cells which can synthesize excessive amounts of a single type of immunoglobulin.

Myeloma cell line Myeloma cells established in cultures.

NAD^+ Nicotinamide adenine dinucleotide (oxidized form).

NADH Nicotinamide adenine dinucleotide (reduced form).

$NADP^+$ Nicotinamide adenine dinucleotide phosphate (oxidized form).

NADPH Nicotinamide adenine dinucleotide phosphate (reduced form).

NTP Nucleoside triphosphate; it is the general term used for ATP, CTP, GTP, ITP, TTP and UTP.

Neurotransmitters Small molecules found at nerve junctions which transmit signals across those junctions.

Neurophiles Organisms which require neutral media for growth.

Nitrogen cycle A cyclic series of chemical reactions carried out by various microorganisms in which molecular nitrogen is reduced to ammonium, the ammonium is converted to nitrite, the nitrite is further oxidized to nitrate, and the nitrate is reduced back to molecular nitrogen.

Nitrogen fixation The conversion of atmospheric nitrogen to a chemically combined form like ammonia and conversion of nitrogen containing organic compounds like amino acids which are essential for growth. Only a limited number of microorganisms can fix gaseous nitrogen.

Nitrogen-fixing bacteria Bacteria that are able to reduce molecular nitrogen to ammonium and then assimilate (fix) the nitrogen into organic compounds. Examples include *Azotobacter*.

Nodulins Proteins, possibly enzymes, present in nodules, functions not yet known.

Nonsense codon A special terminator codon that does not code for any amino acid.

Nonsense mutation A base substitution in DNA that results in a nonsense codon.

Nucleic acid Macromolecules composed of sequences of nucleotide bases. There are two kinds of nucleic acids: (a) DNA which contains the sugar deoxyribose, and (b) RNA which contains the ribose sugar.

Nucleoside A purine or pyrimidine base covalently bound to either a deoxyribose or a ribose sugar moiety but without any phosphate group. The common nucleosides in biological systems are adenosine, cytidine, guanosine, thymidine and uridine.

Nucleotide A nucleoside with one or more phosphate groups esterified mainly at the 5′ position of the sugar moiety. The major nucleotides found in living cells are ATP, ADP, AMP, GTP, CTP, etc.

Nucleus A relatively large spherical body inside a cell which contains the chromosomes.

Oligonucleotides Short segments of DNA or RNA.

Oncogene Cancer producing virus of two types: (a) V-oncogene: a gene derived from a cancer-producing virus whose protein product, when present in a mammalian host cell, confers a transformed (cancerous) phenotype, and (b) C-oncogene: a cellular homologue of a V-oncogene which is normally expressed in non-cancerous cells but may be induced giving rise to the cancer phenotype.

One-step growth curve The growth curve shown by viruses that lyse their host cells.

Operon A group of closely linked genes that code for the synthesis of a group of enzymes which are functionally related as members of one enzyme system (mainly in prokaryotes) or one biosynthetic pathway. It comprises an operator region (switching on or off the activity of the structural genes in synthesizing mRNA), a number of structural genes (equivalent to the number of enzymes in the system) and a regulatory gene.

Organelle A specialized part of a cell that conducts certain functions. Examples are nuclei, chloroplasts, and mitochondria, which represent the source of genetic material, conduct photosynthesis and generate energy respectively.

Organic micropollutant Low-molecular-weight organic substances that are considered hazardous to humans, animals and the environment.

Origin of replication (ori) A sequence of DNA at which replication is initiated in a bacterium or virus.

Osmosis The diffusion of solvent (water) across a membrane from a region of high concentration to a region of low concentration.

Osmotic pressure The pressure required to prevent the influx of water into a solution that is hypertonic. The internal water pressure on the cell membrane and wall due to the influx of water into the cell is because of a hypotonic environment.

Oxidative phosphorylation Coordinated process of substrate oxidation and synthesis of ATP.

Oxidoreductase (OR) An enzyme that catalyses oxidation–reduction reactions by transfer of electrons. ORs are classified into those acting on groups such as CH–OH, CHO or C–O, CH–CH, CH–NH_2, C–NH_2, C–NH, S-groups, heme of donors such as NADH or NADPH, N-compounds, diphenols, hydrogen, as donors and H_2O_2 as acceptor.

Oxygen transfer rate (OTR) The product of volumetric oxygen transfer rate ($k_L a$) and the oxygen concentration driving force (C^*–C) [$ML^{-3}T^{-1}$], where k_L is the mass transfer coefficient based on liquid phase resistance to mass transfer [LT^{-1}], a is the air bubble surface area per unit volume [L^{-1}] and C^* and C are oxygen solubility and dissolved oxygen concentration respectively. All the terms of OTR refer to the time average values of a dynamic situation.

Oxygen uptake rate The actual rate at which volumetric oxygen utilization takes place and is given by $(\mu/Y_{O_2})x$ where, μ is specific growth rate of the concerned cells, h^{-1}, Y_{O_2} = oxygen uptake per unit cell, $g\,g^{-1}$, and x = concentration of cells $g\,ml^{-1}$.

Parasite A living organism deriving its nutrition from another living organism.

Passive immunity Disease resistance in a person or animal due to the injection of antibodies from another person or animal. It is usually short-lasting.

Pasteurization The process of mild heating to kill particular spoilage organisms or pathogens.

Pathogen A disease-producing agent. The term is usually restricted to a living organism such as a bacterium or virus.

Pentose Sugar with five carbon atoms.

Peptide Two or more amino acids joined by peptide bond (–CONH–). A polymer of numerous amino acids is called a polypeptide. Polypeptides may also be grouped by their functions, such as 'neuroactive' polypeptide.

Periplasmic space The region between the cytoplasmic membrane and the outer membrane of the cell wall in a Gram-negative bacteria.

Peroxidase An enzyme that breaks down hydrogen peroxide; $H_2O_2 + NADH + H^+ \rightarrow 2H_2O + NAD^+$.

pH A measure of the acidity or basicity of a solution on a scale of 0 (acidic) to 14 (basic). For example, lemon juice has a pH of 2.2 (acidic), water has a pH of 7.0 (neutral) and a solution of baking soda has a pH of 8.5 (basic)

pH auxostat A continuous culture system in which the controlled parameter is the pH of the culture. Kinetically behaviour of the system can be considered identical with that of a turbidostat.

Phenol coefficient A standard of comparison for the effectiveness of disinfecting agent for the same length of time on the same organism under identifical conditions.

Phenotype The observable structural and functional characteristics of an organism determined by its genotype and modulated by its environment.

Phosphorylation The addition of a phosphate group to an organic molecule.

Photoautotroph An organism that uses light for its energy source and carbon dioxide (CO_2) as its carbon source.

Photophosphorylation The production of ATP by photosynthesis.

Phosphate oxygen ratio (P:O ratio) The ratio of inorganic phosphate (esterified) to the amount of oxygen consumed, g mol P (g atom O)$^{-1}$.

Photosynthesis The biological conversion of light energy into chemical energy (ATP) that involves one of the chlorophylls in plants and many microorganisms.

Physiology Study of the metabolic pathways governing the cell's growth and turnover of waste products.

Plasma The non-cellular liquid fraction of blood. In vertebrates, it contains many important proteins (e.g. fibrinogen, responsible for clotting).

Plasmids Extrachromosomal, autonomous, self-replicating, circular segments of DNA, as well as some viruses used as 'vectors' for cloning DNA in bacterial 'host' cells.

Ploidy The number of sets of chromosomes present in an organism, for example, haploid (one) or diploid (two).

Polar molecule A molecule with an unequal distribution of charges.

Polymer A linear or branched molecule of large numbers of monomers (amino acids, sugars, nucleotides) with repeating (polysaccharides) or non-repeating (protein, nucleic acids) subunits.

Polypeptide A long peptide chain consisting of many amino acids.

Polysaccharide A polymer of sugars.

Pour plate method A method of inoculating a solid nutrient medium by mixing bacteria in the melted medium and pouring the medium into a petri plate to solidify.

Primary treatment Removal of most easily separable contaminates from waste water.

Privileged tissue Body tissue to which there is no immune response.

Prokaryote An unicellular organism lacking membrane-bound nucleus.

Proinsulin A precursor protein of insulin.

Promoter A DNA sequence of a gene that controls the initiation of 'transcription'. It is a region of the DNA involved in binding of RNA polymers for the initiation of transcription.

Propionic acid fermentation A fermentation in which the major waste products are propionic acid and carbon dioxide. Bacteria such as *Propionibacterium* carry out this type of fermentation.

Protease Protein digesting enzyme.

Protein A polypeptide consisting of amino acids. In their biologically active states, proteins function as catalysts in metabolism and to some extent as structural substances of cells in tissues.

Proton motive force (pmf) An energized state of membrane due to electrical potentials and proton gradients (pH differences). The potential created across a membrane when proton (hydrogen ions) are concentrated on one side of the membrane can be written as: $p = \mu^{(-2.3RT\Delta pH)/F}$, where μ = membrane potential, R = gas constant, T = absolute temperature, ΔpH = pH difference across the membrane, and F = faraday constant.

Protoplast Spherical osmotically sensitive structures formed when cells are suspended in isotonic solution and their cell walls are totally removed by the action of lytic enzymes.

Protoplast fusion The joining of two cells to achieve desired results, such as increased viability of antibiotic producing cells.

Protoza Diverse phylum of eukaryotic organisms with their structure varying from simple single cells to colonial forms. Nutrition may be derived by phagotrophic or autotrophic means. Some protozoa are pathogenic.

Pyrogenicity The tendency of some bacterial cells or parts thereof to cause inflammatory responses in the body. Presence of the phenomenon may negate the usefulness of pharmaceutical products used in health care.

Recombinant DNA (rDNA) *In vitro* hybridization of DNA produced by joining pieces of DNA (genes) from different organisms together.

Recombination Formation of a new association of genes of DNA sequences from different parental origins.

Regeneration The laboratory process of growing a whole plant from a single cell or small clump of cells.

Regulatory gene A gene, usually situated upstream of a bacterial operon, which codes for a protein (a repressor) having the ability to induce or repress the transcription of the structural genes within the operon.

Regulatory sequence A DNA sequence involved in regulating the expression of a gene.

Replication The synthesis of a new DNA from existing DNA by duplication process and the formation of new cells by division.

Repression Binding of a protein (repressor) to the operator in an operon thereby preventing synthesis of mRNA and consequently synthesis of protein.

Repressor A protein that binds to the operator site to prevent transcription.

Residence time (retention time) The time a molecule or a cell resides in a bioreactor during a continuous process.

Resistance gene Gene that provides resistance to an envionmental stress such as an antibiotic or other biologically active chemical compounds.

Resistor A device designed to limit the flow of electrons in an electric circuit by a definite amount resulting in a limited current or a voltage drop.

Respiration The oxidative breakdown and release of energy from nutrient molecules by reaction with molecular oxygen (aerobic respiration) or inorganic molecules such as nitrate (anaerobic respiration).

Respiratory quotient (RQ) The ratio of carbon dioxide produced to the amount of oxygen consumed $(\text{g mol } CO_2)(\text{g mol } O_2)^{-1}$.

Restriction endonuclease A site-specific endodeoxyribonuclease which causes cleavage of both strands of DNA at points signaled by two-fold symmetry of base sequences about an axis; used in genetic engineering to splice genes.

Reverse transcriptase An enzyme found in retroviruses that can synthesize a single complementary strand of DNA from a mRNA sequence. It is used in genetic engineering techniques to produce specific cDNA molecules from a purified preparation of a specific mRNA.

Rheogram Fingerprint of rheological relationship, e.g. shear stress versus shear rate.

Rheology Science of deformation and flow of matter.

Rheopectic fluid A fluid when subjected to a constant shear stress, exhibits an apparent viscosity that decreases with time.

Ribonucleic acid A polyribonucleotide of a specific sequence linked by 3′-5′-phosphodiester linkages.

Ri-plasmid Plasmid from *Agrobacterium rhizogenes* used as plant vector.

Saccharification Hydrolytic degradation of polysaccharides to sugars.

Saprophyte Organism that utilizes organic matter in solution from dead or decaying plant or animal tissues.

Scanning electron microscope An electron microscope which provides three dimensional view of the specimen magnified about 10,000 times.

Secondary treatment Biological degradation of the organic matter in waste water following primary treatment.

Selection A process by which cells (plant, animal or microorganism) are chosen for specific characteristics. It describes the use of special conditions to allow survival only of cells with a particular phenotype.

Semiconductor A material such as silicon or germanium with electrical conductivities intermediate between good conductors such as copper wire and insulators such as glass.

Semiconductor device An electronic device that uses a semiconductor to limit or direct the flow of electrons. Examples are: transistors, diodes, and integrated circuits.

Serum The liquid containing immunoglobulins that remains after blood plasma is collected.

Sequencing (protein, nucleic acids) An analytical procedure for the determination of the order of amino acids in a polypeptide chain or of nucleotides in a DNA or RNA molecule.

Shear Movement of a layer relative to an adjacent parallel layer.

Shear rate Rate of change of shear, i.e. velocity gradient, $(du/dy)/v$ $[T^{-1}]$.

Shear stress Component of stress parallel or tangential to the area under consideration $[ML^{-1}T^{-2}]$.

Shuttle vector A vector molecule that is able to replicate in two different host organisms and can therefore be used to 'shuttle' genes form one to the other.

Single-cell protein Microbial cells or protein extracts of cells grown in large quantities for use as human or animal protein supplements.

Slime layer Covering of gelatinous materials adhereing to cell wall of microorganisms.

Slimes Aggregations of microbial cells generally amenable to biological control. Such aggregations may pose serious envionmental and industrial problems.

Sludge Precipitated, mechanically or biologically separated solid matter produced during water and or sewage treatment or industrial processes. Such solids may be amenable to biological control.

Somaclonal variation Genetic variation produced from the culture of plant cells from a pure breeding strain; the biological promoter of such a variation is not known.

Speciality chemicals Chemicals, usually produced in small volumes, that sell for more than 50c per kg. (*See also* Commodity chemicals.)

Species A taxonomic subdivision of a genus. A group of closely related, morphologically similar individuals which are actually or potentially interbred.

Specific enzyme activity Amount of substrate rendered into product per unit dry weight of enzyme protein per unit time.

Specific growth rate coefficient (μ) The rate of change of microbial cell mass concentration per unit weight of cell mass: $\mu = \frac{1}{x}\frac{dx}{dt}$ (where x = cell mass concentration $g\,l^{-1}$). (*See also* dilution rate.)

Specificity The specific compatibility of certain substrates to certain enzyme.

Spectrometer An instrument used to analyse the structure of compounds on the basis of their light absorption properties.

Splicing The procedure by which introns are removed from precursor RNA molecules and adjacent exon sequences are joined together.

Spontaneous generation: The idea that life could arise spontaneously from non-living matter.

Spore Minute, thick-walled, resistant and dormant cellular form that is devoid of metabolic activity and that can give rise to a vegetative cell upon germination.

Sporulation The germination of a spore by a bacterium (via morphological conversion) or by a yeast (via product of meiosis).

Starch A polymeric substance of glucose molecules, component of many terrestrial and aquatic plants used by some organisms as a means of energy storage; starch is broken down by enzymes (amylases) to yield glucose, which can be used as a feedstock for chemical or energy production.

Stationary phase The growth phase of a microbial culture in which there is little or no growth because of environmental or biological limitations. In many cases it is a phase of product formation for secondary metabolites.

Sterile Completely free of viable microorganisms.

Sterilization The killing of all microorganisms and infectious agents, such as prions and viroids.

Steroid A group of organic compounds some of which act as hormones to stimulate cell growth in higher animals, humans and in plants.

Sticky ends The staggered ends of complementary sequences of DNA which result from cleavage by a restriction enzyme.

Strain A group of organisms of the same species having distinctive characteristics but not usually considered a separate breed or variety. A genetically homogenous population of organisms at a subspecies level that can be differentiated by a biochemical, pathogenic, or other taxonomic features.

Structural gene A gene the codes for an enzyme.

Substrate A chemical substance acted upon by an enzyme or a cell to give rise to a product.

Substrate-level phosphorylation The synthesis of ATP by direct transfer of a high-energy phosphate group from an intermediate metabolic compound to ADP.

Subunit vaccine A vaccine that contains only portions of a surface molecule of a pathogen. Subunit vaccines can be prepared by using rDNA technology to produce all or part of the surface protein molecule or by artificial (chemical) synthesis or short peptides.

Sulphate-reducing bacteria Bacteria that use sulphate as an electron acceptor during anaerobic respiration. Examples include *Desulfovibrio* and *Desulfotomaculum*.

Sulphur-oxidizing bacteria Bacteria that use reduced forms of sulphur, such as hydrogen sulphide, elemental sulphur, and thiosulphate, as a source of reducing electrons of energy. Examples include the photosynthetic bacteria, such as *Chromatium* and *Chlorobium* and the aerobic *Beggiatoa* and *Thiothrix*.

Surface-active agent (surfactant) Any compound that decreases the tension between molecules lying on the surface of a liquid.

Symbiosis The cooperative living together of two dissimilar organisms in mutually beneficial relationships.

Synergistic effect The principle whereby the effectiveness of two drugs used simultaneously is greater than that of either drug used alone.

Taxonomy Classification of organisms.

T cell Stem cell processed in the thymus gland that is responsible for cellular immunity.

tDNA Transfer DNA; that part of Ri or Ti plasmids which is transferred to the plant chromosome.

Template The single nucleic acid strand which is copied during replication or transcription.

Tension pressure Force normal to the surface on which it acts $[ML^{-1}T^{-2}]$.

Terminator A sequence of DNA just downstream of the coding segment of a gene which is recognized by polymerase as a signal to stop synthesizing mRNA.

Tertiary structure In a protein, the final three-dimensional structure is formed from the folding of the polypeptide. The final folding is determined by the disulphide bonds, the hydrogen bonds, and the hydrophobic and hydrophilic interactions of the amino acids with water and amongst themselves.

Tertiary treatment Physical and chemical treatment of waste water to remove all BOD, nitrogen and phosphorus, following secondary treatment.

Therapeutics Pharmarceutical products used in the treatment of disease.

Thermal death point (TDP) The temperature required to kill all the bacteria in a liquid culture in 10 minutes at pH 7.

Thermal death time (TDT) The length of time required to kill all bacteria in a liquid culture at a given temperature.

Thermophile An organism having a growth optimum at temperatures between 45 and 70°C.

Thermophilic Of a microorganism that is capable of surviving at elevated temperatures; this ability is important to make such organisms more compatible to industrial biotechnology processes.

Thixotropic fluid Fluid when subjected to a constant shear stress, exhibits an apparent viscosity that increases with time.

Thromobolytic enzymes Enzymes such as streptokinase and urokinase that initiate the dissolution of blood clots.

Ti plasmid Plasmid from *Agrobacterium tumefaciens* used as a plant vector.

Tissue Collection of cells forming a structure.

Tissue culture (*See* Cell culture.)

Torque The rotary motion of a body around its axis. It equals the rate of increase of the moment of momentum, which is the product of the momentum (mass × velocity) and the normal distance from the axis. $G_T = [ML^2T^{-2}]$.

Total carbon (TC) The total amount of carbon in a chemical compound expressed in moles or weight.

Total organic carbon (TOC) The total amount of organic carbon in a mixture.

Total oxygen demand (TOD) The total amount of molecular oxygen consumed in the combustion of oxygen-demanding substances at about 900°C.

Totipotency The capacity of cells of a higher organism to differentiate into an entire organism. A totipotent cell contains all the genetic information necessary for complete development.

Toxocity The ability of a substance to produce a harmful effect on an organism by physical contact, ingestion or inhalation.

Toxin A substance produced in some cases by disease-causing microorganisms which is toxic to other living cells.

Transamination The transfer of an amino group from an amino acid to an organic acid.

Transcription The synthesis of messenger RNA on a DNA template; the resulting RNA sequence is complementary to the DNA sequence. This is the first step in gene expression. (*See also* Translation.)

Transduction The transfer of hereditary material from one cellular organism to another by a virus, with subsequent recombination of the hereditary material with the recipient's genome and the transformation of the recipient.

Transferase An enzyme that catalyses reactions in which a group is transferred from one compound to another. Such transferred groups are C_1 aldehyde or ketonic residues, acyl, glycosyl, alkyl, nitrogen, phosphorus and sulphur containing groups.

Transfer RNA (tRNA) A low molecular weight RNA containing about 70–90 nucleotides that carries and matches a specific amino acid to its correct codon on a mRNA during protein synthesis.

Transformation The introduction of new genetic information into a cell using naked DNA.

Transistor An active component of an electrical circuit consisting of semiconductor materials to which at least three electrical contacts are made so that it acts as an amplifier, detector or switch.

Translation The process in which the genetic code contained in the nucleotide base sequence of the mRNA directs the synthesis of a specific order of amino acids to produce a protein. This is the second step in gene expression. (*See also* t-RNA).

Translocation Movement of a gene from one chromosomal locus to another.

Transmission electron microscope An electron microscope that provides high magnification of thin sections of a specimen.

Transportable elements Segments of DNA which moves from one location to another among or within chromosomes possibly in a predetermined fashion, causing genetic change. This may be useful as a vector for manipulating DNA.

Transposon A DNA element that can be randomly inserted into plasmids or the bacterial chromosome independent of the host cell recombination system. These elements carry genes that confer new phenotypic properties on the host cell, such as resistance to antibodies.

Trickling filter A method of secondary sewage treatment.

TCA cycle (tricarboxylic acid cycle) Metabolic pathway involving the oxidative breakdown of acetyl CoA to carbon dioxide in the presence of oxygen.

Trihalomethanes (THMs) Organic micropollutants and potential carcinogens consisting of three halide elements attached to a single carbon atom. These compounds may be biologically destroyed during water treatment and purification.

Turbidostat A continuous culture system in which the set parameter is cell concentration as reflected by the turbidity of the system.

Turgor pressure Hydrostatic pressure within a cell providing rigidity $[ML^{-2}]$.

Turnover number Number of molecules converted to product by each catalyst molecule (enzyme) per unit time.

Yield coefficient (Y) The amount of a product of primary interest generated by the system; namely, the cell biomass or metabolite or ATP or even wastes per unit amount of culture input. Thus, $Y_{x/s}$ is the yield coefficient of a cell biomass based on substrate, S; or $Y_{p/x}$ is the yield coefficient of product (metabolite) based on cell mass, X.

Vaccine A suspension of attenuated or killed bacteria or virus or portions thereof, injected to produce active immunity.

Vacuole Droplet in a cell often containing reserve food material.

Vector DNA molecule used to introduce foreign DNA into host cells. Vectors include plasmids, bacteriophages (virus) and other forms of DNA. A vector must be capable of replicating autonomously and have cloning sites for the introduction of foreign DNA.

Vegetative cell Cells involved with obtaining nutrients, as opposed to reproduction or resting.

Velocity gradient Derivative of velocity of a fluid element with respect to a space coordinate, i.e. shear rate, $(du/dy)/v$ $[T^{-1}]$.

Velocity profile Velocity distribution normal to the direction of flow.

Viable Of a state in which a cell is capable of growth.

Virus Any of a large group of submicroscopic agents infecting plants, animals or bacteria. They are unable to reproduce outside the tissue of the host. A fully formed virus consists of nucleic acid (DNA or RNA) surrounded by a protein or a combined protein and lipid coat.

Viscoelasticity Partial elastic recovery on removal of a deformed shear stress.

Viscosity A measure of a liquid's resistance to flow.

Voidage Space occupied by gas and or liquid in a fixed- or fluidized-bed transfer unit or reactor.

Volumetric mass transfer coefficient, $k_L a$ The proportionality coefficient reflecting both molecular diffusion, turbulent mass transfer and specific area for mass transfer. Many factors that affect $k_L a$ also affect the solubility of oxygen as substrate concentration affects both fluid viscosity and oxygen solubility.

V factor NAD or NADP.

Wild type The most frequently encountered phenotype in natural breeding populations.

Wort Unfermented extract of malt or other grains.

Yeast A unicellular, commonly saprophytic, fungus of the family Saccharomycetaceae which is used especially in the making of alcoholic liquors and as a leavening agent in baking. This is a plasmid-forming fungus which usually multiplies asexually by budding.

Zygote The cell obtained as a result of complete or partial fusion of genomes or parts thereof.